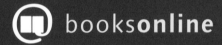

Read SAP PRESS online also

With booksonline we offer you online access to leading SAP experts' knowledge. Whether you use it as a beneficial supplement or as an alternative to the printed book – with booksonline you can:

• Access any book at any time
• Quickly look up and find what you need
• Compile your own SAP library

Your advantage as the reader of this book

Register your book on our website and obtain an exclusive and free test access to its online version. You're convinced you like the online book? Then you can purchase it at a preferential price!

And here's how to make use of your advantage

1. Visit www.sap-press.com
2. Click on the link for SAP PRESS booksonline
3. Enter your free trial license key
4. Test-drive your online book with full access for a limited time!

Your personal **license key** for your test access including the preferential offer

tije-6rfz-xgsy-uapd

Variant Configuration with SAP®

SAP PRESS

SAP PRESS is a joint initiative of SAP and Galileo Press. The know-how offered by SAP specialists combined with the expertise of the Galileo Press publishing house offers the reader expert books in the field. SAP PRESS features first-hand information and expert advice, and provides useful skills for professional decision-making.

SAP PRESS offers a variety of books on technical and business related topics for the SAP user. For further information, please visit our website: *www.sap-press.com*.

Ferenc Gulyássy, Marc Hoppe, Martin Isermann, Oliver Köhler
Materials Planning with SAP
2010, 564 pp.
978-1-59229-259-2

Marc Hoppe
Inventory Optimization with SAP, Second Edition
2008, 705 pp.
978-1-59229-205-9

Martin Murray
SAP WM: Functionality and Technical Configuration
2008, 504 pp.
978-1-59229-133-5

Michael Hölzer, Michael Schramm
Quality Management with SAP, Second Edition
2009, 632 pp.
978-1-59229-262-2

Uwe Blumöhr, Manfred Münch, and Marin Ukalovic

Variant Configuration with SAP®

Galileo Press

Bonn • Boston

Galileo Press is named after the Italian physicist, mathematician and philosopher Galileo Galilei (1564–1642). He is known as one of the founders of modern science and an advocate of our contemporary, heliocentric worldview. His words *Eppur se muove* (And yet it moves) have become legendary. The Galileo Press logo depicts Jupiter orbited by the four Galilean moons, which were discovered by Galileo in 1610.

Editors Patricia Kremer, Frank Paschen
English Edition Editor Meg Dunkerley
Translation Lemoine International, Inc., Salt Lake City, UT
Copyeditor Ruth Saavedra
Cover Design Jill Winitzer
Photo Credit Getty Images/Aaron Graubart
Layout Design Vera Brauner
Production Editor Kelly O'Callaghan
Assistant Production Editor Graham Geary
Typesetting Publishers' Design and Production Services, Inc.
Printed and bound in Canada

ISBN 978-1-59229-283-7

© 2010 by Galileo Press Inc., Boston (MA)

1st Edition 2010

1st German edition published 2009 by Galileo Press, Bonn, Germany

Library of Congress Cataloging-in-Publication Data
Blumöhr, Uwe.
   [Variantenkonfiguration mit SAP. English]
   Variant configuration with SAP / Uwe Blumöhr, Manfred Münch, Marin
Ukalovic. — 1st ed.
        p. cm.
   Includes bibliographical references and index.
   ISBN-13: 978-1-59229-283-7 (alk. paper)
   ISBN-10: 1-59229-283-6 (alk. paper)
   1. Computer integrated manufacturing systems.   2. Configuration
management.   3. Product management.   4. SAP ERP.   I. Münch, Manfred, 1971–
II. Ukalovic, Marin.   III. Title.
   TS155.63.B5713 2010
   670.285—dc22   2009032619

For Lisa, Max, Katrin, and Sonja

Contents at a Glance

Contents

Foreword

At first glance, Variant Configuration in SAP seems to be a purely technical topic for experts, and that is correct because it requires knowledge from IT areas, such as artificial intelligence, conceptual modeling, and semantic services. But this topic goes far beyond the narrower area of IT. It equally involves areas such as supply chain management (SCM), product lifecycle management, and customer relationship management (CRM), and can therefore only be processed in a multidisciplinary fashion. Computer scientists, engineers, and business economists working for SAP and SAP customers have jointly faced the challenge of Variant Configuration for more than 15 years. This intensive and continuous collaboration has an enterprise-strategic purpose that concerns both large international enterprises and medium-sized companies: Customer and production processes with configuration variants have become the core process of high-tech enterprises, whether at leading computer manufacturers such as IBM and HP, at automobile manufacturers such as BMW, Daimler, and Volkswagen, and at medium-sized furniture manufacturers such as the Hüls group of companies. Today, this configuration process is similarly critical to business almost everywhere: in plant construction, mechanical engineering, and small and medium-sized businesses as is discussed in this book.

Here, like in many other areas, the offer for individual customer requirements exceeds the options of classic product description on material coding and bills of material and corresponding storage on an article basis. The variety of variants is so high that totally new options of product presentation and specification had to be found with which to approach customers. Today, it seems that virtually everyone has already used a configurator on the Internet—for instance, to configure a car to your personal preferences—either only for information or for a real purchase decision. These configurators present the provider's wide range of options in a way in which customer-specific interaction, brand positioning, and pricing have been combined in an intelligent product presentation. On the one hand, such variant configurators can be found on the Internet, but on the other hand, they are also available on laptops for mobile field service and embedded in CAD tools.

Of course, this was still a long way off when in 1994 SAP developed the first specification and implementation of the SAP Sales Configuration Engine in cooperation with an American customer at an early stage of the SAP R/3 system. Back then, the Internet was only beginning to loom on the horizon; SCM and CRM were terms still to be defined. At that time, however, it was already clear that a business solution for product configuration involved more than just a sales configurator. The new configurator was supposed to provide two things from the outset: It was supposed to provide SAP customers a high degree of freedom in user interface design for mobile sales processes, and it was supposed to be integrated with processes of SAP Enterprise Resource Planning, that is, sales order management, materials management, and production. For SAP as a manufacturer of standard software this double requirement of free designability and deep integration posed a major challenge.

Today in times of service-oriented architecture, clearly defined component services, and multichannel-multidevice processes, younger colleagues would probably hardly understand where the actual problems were. But even today the challenges go far deeper than might superficially seem to be the case. Because the variant problem starts with sales configuration in sales and distribution and then continues throughout the entire value-added process, everyone will agree that in sales and distribution nothing should be promised that cannot be provided in production later on. If a customer selects a certain specification from a myriad of possible combinations of options and receives a quotation for this configuration, it must be ensured that the customer obtains this configuration exactly as promised. Each customer configuration is a specific variant selected from millions of different possible variants. From the business and logistics viewpoint, it is clear that the variant richness cannot be stockpiled. However, it is only possible to a limited extent to satisfy each sales order individually and independently of one another through make-to-order production. From a business perspective, the goal is to link the profitability and scalability of mass production with an optimal customer individualization. This problem poses a real challenge. Brilliant enterprises have advanced the solution of this problem to create a high customer value and an increase of their own profitability at the same time.

For a scaling, customer-specific production you must ensure that all essential restrictions of feasibility in production have already been considered in the sales process. For parts provisioning in requirements planning and in production, you must supply exactly those components that correspond to the specification of the sales order. A missing or wrong part can bring the entire production to a stop or at least result in expensive standstills.

So the configuration knowledge must be passed from the sales order up to production and be considered particularly for the bill of materials explosion in requirements planning and material flow according to customer requirements. Here, the production planning and control must handle high numbers of very different customer variants in a manageable combination of standard processes and individual tasks. This is a problem of complexity management that can be solved through the close cooperation of all experts involved in the process. Without IT, such processes cannot be mastered in larger scopes.

However, classic IT is not enough here. It increasingly involves the question of which language the different expert areas can use to communicate with one another, that is, the engineers in construction and production on the one hand and the employees in sales and distribution and marketing on the other hand. How can the enterprise as a whole present itself to the customer and speak the language of the customer at the same time? The customer searches for a solution to a requirement that he describes in his language. A translation must be provided here: from the language of the customer and his problem into the language of the solution partner and his internal processes and services. The world of customer requirements is dynamic, so how can you determine future requirements and meet them tomorrow by providing a technical implementation? How can you pursue technical innovations that ultimately create the demand of tomorrow? These are questions that require an effective knowledge management, which, in turn, is only possible using extended methods of information management.

Variant Configuration really is a fascinating topic, and I have greatly enjoyed the related global and interdisciplinary cooperation with the best colleagues and experts. In an ecosystem of customers and partners around and with SAP, many colleagues have made sustainable contributions. Together they all have made a piece of business innovation history. And the story continues.

Dr. Peter Zencke
Executive Board Member for Research and Development at SAP AG, 1993–2008

This introduction presents the goals and the structure of this book. You'll find it easy to go to the most interesting and relevant chapters after reviewing this introduction.

Introduction

Although Variant Configuration has been one of the fully advanced functions of the SAP business process solutions for many years, the interest in this topic has increased drastically lately. Many enterprises, particularly from the production industry, can benefit from the use of Variant Configuration. Especially in Central Europe and the United States, the number of uses of this function has increased considerably.

It's assumed that this development is based on the fact that the implementation of SAP business process solutions has largely been completed in many enterprises. Financial accounting and controlling are usually the first in line for the implementation of an SAP system. Areas such as material management, purchasing, and consistent logistics, are then frequently implemented in a second wave of implementation projects. Applications that can be used universally, including Variant Configuration, are often used at later stages. This similarly applies to the project management solution. Both Variant Configuration and project management are considered complex applications that are only implemented after the classic core functions of SAP business process solutions — so to speak, the voluntary part after the successfully performed mandatory part.

For example, the many improvement potentials that result from an implementation of Variant Configuration are often only recognized after several years of successful operation of an SAP solution. Many enterprises already use configuration solutions. These range from simple Microsoft Office programs, to database applications, to complete custom-developed programs or purchased software components that feature the famous "interface to SAP". A reconsideration of these grown structures often results from the necessity to integrate the different enterprise areas with one another.

With this book, we want to make it easier for you to access product configuration with SAP. The topic is complex, and should be considered from different perspectives. Special emphasis is placed on the following three questions:

1. What did SAP have in mind when developing Variant Configuration as part of the SAP solution?

2. What should you know about the functions of Variant Configuration to be able to use it efficiently?

3. How can you learn from others that use Variant Configuration?

Target Audience

This book is intended for everyone who wants to understand Variant Configuration with SAP — for whatever reason. We've broken down the chapters as follows, according to audience group:

▶ *IT managers and decision-makers* should read Chapters 1, 3, 5, 6, 7, 8, 9, 11, and 12 in particular.

▶ *Sales employees* will especially benefit from Chapters 1, 3, 5, 6, 7, 9, and 11.

▶ *Employees in production preparation and sales order processing* should check out Chapters 1, 3, 4, 6, 7, 9, 10, and 11, as well as Chapter 2.

▶ *Managers of implementation projects* on Variant Configuration with SAP are recommended to read Chapters 1, 3, 7, 10, and 11, but particularly Chapters 8 and 9.

▶ *Implementation and process consultants* may be interested in Chapters 1, 3, 5, 6, 7, 8, 9, and 11, as well as Chapters 2, 4, and 10 in particular.

▶ *Users* with expertise in SAP Variant Configuration should read Chapters 1, 6, 7, 9, 11, and 12, and especially Chapters 2, 3, and 5.

You can see that not all chapters are equally important to all readers. How important a specific chapter is to *you* depends on your background and your fields of interest. However, rest assured that you will gain new and sometimes surprising insights from all of the chapters.

The next section provides a brief overview of what you can expect in the individual chapters for better orientation.

Structure of the Book

The book is divided into 12 chapters whose main ideas are briefly presented here.

▶ **Chapter 1, Basic Principles of Variant Configuration**, familiarizes you with the basic concepts of product configuration in general and then describes the terminology and components of Variant Configuration with SAP. It also discusses how

Variant Configuration benefits your business processes. This chapter provides the necessary basic knowledge required for the subsequent chapters.

▶ **Chapter 2, Creating a Product Model for SAP Variant Configuration**, deals with product modeling in SAP ERP and forms the clear focus of this book. It is aimed at everyone who is involved in the provisioning of required master data. Experienced modelers will find in-depth information that goes beyond basic knowledge. However, reading this chapter can't replace the participation in training.

▶ **Chapter 3, Business Processes in SAP ERP**, addresses Variant Configuration in business processes, discussing which business processes are supported and how should you use them. You'll develop in-depth insight into the integration aspects along the supply chain and in planning using Variant Configuration.

▶ **Chapter 4, Customizing SAP ERP for Variant Configuration**, provides answers to the question about which system settings are important for the use of Variant Configuration. It addresses the customizing for Variant Configuration and the classification system, as well as the special settings for business processes that are noteworthy in the context of Variant Configuration.

▶ **Chapter 5, Special Features of Product Configuration in SAP CRM**, teaches about integrating the product configuration with the SAP CRM solution. This chapter starts with an overview of the configuration processes that are supported in the different distribution channels. You then get to know the maintenance environment for product models that are used especially for sales processes in SAP CRM. Finally, you're presented with a presentation of the functions and settings in the user interface that can be used for interactive configuration in SAP CRM.

▶ **Chapter 6, Enhancements in SAP Industry Solution DIMP**, shows that product configuration solutions must often meet industry-specific requirements. The industry solution *Discrete Industries and Mill Products* (DIMP) was selected as an example. It provides variant manufacturers with several additional functions as a part of SAP ERP and is therefore of particular interest.

▶ There are many function-enhancing additional packages for the standard SAP business process solutions that are provided by partner enterprises and other third parties. **Chapter 7, Enhancements and Add-Ons in the SAP Partner Environment**, introduces you to some of the add-ons that focus on product configuration. This chapter provides you with good insight into the wide range of enhancements available.

▶ **Chapter 8, Project Lead Reports on Projects and Project Structures**, shows the perspective of project leads in an implementation project. The authentic report of a project lead at an SAP customer illustrates the challenges of an implementation project for product configuration with SAP. The SAP consulting recommendations for compiling the project team, as well as the structured procedure in implementation, forms an optimal supplement. This chapter is a must-read for everyone who is involved in implementation projects.

▸ For as interesting and informative the theory may be, the benefit of practical use is decisive. **Chapter 9, Customer Reports on the Introduction of SAP Variant Configuration**, deals with this topic and presents six implementations of Variant Configuration with SAP. Here, you'll not only get an impression of the corresponding application and the selected solution approaches, but also benefit from the experiences and estimations of the implementation teams.

▸ **Chapter 10, Challenges in Variant Configuration**, discusses three challenges that frequently occur in real life. These involve the optimization of the runtime behavior of comprehensive configuration problems, the correct handling of changes to the master data and production orders in Variant Configuration, and the handling of complex system configurations.

▸ This book wouldn't be complete if it didn't introduce the very agile user group that is intent on independence for product configuration with SAP. **Chapter 11, Configuration Workgroup**, provides you with information on the organization of the CWG, its pursued interests, and offers that could facilitate and enrich your work.

▸ **Chapter 12, Outlook for SAP Business ByDesign**, presents SAP's new solution for medium-sized businesses that supports lean product configuration processes during the product launch phase. You'll learn how SAP Business ByDesign handles the customer's product specifications.

Our aim is to bring together our expertise on Variant Configuration in these chapters and to provide you with a book that supports you in the different ways of dealing with Variant Configuration. So that you not only benefit from our expertise, but also our experience, this book includes the following icons.

[!] ▸ This icon warns you about frequent errors or problems that you may encounter in your work. Be sure to read the corresponding section thoroughly.

[+] ▸ This icon highlights tips that may make your work easier and helps you to find further information on the current topic.

[Ex] ▸ This icon indicates if we explain and give more information on the current topic based on practical examples.

We hope that this facilitates your work with this book.

Finally, we would like to add that all gender-specific pronouns used in this text are masculine to simplify reading.

Acknowledgments

In the development of this book we received support from many people. This support ranged from advice and remarks, to individual chapters, to providing content of entire sections. We would like to sincerely thank the people listed below in alphabetical order and all those not mentioned here for their support. Without your help we couldn't have completed this book.

▶ **Manfred Brillert** is CIO of the Hüls group of companies, which also includes Hülsta enterprise. He was responsible for the implementation of SAP software in the group of companies, first as a project lead and then as a member of the steering committee. He is the co-author of the section on Variant Configuration at Hülsta, which you can find in Chapter 9.

▶ **Raimond Buchholz** has been a software developer at SAP for more than two decades and has worked on SAP Variant Configuration from its start. He provided valuable remarks for Chapter 1.

▶ **Don Cochran** is VP sales and business development, and **Daniel Naus** and **David Silverman** are principal consultants at eSpline LLC. They provided the section on managing variant configuration and the partner add-on software Avenue Orchestrator in Chapter 7. They also contributed to the success of the English version of Chapter 2.

▶ **Wilfried Dahlhaus** assumes responsibility for an SAP team within the central IT at Hülsta. As a project lead, he has comprehensive knowledge in the area of Variant Configuration. In combination with a graphical configurator, he implemented Variant Configuration at Hülsta and other companies of the Hüls group of companies. He's the author of the section on the Variant Configuration project at Hülsta, which you can find in Chapter 9.

▶ **Markus Deger** is project lead at SAP Deutschland AG & Co. KG. In addition to Variant Configuration projects in SAP ERP, he has led several larger implementation projects including the Internet Pricing and Configurator. He provided the recommendations for performing implementation projects (see Chapter 8).

▶ **Jens Hennecke, Sascha Rauhe**, and **Rainer Förster** are managing directors at the AICOMP Group and contributed the corresponding section about AICOMP on the VCPowerPack topic in Chapter 7.

▶ **Sönke Janßen** is an employee at the central IT at Hauni-Maschinenbau AG. His section in Chapter 9 about the project at Hauni AG and the engineer-to-order process at a plant construction enterprise helped describe Variant Configuration in real life.

- **Lech Kochanowski** is SAP technology manager of IM architecture and innovations at Krones AG and can look back on many years of experience in the Variant Configuration environment. His report on the Krones project in Chapter 9 sheds light on a complex, but rather successful multi-stage product model.

- **Dr. Hans Joachim Langen** is head of the PLM business area at itelligence AG. He and **Dr. Reiner Kader** wrote the section on CAD automation with it.cadpilot in Chapter 7. Dr. Kader is CEO of ACATEC Software GmbH.

- **Henk Meeter** is solution architect in the Product and Solution Configuration Center of Expertise at SAP Custom Development. Based on his vast experience in the solution of complex system configurations, he developed the shelving system example presented in Chapter 10 and proofread the translation of Chapter 9.

- **Thomas Niemann**, global manager of SAP Systems at Getriebebau NORD, is responsible for the global implementation of Variant Configuration. He wrote the project report in Chapter 9.

- **Scott Perdue** is employee of the American variant manufacturer Baldor Electric. By providing a report about the variant configuration project at Baldor Electric for Chapter 9 he did significantly revalue the international version of this reference book.

- **Wolfgang Schildmann** is a consultant for SAP implementations at Siemens IT Solutions and Services, and his comprehensive feedback considerably contributed to Chapters 1 and 10.

- **Steve Schneider** is an employee at a large office furniture enterprise. With regard to Variant Configuration, he assumes responsibility for the implementation and operation of the SAP business process solution within this enterprise. He has engaged in the Variant Configuration topic since 1993 and has worked with many different configurators. Since 1997, however, he has focused on SAP. He wrote the field report on the implementation of SAP Variant Configuration (see Chapter 8).

- **David Silverman** is an employee of eSpline LLC, and **Christophe Faure** is an employee of Fysbee SAS. They provided the section on the partner add-on software ConfigScan Validation Suite in Chapter 7.

- **Sybit GmbH** provided the partner report on the Sybit Model Tester and the Sybit Configuration Visualizer in Chapter 7.

- **top flow GmbH** contributed a section on the top flow framework and on the top flow Variant Engine in Chapter 7.

- **Kirsten Ukalovic** is a freelance translator and manages a small family enterprise. Her German translation of the field report in Chapter 8 and her comprehensive proofreading contributed to the punctual release of the original book version in German.

- **Jan-Clemens Vogt** is a project lead in the SAP environment in CC Information Systems and Business Processes at Felix Schoeller Service. His section in Chapter 9 provided insight and specific information on Variant Configuration at the paper manufacturer Felix Schoeller.

- **Stefan Weisenberger** is solution director of mill products in the Mill Products & Mining industry business unit at SAP AG, and in this role he has contributed to the further development of industry-specific, characteristic-based logistic processes in the paper, metal, and textile industry over many years, particularly in the integration of supply chain planning, manufacturing, and CRM. He provided significant remarks for Chapter 6.

- **Frank Wietzorrek** is a development architect at SAP AG and assumes responsibility in the custom development area. His main focus is on the management of development and implementation projects in the Variant Configuration area. He's responsible for the Order Engineering Workbench. In this role, he made major contributions to the specification, development, and launch of this solution, and he's the co-author of the corresponding section in Chapter 3.

- **Michael Zarges** works at customer management in the Industry Solutions area at SAP AG. Thanks to his support in more than 50 customer projects and cooperation in the product configuration area in SAP CRM he's gained extensive knowledge, based on which he wrote a large part of Chapter 5. Additionally, as a member of the board, he considerably contributed to the success of the Configuration Workgroup.

- **Dr. Peter Zencke** was an Executive Board Member for research and development board at SAP AG from 1993 through 2008. He always strongly supported the product configuration as an elementary part of the business process solutions of SAP. The foreword that he willingly contributed to the book emphasizes his many years of commitment to Variant Configuration with SAP.

Our special thanks go to the team of Galileo Press, in particular, Patricia Kremer and Frank Paschen from the editorial office and Alexandra Müller for proofreading. Their generous support along the long and sometimes difficult way helped us to get through hard times. We also would like to thank those who supported the English version of this book, namely Meg Dunkerley at SAP PRESS, Oliver Manden at Lemoine International, GmbH for providing an excellent translation to English, and Ruth Saavedra for copyediting the English translation.

Uwe Blumöhr, **Manfred Münch**, and **Marin Ukalovic**

Without a doubt, there are things on this planet that the human mind cannot understand. Fortunately, the SAP solutions for product configuration are not among those. This chapter provides the necessary knowledge and the required information to understand the basic principles of variant configuration.

1 Basic Principles of Variant Configuration

If you deal with the topic of variant configuration, you quickly come across specific concepts that on the surface, seem to be self-explanatory. But if you take a closer look, it's apparent that the technical discussions on variant configuration are sometimes frustrating because they are impossible without the common basic knowledge of variant configuration. Without this knowledge, discussions usually end in confusion and misunderstandings.

Quite a few experienced business people suspect a secret science behind variant configuration after their initial impression—a science that is primarily characterized by an abstract, complex terminology.

This chapter describes the basic concepts of variant configuration. You'll learn what's behind the concept of *product configuration*. From there, we'll discuss the specific features of *SAP Variant Configuration* and describe how business processes in your enterprise can benefit from it.

Regardless of why you deal with this topic, in this chapter, you'll quickly gain the confidence to find your way around the technical discussions, and you'll know the right questions to ask in important decision processes.

1.1 What Is Product Configuration?

Before detailing SAP Variant Configuration, let's look at the superordinate topics of product configuration and configuration tasks in general.

1.1.1 Terminology

This book has a very strong focus on the product configuration in logistic processes. The question about a general definition is therefore not only allowed, but inevitably forms the starting point of the discussions.

First, you should note that configuration is not limited to products. Many systems enable you to adapt certain parameters to your requirements. In a *configuration task* you make adjustments to the system parameters to find a suitable setting.

Several concrete examples illustrate what configuration tasks can entail:

- **Setting parameters of a software system**
  This enables a purpose-driven use, for example, in the Customizing of the Enterprise Resource Planning solution of SAP (SAP ERP) or in Business Configuration of SAP Business ByDesign.

- **Calculation of the annual income tax**
  The income statement requires individual specifications on income and costs.

- **Conclusion of an automobile third-party insurance**
  Such a conclusion is based on individual specifications on the car and the drivers. These specifications determine the insurance contributions and are defined together with the insurance contract.

- **Purchase of a new car**
  Prior to purchasing a car, you have to decide on properties such as color and equipment.

- **Tailoring a custom-made suit**
  The tailor measures his customers to make perfectly fitting suits.

- **Maintenance of a cable car**
  Maintenance work generally depends on many factors, such as the operating hours passed, the age, and the operational demand of the object to be maintained.

- **Project plan for holding a conference**
  Among other things, this plan depends on the number of participants expected.

- **Submitting a road construction measure**
  Type and length of the road are characterizing factors here, for example.

- **Supplying an open-plan office with office furniture**
  Such a supply is associated with a number of dependencies, e. g, for adjacent desks.

A Sudoku puzzle also is an example of a configuration task (see Figure 1.1).

Solving a Sudoku puzzle requires you to consistently fill a 9 × 9 grid. Each column, each row, and each box bordered in bold contain the digits from 1 to 9 only one time each.

Configuration tasks are versatile and can occur in many different contexts. From the examples mentioned earlier, the car purchase and the tailor-made suit are obviously associated with the specification of products. The insurance contract is also a product that is individualized by specifications. In the office furniture example, the individual specification of the furniture is not only determined by the properties of an individual table or cabinet, but also by their composition within the open-plan office.

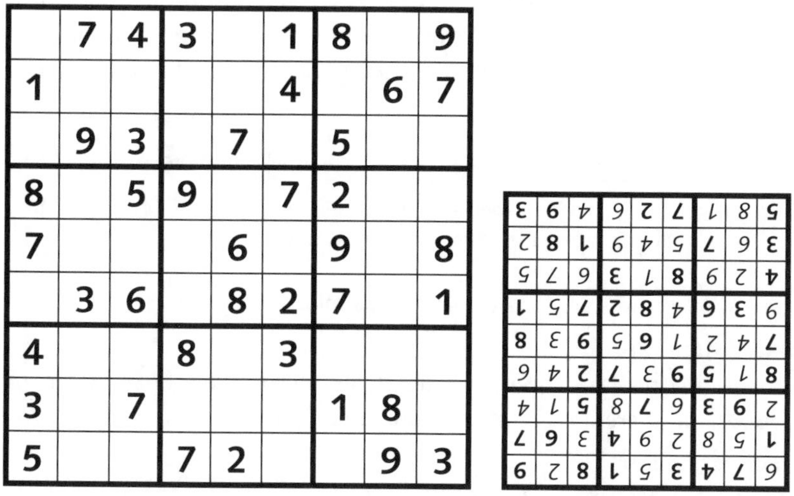

Figure 1.1 Sudoku Puzzle as a Configuration Task, Including Solution

Let's take a closer look at the car purchase and tailor-made suit examples. In both cases, a product is specified according to individual requirements. A car can be described very well using parameters. Many customers inform themselves on the Internet before the purchase and analyze different equipment settings by changing the known parameters.

People who don't buy their clothes off the rack but at renowned tailors are usually unfamiliar with the regular sizes. The purchase of a tailor-made suit requires a face-to-face meeting in which the tailor measures typical parameters, such as length of leg and chest size, and discusses specifics and wishes, which he hasn't anticipated before. Possibly the tailor draws sketches that he discusses with the customer and uses for cutting the cloth later on, for example.

Product Specification [+]

A product specification involves the determination of all requirements made on a product, that is, the features that characterize the product and must be considered for its provisioning.

The form in which the agreed product features are specified can be manifold. In most cases, you will use textual and graphic descriptions, and frequently, you'll use well-defined parameters. A product specification can include the adaptation of the product to individual requirements. But also, standard products usually have an underlying specification that is explicitly formulated.

If adaptations of or deviations from an otherwise identical product are part of the specification, then the specification describes a *product variant* (or *variant* for short). If these adaptations are at least partly described in formal parameters, then the individualized specification includes the determination of these parameters and in turn, the solution of a configuration task.

[+]

Product Configuration

The *product configuration* deals with a special type of configuration tasks. It is about the specification of products that are available in different variants.

The typical procedure of product configuration is as follows:

1. **Formal description**
   The product is described through the formal definition of a set of parameters or product options and therefore becomes a configurable product.

2. **Definition of parameters**
   Based on the formal description, you select individual values for the parameters.

3. **Recording the specification**
   The values that make up the individual appearance of a product are recorded in a specification.

An essential prerequisite for the automated product configuration is the formal description of the parameters. Corresponding software systems always cover the following three aspects:

1. **Modeling**
   The modeling of products formally defines parameters, such as characteristics, rules, and components. This is generally referred to as a *configuration model*.

2. **Configuring**
   The active configuration of a product enables the individual selection of parameter values that are consistent with the underlying model. This step is typically implemented in dialog with the user and is referred to as *interactive configuration*.

3. **Saving**
   Representing or saving the values documents the individual appearance of the product. The value set is the result of an active configuration and is therefore often referred to as the configuration result, or *configuration* for short.

The procedure shown in Figure 1.2 can be generalized. In general, this is referred to as *variant configuration*.

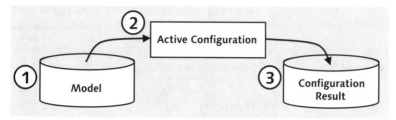

Figure 1.2 Basic Procedure of Variant Configuration

Variant Configuration   **[+]**

Variant configuration deals with the configuration tasks that can be managed with the three-level approach of modeling, configuring, and saving. So variant configuration can be considered as a generalization of product configuration.

Different business functions of the SAP Business Suite use the technologies of variant configuration. In addition to the product configuration, this includes other configuration tasks, for example, the configuration of networks in the project system solution. However, product configuration is by far the most frequently used application of variant configuration. Therefore, this book focuses on the solution of product configuration tasks using SAP software.

Configuration of the Box Product   **[Ex]**

Let's discuss this process based on a physical product:

▶ A configurable product, Box, is supposed to be available in three colors. Accordingly, the model includes a characteristic, Color, for which you determine the three possible attributes, red, yellow, and green.

▶ In the first active configuration, you choose the red color according to the model.

▶ The result of configuration 1 is: Box {Color: red}.

The representation in a model is particularly worthwhile if you use the model not for individual product instances, but for a large set of product instances.

The concept of *mass customization* is a combination of the two terms *mass production* and *individual customization*. Even if the concept of *mass customization* is partly considered as identical to product configuration, it is rather a special aspect of product configuration: In addition to the mass distribution of a configurable product, other considerations can also justify the effort of formal modeling, for instance, the technical complexity of products such as elevators or printing presses. The formal modeling is only worthwhile, however, if the configuration task is to be solved repeatedly in a corresponding frequency. The formalization is the basis for automating different business tasks, for instance, determining a sales price or defining parts required for production.

The software that supports you in solving configuration tasks is often referred to as a *configurator*. Various SAP business software solutions include a configurator.

1.1.2 Elementary Configuration Modules

The central meaning of the model in which the parameters of the configuration tasks are formally described has already been mentioned. This section builds upon these parameters.

Characteristics and values typically occur in configuration tasks that can be modeled. A *characteristic* enables the description of a characterizing property, for example, color. A *characteristic value* determines a possible attribute of this property, for instance, red. The expressiveness of a characteristic is basically determined by its type and value range. For example, a characteristic can be labeled as a listing set with the values, red, yellow, and green, or as a numeric value for determining the length in centimeters without specifying the decimal places.

[+] Characteristic

Characteristics are the variables of the configuration task. In the product configuration, the characteristics are usually used to describe the product options.

What makes the solution of configuration tasks so exciting is that you usually have to meet certain additional conditions. Typically, there are dependencies between characteristics; for example, you can calculate a volume characteristic by multiplying the characteristics, length, width, and height. Logical boundary conditions that apply for the solution of configuration tasks are formulated in *configuration rules*. The way rules for modeling are formulated and processed in the configuration characterizes the capabilities of a configurator. Here, you must pay particular attention to a problem-oriented approach.

Frequently, you can subdivide a system that is the subject of a configuration task. The *component structure* describes the structure of a system and its parts. The *component decomposition* structures the components through a hierarchical division of the system into subsystems. The components can in turn be split into (sub-) components.

[+] Bill of Materials

For configuration tasks that involve products from the production industry, the *bill of materials* (BOM) typically is the carrier of the hierarchical decomposition.

Figure 1.3 shows an example of a BOM.

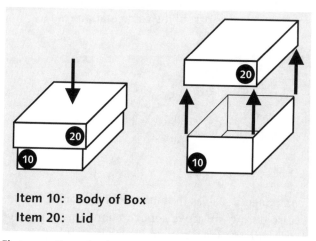

Item 10:   Body of Box
Item 20:   Lid

Figure 1.3   Example of a BOM for a Box Product

The determination of a BOM, which is required for manufacturing a product in a specific configuration, is frequently an essential goal in product configuration. An important instrument for this is the configurable BOM. The component decomposition is parameterized in the *configurable BOM*; specific components are only part of the decomposition under certain circumstances. This way you achieve a general validity. The configurable BOM consists of all possible components.

If you clearly define or predefine all possible components in the configurable BOM and can only select the individual components later based on a concrete configuration, this is referred to as a *super BOM* or *maximum BOM*.

The super BOM in Figure 1.4 shows that the lid of the box can have different colors. For each variant, in the BOM explosion the system determines the component set—including the lid with matching color—according to the configuration.

Item 10: Body of Box	
Item 30: Red Lid	if "color" "red"
Item 40: Yellow Lid	if "color" "yellow"
Item 50: Green Lid	if "color" "green"

Figure 1.4   Example of a Super BOM for the Box Product

Alternatively or supplementary to the principle of the super BOM, the component structure can also allow for enhancements with additional components. These enhancements can be implemented manually or automatically.

Figure 1.5 shows that owing to an individual requirement, you must manually add a golden ribbon to the box that is not included in the model of the super BOM. The BOM that is created this way is then saved as part of the configuration result.

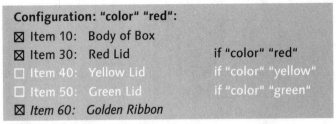

Figure 1.5 Example of a Manual Enhancement of the BOM

The *dynamic instantiation* of components enables the automatic addition of components in the component structure during the active configuration. This is particularly interesting if the dynamically added component itself is configurable — if it is basically known in modeling, but the actual number of occurrences and the actual appearance of its configuration is unknown at the time of modeling. This is illustrated based on the following example.

[Ex]

> **Dynamic Instantiation**
>
> Imagine that an open-plan office is equipped with office furniture. Many desks will be used that are described by parameters for form and style and that are partly connected with one another. Certain geometric connection conditions that can be ensured through configuration rules apply to adjacent desks.
>
> The static planning of desks as components in a super BOM seems pointless. The structuring in a BOM is generally of little use in this case. The dynamic creation of an adjacencies structure that enables you to add desks at free connection positions seems to be a far more natural approach. It's rather a composition problem than a component decomposition.

For product configuration in the production industry, the routing often plays an important role too. The *routing* breaks down a production cycle into a sequence of partly parallel or alternative operations. Analogous to the BOM, you can also apply the principle of the configurable routing or maximum routing here.

1.1.3 Product Configuration in Logistic Scenarios

The logistic processes for the manufacturing of a product basically depend on the business model of sales order processing. The following sections describe production scenarios and distinguish the different procedures from one another.

The theoretical distinction of the process models also reflects in a procedure that is different in terms of software. In real life, the scenarios are often used in a mixed form, and the transitions are rather smooth. Product configuration is not a separate production scenario; rather, it interacts with the individual scenarios more or less. It has a far-reaching influence on the processing steps in the logistic process chain. In particular, the sale and the production are fundamentally influenced by the use of a product configurator.

In the *make to stock* approach (MTS) products are manufactured and temporarily stored independent of concrete sales orders owing to requirement assumptions. The stock decouples the production operation from sales and delivery and consequently accelerates the logistic processing.

If configurable products are specified as product variants, they can be manufactured and stored independently of sales orders. This is a typical procedure in the clothing industry. In the automotive industry, special editions are manufactured as product variants. The selection of a suitable variant plays a decisive role then particularly in sales and delivery. All product parameters relevant for the MTS production must be defined in advance. However, this doesn't exclude that some parameters are only specified when the individual customer requirement is known. For example, you can still change the sound system of a car's special edition retroactively if required. So the MTS production can also refer to not fully specified variants.

Figure 1.6 shows how MTS integrates with the scope of the make-to-order (MTO) scenario; the figure illustrates the subdivision of MTO according to increasing individualization with corresponding product examples.

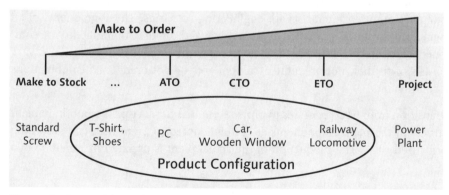

Figure 1.6 Scope of the Make-to-Order Scenario

In the scenario, the production of a product is triggered by a sales order. This is obviously a procedure for very expensive, rarely sold, or specifically defined products. MTO is the business model with the longest tradition.

As the classic scenario for customer specific production MTO is predestinated to be further refined through product configuration. The logistic process is already individualized, and the ordered goods can be tracked individually during the entire process. The product configuration formalizes the degree of individualization of the ordered goods and optimizes the logistic process through increased automation. If you take a closer look, you can differentiate three levels:

▶ **Assemble to order (ATO)**
Here, you are provided with a fixed set of components for the individual composition of the product. These are nonindividualized standard components that can be provided according to the MTS scenario, for example. You can consider ATO as a special type of the configure-to-order scenario in which you only parameterize the combination of standard components, for instance, in the last production level.

▶ **Configure to order (CTO)**
In this approach, you determine the design and composition or the configuration of the product when the customer places an order according to the offered options. The vendor creates the product only after order receipt in accordance with the configuration selected by the customer. In the CTO business model, a clear focus is on the product configuration. The parameterization of the product is clearly visible for the customer in any case.

▶ **Engineer to order (ETO)**
ETO means that the design of the product is at least partially determined only after order receipt. When you specify the product details, further individual decisions must be made that require corresponding product knowledge. The product configuration often interacts with engineering scenarios. The components of an engineering solution can often be individualized, or the combinability of components is subject to formalized rules. However, the engineering solution always consists of further, not predefined components or not predefined combinations of components.

The transition from ETO processes to project-oriented production is smooth. In plant construction, the plant components are often adapted to a specific plant according to formalized rules. The production of the actual plant is usually implemented in a project.

1.1.4 Core Problem of Variant Diversity

Mastering the diversity of configurations that can be generated from a parameterized model represents the actual problem in fulfilling the configuration tasks. The big challenge is to structure the data required for products and BOMs as efficiently as possible. Let's look at two different cases:

▶ A product only occurs in a few versions.

▶ A product is described using parameters.

With regard to variant diversity, these cases considerably differ in their complexity. For a product that is available in very few versions you can easily differentiate between the individual variants. It poses no problems if the datasets that are separate in terms of administration are managed for the few products. In sales, the products receive different product numbers, the prices are defined separately, and all datasets in logistics are managed separately for each product. The means of modern electronic data processing allow for an economic workflow even if redundancies are relatively high. Therefore, many similar products can be managed independently of one another.

If a product is described using parameters, this quickly results in so many variants that each possible limitation is exceeded and that separate datasets are no longer manageable.

Let's return to the sample product, the box. As parameter you've assigned the color characteristic with three values. You can use them to create three differently characterized configurations. If you add a second characteristic, for example, a surface finish with the attributes, glossy, matte, and without, this increases the number of possible configurations to 3 × 3, that is, nine. You obtain the number k of the possible configurations of a product by multiplying the number n_{Mi} of possible values of the characteristics (M1, M2, ..., Mn) with one another. The following therefore applies:

$$k = n_{M1} \times n_{M2} \times n_{M3} \times ... \times n_{Mm} \Longleftrightarrow k = \Pi_{i=1...m} \quad n_{Mi}$$

Product with 5 × 3 Values   [Ex]

So let's assume that a product has five characteristics with three values each. In this case, the number of configurations is:

$$k = 3 \times 3 \times 3 \times 3 \times 3 = 3^5 = 243$$

It is unlikely that each characteristic of a product has exactly the same number of values. To generally estimate the variant diversity, for a number of m characteristics you can nevertheless estimate an average number of attributes for each characteristic with n. For the variant diversity, this results in the following estimation:

$$k \approx n^m$$

n \ m	1	2	3	4	5	...	10	...	15
"linear"	1	2	3	4	5	...	10	...	15
"quadratic"	1	2	9	16	25	...	100	...	225
2	2	4	8	16	32	...	1024	...	32,768
"e ≈ 2.718..."	≈ 2.7	≈ 7.4	≈ 20	≈ 55	≈ 148	...	≈ 22,026	...	≈ 3,269,017
3	3	9	27	81	243	...	59,049	...	14,348,907
5	5	25	125	625	3,125	...	9,765,625	...	30,517,578,125
7	7	49	343	2401	16,807	...	282,475,249	...	4,747,561,509,943

Table 1.1 Number of Configurations for m Characteristics with n Values each

Table 1.1 shows the number of possible configurations (k) that results from m characteristics with n different values each. Three rows are added for comparison: The row marked with "linear" corresponds to a linear increase, the row marked with "quadratic" corresponds to a quadratic increase, and the row marked with "e ≈ 2.718..." shows the increase in a natural growth; e is the *Euler's number*.

If you assume an average of three values per characteristic, the following estimate applies to the number of possible configurations:

$k \approx 3^m$

For five characteristics with three values each, this results in 243 different configurations. For 10 characteristics it results in 59,049, and for 15 characteristics, in more than 14 million (exactly 14,348,907) configurations. If you assume 7 values per characteristic for 15 characteristics, this results in more than 4.7 trillion possible configurations.

To conclude this consideration, let's look at an example from the production industry.

[Ex] **Number of Configurations of the Condenser Product**

On the Internet, a condenser manufacturer offers condensers in 2 designs, 75 capacity levels, 5 tolerance ranges, 15 grid dimensions, 25 DC voltage steps, and 21 AC voltage steps, and with 7 types of dielectric fluid. 41.3 million configurations result from these seven product characteristics:

$k = 2 \times 75 \times 5 \times 15 \times 25 \times 21 \times 7 = 41,343,750$

Therefore, the number of possible configurations increases exponentially with the number of characteristics. Problems with linearly increasing complexity are considered as highly manageable in algorithmics. In case of a quadratic increase, this may be considered a challenge. Problems with exponential growth behavior are considered very difficult to control in terms of algorithm.

The high number of theoretically possible configurations can convey the impression of monotonous pools of numbers and stupefying processing of large quantities. However, the actual appeal of configuration tasks is that not all possible configurations actually make sense. Usually, rules and restrictions apply for the combination of characteristic values. The configuration rules generally restrict the number of possible configurations to a considerably smaller number of consistent configurations.

Let's return to our examples: The possible capacity levels of a condenser depend on its design and the desired operating voltage. Only approximately 20,000 out of the 41.3 million configurations can actually be manufactured. In a Sudoku puzzle about 2×10^{77} configurations are possible. According to the formula $k \approx n^m$ the number with 77(!) zeros results as 9^{81} from 81 characteristics with 9 values each. This estimation is exact for the Sudoku puzzle. However, the Sudoku rules restrict the number of possible combinations to $6.6 \times 10^{21} = 66,000,000,000,000,000,000,000$ — with 21 zeros this is still an impressive number.

1.1.5 Procedural and Declarative Approaches

The rules and restrictions that apply to the combinability of characteristic values adds the extra something to variant configuration. For solving a configuration task, the handling of the rules is of central importance whether you can find a suitable configuration at all, which effort must be made, and whether the proceeding can be understood.

Two basically different approaches are the procedural and the declarative formulation of rules. You require a basic understanding of the character of these two approaches to correctly estimate the complexity of a configuration task and to be able to tackle it using suitable means.

Let's start with a simple example: For a rectangular object like the previously introduced box, there is a dependency between the base area of the box, its height, and its volume. The following formula shows this dependency:

Volume = base area × height

It seems obvious to interpret this dependency as a simple rule: If the base area and height are known, then you can calculate the volume accordingly. The dependency is evaluated in a specific direction. A simple rule triggers an action if certain conditions are met. If the conditions are not met, the action is not implemented. Conditions can be given implicitly — as in the example of the box volume — or formulated explicitly — as in an "if... then..." notation.

Let's consider another dependency: The base area of the box depends on the length and width of the box:

Base area = length × width

Interpreted as a simple rule, this means that if length and width are known, then you can calculate the base area. Because you require the base area to determine the volume, you must evaluate this rule before you can calculate the volume.

Procedural Approach

In the *procedural approach* for processing configuration rules, you define a set of simple rules whose successive processing is carried out in a defined sequence. The model, that is, the modeler of the rules, specifies the process of the steps and their sequence.

During the interactive configuration the processing of the rules is carried out in interplay with the user inputs. A configuration step includes the selection of one (or more) characteristic values and the processing of the set of rules. Generally, each user input triggers the complete processing of the procedural set of rules according to the predefined process. You must reset the information that was determined from the procedural set of rules in the previous step before you reprocess the rules.

[Ex]

Resetting the Values When Executing a Configuration Step

Let's assume that you selected 8 inches for the length, width, and height when you configured the box. As a result, you determined a base area of 64 square inches and a volume of 512 cubic inches. If you delete the value for the width in the next configuration step, the base area and the volume of the box are undefined again. The values that were previously determined procedurally cannot simply be kept.

When you formulate the procedural set of rules, you implicitly make assumptions about the process and the value assignment sequence of the interactive configuration. This way, the previously formulated rules for the box exclude a user wanting to determine the length, width, and height of the box based on a desired volume.

Forward Chaining in a Production Rule System

In contrast to the strictly procedural approach, the processing sequence of rules can often be determined automatically. If the conditions for executing a rule are met, the rule is executed. The result of the execution can influence the conditions for further rules that are then executed if required. In a *production rule system*, the rules are connected to one another using *forward chaining* (see Figure 1.7). In contrast to the strictly procedural approach, you don't necessarily need to execute the rules for each configuration step again.

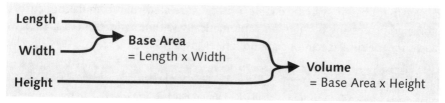

Figure 1.7   Forward Chaining Based on the Box Volume Example

Declarative Approach

In the *declarative approach* the focus is on the description of the problem. Dependencies are formulated in *constraints*, for example. Here, you don't have to determine a general evaluation direction. This means that frequently you can evaluate a dependency in different directions; the declarative formulation usually allows for different evaluations of a dependency. You can provide additional information for the types of desired evaluations.

Let's illustrate this situation with an example. The following restriction shall be made for the dimension of the boxes:

Length ≥ width

From this dependency, you can infer various evaluations:

▶ Restrict the length to a minimum dimension if the width is known.

▶ Restrict the width to a maximum dimension if the length is known.

▶ Notify an inconsistency if the length is not greater than or equal to the width.

In contrast to the formulation in the simple rules, the declarative notation in constraints draws the attention away from the algorithmic processing sequence toward a description of the problem relations.

In the processing of constraints the evaluations are connected in the form of forward chaining — just like in a production rule system. According to this method, the processing sequence is determined automatically. For each *inference* (for instance, of a characteristic value), information is stored about the justification, that is, the reason of the derivation. In this way, you can explain the automatically determined inferences. Via *backward chaining* you can track the cause over any number of steps.

Far more important than the ability to explain is the ability to discard inferences via the principle of justifications as soon as the cause of the inference is no longer given. For example, if a manually selected characteristic value is discarded, the system also resets all inferences that were determined based on the original assigned value. Constraint-based systems enable a problem-related rule formulation, ensure an automated processing of rules, and automatically withdraw inferences as soon as the causes for the rule executions are no longer given.

[+] **Declarative Rules**

Because declarative rules formulate a clearly defined context irrespective of the processing sequence of other rules, in comparison to procedural sets of rules, they are considerably more maintenance friendly in case of changes.

Product configuration with SAP software enables the formulation of simple rules, the procedural execution of rules, the formulation of production rules, and constraint-based rule processing. This versatility provides the modeler with powerfull tools. Their targeted usage, in turn, calls for a sensible use of the individual tools and knowledge of possible interactions, for instance, with regard to the runtime behavior, if they are used in combination.

1.2 What Is SAP Variant Configuration?

Now let's take a closer look at SAP Variant Configuration and its areas of use. Without a doubt, the main area of use is the product configuration for which we will create a little *Hello World* example from scratch. Furthermore, you learn about the different components of SAP Variant Configuration and the business process solutions in which they are used.

1.2.1 Product Configuration Using Variant Configuration

Configuration tasks occur in several business processes. SAP Variant Configuration is primarily used to support logistic processes for multivariant products and is an integral part of *SAP ERP Central Component* (ECC). ECC refers to the component that was called *R/3* in previous software versions. The variant configurator (Variant Configuration) is part of the logistics module and can be abbreviated as *LO-VC*.

In SAP Customer Relationship Management (SAP CRM) sales-related processes are often implemented without direct access to an ERP system, for instance, for a sale via a call center, in online catalogs for business-to-business (B2B) and business to consumer (B2C) applications , or if the sales employees work in the field. The *Internet Pricing and Configurator* (IPC) includes a configurator that can process configuration tasks even without access to an ECC system. From the perspective of the ECC processes, the functional scope of Variant Configuration is extended to a wide range of CRM processes using IPC. For SAP CRM scenarios, which generally don't use ECC functions, using the IPC opens up configuration tasks that go — in some cases — far beyond the functional scope of LO-VC.

1.2.2 Further Areas of Use

In addition to its use in product configuration, LO-VC is integrated with further applications that can be used independent of or in combination with product configuration. Similar to products, you can configure the following objects:

- **Standard networks**
  A standard network describes a sequence of recurring processes and can be used as a template for creating project networks. When you generate a project network, the system determines a suitable variant from a configurable standard network in an interactively processed configuration task.

- **General maintenance task lists**
  A general maintenance task list is a sequence of operations for implementing maintenance work in technical plants. A suitable variant is determined from an interactively processed configuration task based on a configurable general maintenance task list.

- **Model service specifications**
  A model service specification is a combination of frequently required services and can be used as a template for procurement transactions. When you create a model service specification, the system determines a suitable variant from a configurable model service specification in an interactively processed configuration task.

The product configuration is by far the most frequent area of use of SAP Variant Configuration.

1.2.3 "Hello World" Example

What are the minimum requirements for interactively configuring a product in SAP ERP? A product, a characteristic, and the assignment instruments, *variant class* and *configuration profile*, are sufficient. Via the variant class you combine characteristics according to the object-oriented approach and assign them to the product. In the configuration profile you make basic settings for the interactive configuration. Figure 1.8 shows the master data elements for the interactive product configuration that are required as a minimum.

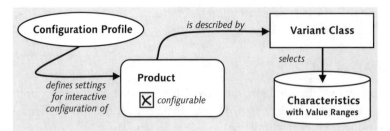

Figure 1.8 Master Data Elements that are Required as a Minimum

Based on the previously used box example, we will now create the master data elements in the system. For the box product you can select from three colors in an interactive configuration.

Creating a Configurable Product

You create the product using Transaction MM01 (Create Material). In SAP ERP, the concept of a *material* is used for all goods that are the subject of a business activity and is therefore a synonym for *product*. You can find the transaction under the menu path SAP MENU • LOGISTICS • MATERIALS MANAGEMENT • MATERIAL MASTER • MATERIAL • CREATE (GENERAL) • IMMEDIATELY.

In the initial screen you make the following specifications. You define the material number in the MATERIAL input field; in this example, it is defined as BOX. You select MECHANICAL ENGINEERING, for example, as the INDUSTRY SECTOR and CONFIGURABLE MATERIAL as the MATERIAL TYPE. In any case, you should select a material type for which variant configuration is permitted in customizing.

[+]
> **Material Type KMAT**
>
> For the material type CONFIGURABLE MATERIAL, which is referred to as KMAT within the system, variant configuration is not only permitted, but is even defined as mandatory.
>
> The term *KMAT* has established itself in the terminology of product configuration and is internationally used as a brand and common noun (such as Band-Aid, Kleenex, or Jeep).

In the BASIC DATA 1 tab you specify the material description, in this example, BOX, and the BASE UNIT OF MEASURE, for instance, PIECE(S).

In the BASIC DATA 2 tab, You select the checkbox for MATERIAL IS CONFIGURABLE; for materials with the CONFIGURABLE MATERIAL material type this checkbox is automatically selected (see Figure 1.9). You can save your entries in the database by clicking on the SAVE icon.

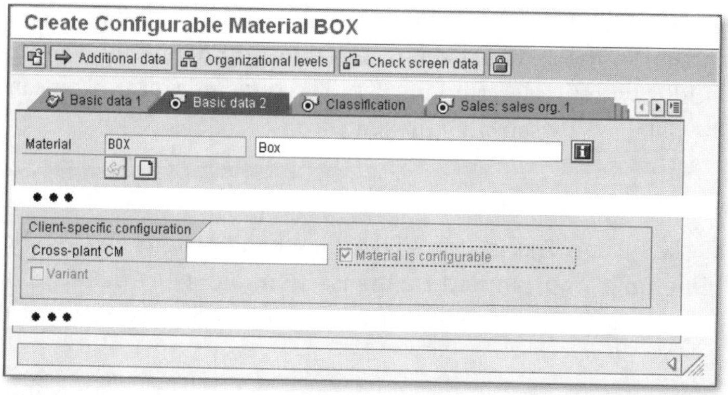

Figure 1.9 Master Data Record of the Box Product

Creating a Characteristic

You create the characteristic using Transaction CT04 (Characteristics). You can find the transaction under the menu path SAP MENU • CROSS-APPLICATION COMPONENTS • CLASSIFICATION SYSTEM • MASTER DATA • CHARACTERISTICS.

In the initial screen, you can specify the characteristic name as COLOR in the CHARACTERISTIC input field and create the characteristic by clicking on the CREATE icon.

In the BASIC DATA tab, you define the following properties of the characteristic: COLOR as the DESCRIPTION, CHARACTER FORMAT as the DATA TYPE, 10 as the NUMBER OF CHARS, and SINGLE VALUE and RESTRICTABLE, for the VALUE ASSIGNMENT. Figure 1.10 shows these settings.

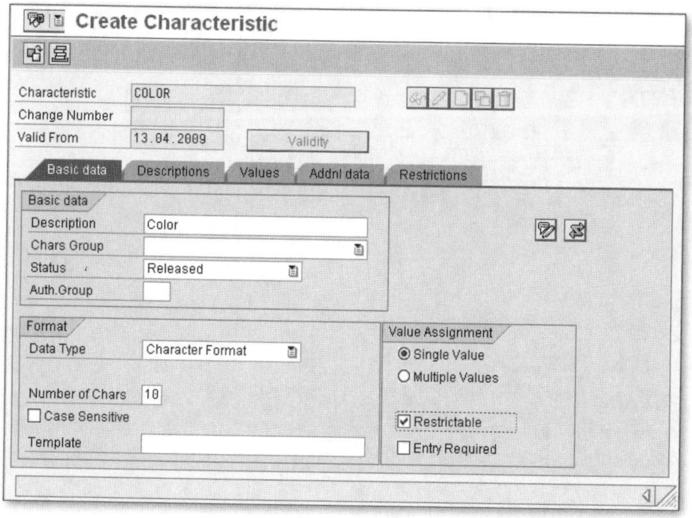

Figure 1.10 Master Data Record of the Color Characteristic

Now, select the Values tab and define three single values as the value range. Here, you also specify a description for each characteristic value. You define the values: RED, YELLOW, and GREEN with the descriptions RED, YELLOW, and GREEN. You can save your entries in the database by selecting the SAVE icon.

Creating the Variant Class

You create the variant class using Transaction CL02 (Class) and assign the COLOR characteristic at the same time. You can find the transaction under the menu path SAP MENU • CROSS-APPLICATION COMPONENTS • CLASSIFICATION SYSTEM • MASTER DATA • CLASSES.

In the initial screen, you specify the CLASS as BOXES and select CLASS TYPE 300. Click on the CREATE icon to create the class. The selected class type must be permitted in variant configuration; class type 300 is permitted in the default setting of variant configuration. All configurable products that can be described using the characteristics combined in this class can be assigned to the class.

In the BASIC DATA tab, you specify BOXES as the DESCRIPTION for the class. In the characteristics tab (CHAR.), you assign COLOR as the characteristic (CHAR.). You complete the class maintenance by clicking on the SAVE icon (see Figure 1.11).

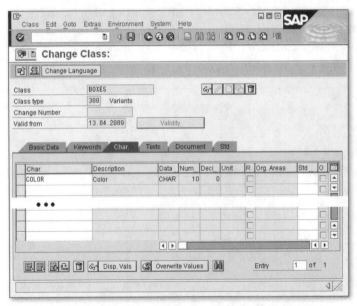

Figure 1.11 Master Data Record of the Boxes Class—Characteristics

Completing and Interactively Configuring the Configuration Model

You now have the basic elements available for the configurable box. You can implement the still missing assignment of the product to the variant class and the creation of the configuration profile using Transaction PMEVC (Product Modeling Environment for Variant Configuration or PME VC) (see Figure 1.12). In this transaction, you can also test the interactive configuration of the box (see Figure 1.13).

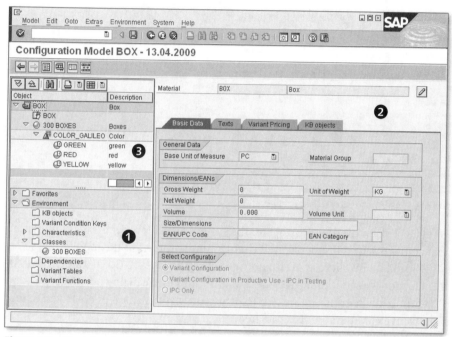

Figure 1.12 Modeling Environment for Variant Configuration

You can find the transaction under the menu path SAP Menu • Logistics • Central Functions • Variant Configuration • Modeling Environment for Variant Configuration. You can access the modeling environment via Material BOX and Class Type 300.

When you click on the Test icon ([Shift] + [F8]) in the menu bar, the system displays the message "Simulation with material BOX requires class assignment." To be able to make the class assignment, you must load the class into the worklist of the modeling environment ❶ that is displayed in the lower-left area of the screen. There, you follow the Environment • Classes path and right-click to select the Add Single Class... function. After you've inserted the BOXES class in the worklist, you can have the system display its details in the right area of the screen ❷ by double-clicking on the class. To assign the product to the class, select the class under the menu path

Environment • Classes and drag it to the BOX product displayed below Object in model tree ❸ in the area above. Then the class appears in the model tree as a subentry of the BOX product.

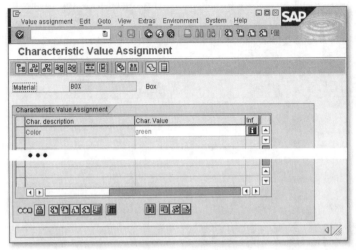

Figure 1.13 Testing the Interactive Configuration

When you click on the Test icon again, the system displays the message "Simulation with material BOX requires a released configuration profile." In the model tree, select the BOX product by clicking on it and right-click the Create Configuration Profile… function. Accept the suggested default settings in the dialog box. Now click on the Test function key and the system informs you that the changes are saved before the interactive configuration is simulated.

Figure 1.12 shows the completely maintained configuration model for the BOX product in the Product Modeling Environment for Variant Configuration. Figure 1.13 illustrates the characteristic value assignment for the BOX product. In the test of the interactive configuration you can select a value for the assigned characteristic, Color, in the Char. Value column. The value help for the Char. Value, which is activated from the input field using the [F4] function key, lists the permitted characteristic values.

1.2.4 Variant Configuration LO-VC

Variant Configuration is programmed in the SAP proprietary programming language ABAP, which focuses on business database applications in large computing installations or in *back-end systems*. Even if such systems could run on small PC or laptop devices today, the minimum software infrastructure still has a considerable scope that is typically managed by IT departments in a data center.

Classification System as the Basis

The basis of Variant Configuration is the *classification system*, which is also programmed in ABAP. It provides a set of global characteristics that can be combined in classes and used for different applications. The abstraction in classes is supplemented by inheritance in *class hierarchies* and *class networks*. A variant class is substantiated by assigning configurable products. While configuring a product, for instance, for a sales order the characteristics of the variant class can get values assigned.

> The classification system itself already allows for the formulation of object dependencies; for example, you can use a precondition for a characteristic value to exclude the corresponding value as long as the condition is not met (for example, that a second characteristic adopts a specific value). Therefore, the classification system includes the basic elements that are required to solve a configuration task.

[+]

Configuration-Specific Modules

Variant Configuration as an integral and inseparable part of the logistics module provides further elements that are required for product configuration:

► **Configuration indicator**
The configuration indicator in the material master characterizes a product as configurable.

► **Configuration rules**
Configuration rules are expressed via object dependencies in SAP Variant Configuration. A dependency formulates a configuration rule.

► **Configuration profile**
The configuration profile allows for settings that are supposed to be valid for a configuration task and the assignment of object dependencies. In addition to the preconditions and selection conditions of the classification system, further dependency types are added via the configuration profile.

  ► *Procedures*
  Procedures enable the processing of algorithmic specifications in a predefined sequence.

  ► *Constraints and constraint nets*
  Constraints and constraint nets are used for the declarative formulation of configuration rules whose processing sequence is determined automatically.

► **Variant tables**
In variant tables you can formulate dependencies between characteristics by listing value combinations row by row.

Variant Tables and Variant Functions

Let's take a closer look at the variant tables. Using the variant table of Table 1.2 enables you to determine the values for the characteristics, WIDTH, HEIGHT, and LENGTH, from the value of the GEOMETRY TYPE characteristic. So, the table column with the GEOMETRY TYPE characteristic is a key column.

Geometry Type	Width	Height	Length
Small cube	2 in	2 in	2 in
Medium cube	8 in	8 in	8 in
Large cube	20 in	20 in	20 in
Jewelry case	2 in	1 in	4 in
Cigar box	4 in	2 in	10 in
Book box	8 in	3 in	16 in
Shoe box	12 in	8 in	20 in

Table 1.2 Example of a Variant Table—Here With Key Column

In procedures or constraints you can refer to variant tables. Analogous to variant tables, you can also define *variant functions*. The relation between input and output characteristics can be formulated as a program, that is, as an ABAP function module. You can then call the variant function in a procedure, for example.

Bill of Materials

Another elementary part of Variant Configuration is the *variant bill of materials (BOM)*. The BOM of a configurable material enables the automatic deactivation of BOM items. If a selection condition is linked to an item, then the item only becomes effective if the selection condition is met. Similarly, you can deactivate entries of the *variant routing* via selection conditions. You can calculate certain fields, such as the BOM item quantity or times for the operation. For this purpose, you assign procedures to a BOM item or an entry in the routing.

Class Nodes

In a variant BOM, you can also add *class nodes* in a BOM item. A class node combines a set of materials, of which exactly one material needs to be selected for the item in order to generate a physical BOM. In the BOM explosion, the material is selected from a class node by checking the characteristic values that are assigned to the configuration against the characteristic values that are assigned to the material at the class node. So a class node replaces multiple alternative BOM items and the associated selection conditions.

Model Concept

Variant Configuration uses many atomic *master data elements* that are maintained in different transactions. They are often subject to changes that are independent of one another. The total of atomic master data elements that are available at the time of solving a configuration task makes up the configuration model that is valid for a configurable product. This approach is also reflected in the master data maintenance. For each master data element there is usually a separate maintenance transaction. PME VC — that is, Transaction PMEVC (Product Modeling Environment for Variant Configuration) — supplements the atomic maintenance transactions by combining maintenance operations or master data elements.

All master data elements that are specific to Variant Configuration and required for the **[+]** interactive configuration of a product can be processed using Transaction PMEVC.

Main Tasks

Two main tasks can be assigned to Variant Configuration:

▶ **High-level configuration**
The high-level configuration usually involves interactively executed value assignment during configuration of a configurable product in the sales order item or the planned or production order. Here you manually select the product options step by step. The system executes rules, implements further inferences, and checks the configuration for completeness and correctness.

▶ **Low-level configuration**
Low-level configuration refers to the explosion of the BOMs and routings that run in the background without user interaction. In Material Resource Planning, for example, the system determines the material requirements of all sales orders that are available in the system for the next day. Here, requirements from variant BOMs are also to be determined based on the configuration result defined for the order.

There are also several additional tasks; for example, when you print a sales order, the system may have to provide language-dependent texts for the configuration result defined in the order item.

Reference Characteristics

Reference characteristics enable an in-depth interaction of the configuration with the logistic process. A reference characteristic is a characteristic that serves as an interface to the embedding application. Via a reference characteristic, the configuration has read and partial write access to the context of the logistic application, for instance, to data fields in the sales order or the BOM. Again, let's illustrate this situation with an example.

[Ex]

Reference Characteristic for Variant-Dependent Surcharges

In the configuration, you can determine a set of so-called variant condition keys using object dependencies. These keys, in turn, are transferred in a dependency to the SDCOM-VKOND data field of the sales order item via a reference characteristic. The variant condition keys entered in this field are used by pricing to identify the corresponding price conditions (with assigned sales prices). Found price conditions are then used for price calculation. This way, you can determine characteristic-based surcharges from the configuration, although the pricing is a logistic function that runs outside Variant Configuration.

Tools

Variant Configuration provides a comprehensive range of tools whose capabilities and versatility are often underestimated. In addition to the object-oriented approach for describing the configuration parameter, the procedural methods for inferring parameter values are supplemented by declaratively formulated rules. This combination allows for a proper handling of configuration tasks even across complex component structures.

[!]

Usage of Tools

The targeted usage of tools and particularly the sensible interaction of the instruments require profound knowledge and sufficient experience. Improper usage can result in undesirable side effects.

The use of best practices and benefiting from the experiences of others is promoted by the *Configuration Workgroup* (CWG), the user group for product configuration with SAP software. For further information on the CWG refer to Chapter 11, Configuration Workgroup, and its web site, *www.configuration-workgroup.com*.

1.2.5 Internet Pricing and Configurator (IPC)

The name *Internet Pricing and Configurator* includes several software components that were developed in the Java programming language. Essentially, this includes the following components:

- **Sales Configuration Engine (SCE)**
  The Sales Configuration Engine is a configuration tool you can use to solve interactive configuration tasks on the basis of product models.

- **Sales Pricing Engine (SPE) and Transaction Tax Engine (TTE)**
  The Sales Pricing Engine and the Transaction Tax Engine implement the pricing and the calculation of transaction taxes on the basis of the *condition technique*. A prerequisite is that the prices and taxes are modeled in corresponding condition records on a database that the IPC can access. The following sections don't further describe the pricing and tax calculation.

- **Product Modeling Environment (PME)**
  The Product Modeling Environment is a modeling tool that enables the definition of product models for product configuration. To disambiguate the PME from PME VC, it is also referred to as *Java PME*.

The IPC includes several further components, for example, components of graphical user interfaces for the interactive configuration.

In 1999 the IPC was the first CRM component available at SAP, even though it had a different architecture and didn't include all of the parts that are available today. Although it is not a complete software solution, there were times when you could purchase the IPC separately and integrate it with a project-specific solution. Two motives in particular resulted in the development of the first IPC component, the Sales Configuration Engine, in the mid 1990s:

- **Decoupling from a complex backend system**
  True to the motto *model once—configure anywhere*, efforts were made to solve a configuration task "offline," that is, without direct access to complex backend software. *Sales Force Automation* requested a better support of field sales employees. The overwhelming trend toward the Internet could be quickly accommodated by providing a Java-based software.

- **Creation of the Configuration Workgroup**
  Early in the 1990s, a small number of users of SAP Variant Configuration teamed up in the Configuration Workgroup to expand the configuration functionality from a pure product configuration to system configurations (see Chapter 11, Configuration Workgroup). Treatable component structures were no longer supposed to be restricted to predefined super BOMs. The aim was to be able to freely and

dynamically add new components as from a construction kit during system configuration. At this point, remember the example of the open-plan office that is supposed to be equipped with configurable office furniture that can be connected with one another.

Even if the original development of the SCE focused on system configurations, the business requirement to use existing product configuration models also on the Internet and in the field was considerably higher. SAP CRM and the IPC have been established as integral parts of the SAP CRM solution. The product configuration functionality provided by the IPC can be used in the following operation modes:

► **LO-VC-compatible mode**
The product modeling and the sales order processing are carried out using SAP ERP; the sales processing is done at least partly using CRM scenarios, for instance, Internet, field service, or call center.

► **Product configuration only in SAP CRM**
Here, the product modeling and the sales order processing are carried out without using SAP ERP. For the product modeling, you use the IPC component, PME, integrated with SAP CRM.

► **Advanced mode**
The support of complex system configurations is not integrated with SAP Business Suite and therefore requires a custom development project. The modeling can be implemented either in SAP ERP or using Java PME. With both tools an advanced maintenance mode must be activated in order to use advanced mode modeling features.

In contrast to the atomic master data management of Variant Configuration, the configuration models of the IPC are maintained in knowledge bases (KB). A *knowledge base* is sometimes also referred to as a knowledge-base object (KBO). It includes all master data of one or, in exceptional cases, multiple configurable products. Different validity statuses are kept in runtime versions, whereas a runtime version is always complete and includes a validity date that indicates the validity start date. If the master data maintenance is carried out in SAP ERP, the system creates a runtime version as a "snapshot" of the atomic master data elements in SAP ERP. If the modeling is done using the Java PME, the modeling process directly works on the respective runtime version.

The compatible mode on the basis of LO-VC models is the operation mode of the IPC product configurator that is used most frequently. In comparison to LO-VC, you must consider some specifics if you configure using the IPC:

▶ **Certain reference characteristics cannot be used as interfaces**
Owing to the changed context in which the configuration is carried out, you cannot use certain object characteristics as interfaces. Data elements that the context doesn't know can neither be read nor written on.

▶ **Programming in Java**
Variant functions must be programmed in Java for the use in the IPC. Variant functions implemented in ABAP in the backend cannot be executed if there is no access to the corresponding backend. Even if access is available, a wrapping in Java has to be implemented.

▶ **Older and special functions**
Functions considered obsolete and some special functions of SAP ERP are not reproduced by the IPC. Examples are the Action dependency type or characteristics with value hierarchies.

The SAP online help under *http://help.sap.com* provides a detailed description of the specific features that must be observed for the parallel operation of Variant Configuration and the IPC. You can find the documentation, which is often referred to as *delta list*, in the ERP MASTER DATA AND CONFIGURATION ENGINE section.

To avoid the parallel operation using two configuration tools for a configurable product and a product model, in SAP ERP you can use the IPC under the *one configurator* motto also within ECC as a tool for the high-level configuration. This option to use the IPC also for order processing in the ECC backend for a configurable product is always available if the IPC must be used for the same configurable product in an SAP CRM application anyway. Only in these cases does a simplification emerge in comparison with the use of Variant Configuration: The IPC is an active part of the solution in any case, and if changes are made to the product model, the runtime behavior is tested using the IPC only.

SAP CRM and SAP ERP are based on the SAP NetWeaver technology platform, which allows for the integration of Java programs via the *Virtual Machine Container* (VMC). Based on this architecture, the core functions of the SCE were integrated with the Application Platform AP 7.00 under the name *AP Configuration Engine*. The Application Platform forms the basis for the backend components, CRM 5.0 and ECC 6.0 (see Figure 1.14). The user interface for the interactive configuration is provided as a web application.

Many configuration applications use the AP Configuration Engine, particularly for sales via the Internet or call centers. Only for the configuration in the field you will still use the complete SCE including its graphical user interface.

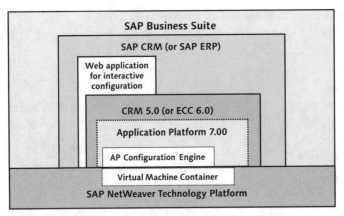

Figure 1.14 AP Configuration Engine in SAP Business Suite

[+] **IPC Product Configurator**

Usually, in all scenarios in which you apply the SCE or the AP Configuration Engine, you use the general term *IPC Product Configurator*.

1.3 How Do Business Processes Benefit from Variant Configuration?

Those who use SAP Variant Configuration as a core module of the business process solution for product configuration make a future-oriented investment in the infrastructure of the enterprise. However, investments in infrastructure measures must be accompanied by business benefits. The following section briefly discusses the factors that can turn the use of Variant Configuration into a profitable investment. We also reference concrete examples.

1.3.1 Prerequisite for the Usage of Variant Configuration

An essential prerequisite for the usage of product configuration is a product structure which can be described using formal parameters and which features a certain complexity. If you only sell vanilla ice cream with chocolate topping, product configuration will hardly bring any benefit.

Initially, it is not important whether the product parameters only describe different properties of certain standard products or whether they have major influence on the design of a special product instance. The following example shows that the search and configuration are often interconnected.

> **Combination of Search and Configuration** [Ex]
>
> Let's take the example of the "white goods": Fridges are usually not manufactured individually for customers, but are provided in different sizes and colors and with different features. Nevertheless, properties such as the width, height, energy efficiency category, or freezer compartment quality play a decisive role in the selection of a fridge. Finding the appropriate fridge from a large set of differently equipped devices can be approached as a configuration task.

You often structure a product range according to descriptive properties and then determine the final attribute of a product in the configuration. When you buy a car, you first select the brand and the model. Then you specify model-specific options in a configuration task. Some enterprises decide to map their entire product range in a comprehensive configuration model. Sometimes, the configuration model then becomes very comprehensive.

Occasionally, the technical parameters that characterize a product group do not meet the criteria according to which the product is used.

> **Transition from Technical Parameters to Application Criteria** [Ex]
>
> A manufacturer of carbide cutting tools, such as drill bits, offers thousands of individual products. To simplify the selection, the usage scenario of the tools was modeled as a configuration task based on application-oriented parameters. After the customer has answered some questions, for example, about the material that is supposed to be processed using the tool, the system determines a small selection of suitable cutting tools.

1.3.2 Factors for the Usage of Variant Configuration

The formal description of product parameters forms the basis for the *automation* of frequently recurring processes that require a vast expertise if they are executed manually. Specialists who are engaged in the sale, planning, production, or purchasing of parameterized products are supposed to be relieved from these time-consuming routine tasks. System users with limited product knowledge are supposed to be supported by the system to execute the process steps correctly.

The goal of *cost reduction* is closely linked with automation. A comprehensive implementation of Variant Configuration usually requires considerable investments. An incremental implementation, for instance, for specific product lines, and careful value assessment and gradual optimization have proven themselves in many cases. It is very useful to follow best practice approaches — as far as they can be transferred to the actual application case. Nevertheless, you are required to continually make adaptations to concrete product structures.

The greatest benefit by far in a successfully implemented usage of product configuration is the *increase in productivity* in the entire enterprise. This is revealed in the reduction of lead times, for example, or in error-free quotations and sales orders, shorter response times, minimization of the complaint rate, and the increase of customer satisfaction. Let's briefly summarize the benefits that Variant Configuration offers for enterprises:

- Automation of recurring processes
- Documentation of expert knowledge
- Cost reduction
- Increase in productivity

Furthermore, external factors make the use of variant configuration appealing. The increasing desire for *individualization* of products requires manufacturing enterprises to implement individual, customized customer requests. Dealing with this task is of high significance to maintain market position, unique attributes, and not least the price level in international competition. You must ensure a high *product complexity and variant diversity* both during order acquisition and in order fulfillment.

The option to select from a variety of products is often not sufficient. Customers want to order customized products and system solutions beyond the standard offer. Here, they accept only minor losses with regard to price and delivery time. The *quality and service* of these products must correspond to the quality and service of the products from mass production. This means vendors must respond individually to customer inquiries and be able to submit quotations that are technically clarified and definite with regard to calculation. They must ensure that they can manufacture with optimized costs and quality and deliver on time.

Let's briefly summarize the requirements that enterprises must face in production:

- Product variety and complexity
- Individual design
- Unchanged service and quality

These requirements for individualization and high variability are opposed to the enterprise-internal need for standardization and modularization.

Product configuration makes a considerable contribution to solve this ostensible conflict of goals. It enables you to control the high number of variants in a manageable set of master data. Not every variant is independent and fully defined in the logistic systems. Rather, they are mapped based on a parameterized configuration model.

Thanks to the versatile combinability of standard components, you can manage a high level of product complexity.

The formal recording and documentation of expert knowledge additionally enables a person-independent use of knowledge that is provided by "old timers." Such strategic considerations can be essential for an enterprise to survive, but you must also consider this aspect from the other side: Only if product experts recognize benefits in the use of a product configuration, will they give the necessary support for an implementation. Because product experts are relieved from routine tasks, they can accelerate the innovation cycle. The systematic mapping in product configuration facilitates the systematic development of the product portfolio.

Interactive product configuration can supplement the classic product catalog and consequently bring the customer closer to the product selection. This ranges from pure product information to self-service scenarios. Many car buyers use the car configurators that are available on the Internet to inform themselves about the currently available extras. Like so many other vendors, the manufacturer of carbide cutting tools, mentioned in Section 1.3.1, Prerequisite for the Usage of Variant Configuration, has connected its online product configurator to electronic purchase-order processing.

1.3.3 Exemplary Consideration on the Master Data Volume

As you can see, the mere reduction of master data by using a classification and variant configuration procedure that is based on it only represents a minor aspect of the potential benefit. Nevertheless, you can avoid a great deal of maintenance effort with by applying product configuration, whereas you must also take into account the increased maintenance effort of the product model, of course.

To give you an idea of the effects of the implementation of product configuration, the following illustrates the dimension of saved master data based on a real-life example.

As described in Section 1.1.4, Core Problem of Variant Diversity, the product range of the condenser example that can be described with only seven characteristics spans a scope of more than 40 million variants. You can exclude most of them owing to invalid combinations of characteristic values.

[Ex] **Data Volume With and Without Configuration**

Let's assume that just under 1% of the 40 million variants, that is, clearly below 400,000 variants, constitutes a feasible condenser product, and only 1/40th of it, that is, approximately 10,000 variants, represents an actually occurring product version, for which you require a product master record in case each product version is modeled by a separate product. For each product you must provide further master data, for instance, the sales price or the BOM.

If product configuration is used, you would require only one single product master record, a rather small configuration model, and a variant BOM.

Figure 1.15 shows a comparison of the master data volume for this example: The gray area corresponds to the master data volume in the case of an explicit definition of each valid variant, and the white dot in the middle corresponds to the master data volume when you use product configuration.

Figure 1.15 Volume With and Without Configuration (White/Gray)

Without doubt, the explicitly defined master data volume is controllable in this example. However, it is difficult to differentiate the many different products, and changes to the product portfolio usually require complex mass changes to the master data.

1.4 Summary

This basic principle chapter approached variant configuration with SAP in three steps. In the first step, you gained the basic understanding of product configuration tasks. So you should be familiar with important terminology and essential configuration elements, such as characteristics, configuration rules, and component structures. You should also be able to classify the product configuration in the logistic scenarios and understand the core problem of variant diversity that grows exponentially with the number of product options. And you should be able to easily and clearly differentiate procedural and declarative approaches.

In the second step, you learned about the SAP-specific concepts and learned that Variant Configuration LO-VC is a core component of SAP ERP. You also got to know the components of the Internet Pricing and Configurator (IPC), and learned that the IPC enables CRM scenarios with product configuration and without access to SAP ERP. The Hello World example of the configurable box familiarized you with the handling of a configurable material (KMAT). So you can reproduce it in an SAP ERP system at any time and you understand that SAP Variant Configuration provides you with a comprehensive set of tools you can use to solve even the most complex configuration tasks. And you learned that the implementation of SAP Variant Configuration is a worthwhile investment in your enterprise workflows if you utilize the best practices that have been developed and tested. In addition, you learned about the SAP user group, the Configuration Workgroup, as a central syndicate of all those who use SAP Variant Configuration.

In the third step, you read about why you use SAP Variant Configuration as a central solution for product configuration tasks. Hopefully, it became clear to you that the central problem of managing a large number of product variants is only one of many reasons. With this understanding, it should be easier for you to assess the future-oriented investment in a business process solution including SAP Variant Configuration. You can now competently participate in decision processes and general tasks that must be performed in a project in the area of SAP Variant Configuration.

Carefully reading the following chapters of this book will enable you as a competent product configuration expert to focus on specific topics. You can select these specific topics according to your personal requirements. Equipped with the necessary basic knowledge, it will be easy for you to deepen this knowledge and use it in practice. We hope you enjoy it.

This chapter introduces you to the essential aspects for a complete product model in Variant Configuration. Those who are new to this topic will get a good overview, and experienced users will enhance their knowledge and find valuable material to reference.

2 Creating a Product Model for SAP Variant Configuration

This chapter introduces all modeling steps in Variant Configuration in detail. As you can see, this chapter is by far the most comprehensive chapter in the book — even though the descriptions are restricted to keep things in proportion. As a result, the chapter is not exhaustive in any respect, but will provide you with enough information for you to understand the modeling steps.

The chapter begins with an overview of the topic and introduces the individual steps. These steps are then discussed in detail in the individual sections. The result is a complete Variant Configuration model that consists of all essential elements. We'll cover the tools from the classification system that meet the requirements of a value assignment interface; we'll introduce the class node as a possible element for configurable Bills of Materials (BOMs). You'll also learn that the configurable product requires a material master, BOMs, and routings. In addition, you'll learn about a control tool for the configuration: the configuration profile. Object dependencies make up a large part of the descriptions in this chapter, which are required to control the configuration in the value assignment interface (mainly in sales and distribution) and to ensure that the product is manufactured or procured as requested. Finally, we'll cover the model of the configurable product from the view of pricing, costing, material variant matching, and the IPC.

2.1 Overview of the Modeling and Integration of Variant Configuration

Let's start with a discussion of when you should use Variant Configuration. We'll assume that your enterprise manufactures and distributes products in different variants. This doesn't necessarily mean you must use Variant Configuration. If the number of variants is manageable, you don't have to use Variant Configuration. The following sections discuss the two options: production without and with Variant Configuration.

2.1.1 Multivariant Product without Variant Configuration

If the number of variants is manageable, you map each of the individual variants with a separate material master. The classification enables you to describe the variant to facilitate searches and evaluations. For planning, production, costing, and further tasks, you require a BOM and a routing for each variant. You can simplify the creation of BOMs for similar variants by combining them in variant BOMs. This procedure is also possible for the creation of routings.

2.1.2 Multivariant Product with Variant Configuration

If the number of possible variants is too large, you can no longer use the procedure described above. A frequently encountered example of Variant Configuration is buying a car.

[Ex] **Need for Variant Configuration**

If you look at a catalog for a model and expand all options regarding engine, chassis, color, extras, and so on, the number of possible combinations easily reaches millions and billions; this is a point where you can no longer use one material master for each variant. In this case, you combine all millions and billions of variants in one material number.

If you want to order and buy such a product (a car in this example), you'll require the material number of the multivariant model. To clarify which of the possible variants you want, the dealer requires the properties that your variant is supposed to have (for example, engine, chassis, extras, and so on). For this purpose, the sales employee (or the customer if the Internet is used for the ordering process, for example) must be provided with a tool that fully supports him in describing his variant by the corresponding properties.

All necessary information on costs, prices, times, dependencies, and so on must be available here. Furthermore, this tool should also include an immediate technical feasibility check. This tool is Variant Configuration.

Variant Configuration must ensure that a sales order or purchase order with such a multivariant product is generally complete and consistent when it is saved. Using a multivariant and configurable product should be as easy for the user as using a product without Variant Configuration. You must also ensure a manageable production or procurement process.

Traditionally, the production is customer-specific in this case. That means the production or procurement process is triggered by the sales order. For this purpose, the model of the configurable product should automatically provide all required information, for example:

- ▸ Determine all required parts
- ▸ Trigger all resulting requirements
- ▸ Create procurement elements
  - ▹ Production order including the appropriate processes and components, for example
  - ▹ Purchase requisition and purchase orders for external procurement, for example

This is the traditional way. Of course, other procedures should also be supported. It doesn't necessarily have to be customer-specific production. You can also combine it with make-to-stock production (key word: material variants, make-to-stock production for configurable materials). The model of the configurable product doesn't have to include the full range of variants. It often makes sense to use incomplete models. This applies particularly to engineering-to-order (ETO) processes.

The model of a configurable product contains the following master data (see Figure 2.1):

- ▸ **Material master**
  The focus is on the material master, which contains the information that the material is configurable. In addition, the appropriate control parameters ensure that the material is managed properly in the supply chain processes. Most of the other objects refer directly to this material master.

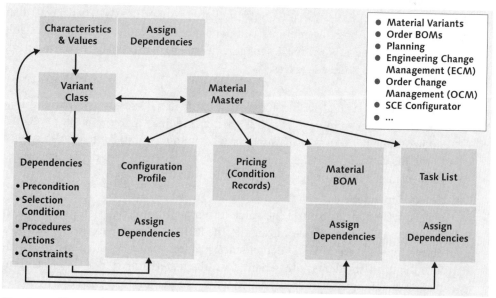

Figure 2.1 Objects of the Variant Model and Their Dependencies

► **Characteristics**

Characteristics are created to map the different variants of the configurable product. You create such a characteristic for each property of the configurable product for which multiple variants are supposed to be available. The characteristics are created without prior reference to the material master.

► **Variant class**

The characteristics are collected in a so-called *variant class*. Variant classes are specific classes that allow you to configure sales orders, for example, via the characteristics that are assigned to the class. The variant class is directly linked to the material master via the classification.

► **Configuration profile**

You create a configuration profile for the configurable material master. Configuration profiles are always object-specific. They contain essential control settings for the configuration process and object dependencies for the sales configuration (also called interactive configuration or high-level configuration).

► **Pricing**

In the standard version, pricing is set specifically for each material. In the default setting, the pricing condition records that are specifically available for Variant Configuration (so-called *variant conditions*) are material-specific, too.

► **Material BOM**

The material BOM generally also requires a material master. Regarding content, the material BOMs in Variant Configuration are super BOMs. That means that they contain all possible required components. In addition to using the default options, a material BOM for a configurable material also enables you to assign classes and object dependencies.

► **Routing**

The routing is created with reference to the material master as well. The link "material master – routing" is not as strict as the link "material master – BOM." Similar to the BOM, it is also a super task list with regard to content. Also similar to the BOM, you can assign object dependencies here. These dependencies are then evaluated for a configurable material.

► **Object dependencies**

Object dependencies control the value assignment in the Sales and Distribution document (SD document), for example. The BOM and routing explosion also uses object dependencies. Object dependencies address characteristics and their values. You can also address the variant class, the material master, and objects from the BOM. You can assign object dependencies to the following levels:

- Characteristics and values
- Configuration profiles
- BOM and routing elements

The Variant Configuration model is supplemented by additional master data and settings. These include material variants, order BOMs, planning profiles, planning tables, and other planning settings.

The Engineering Change Management settings also belong to the Variant Configuration model. Here, you can schedule the change of most of the model data of the Variant Configuration model including history in the so-called change management for master data, Engineering Change Management (ECM). Order Change Management (OCM) enables you to include changes to model data and changes to customer requirements in the production process during the process, especially in the case of long-term continuous production.

The properties of this master data (see also Figure 2.1) cause mandatory dependencies regarding the maintenance sequence of the individual master data of the variant model.

1. **Characteristic maintenance**
   The characteristic maintenance doesn't require model maintenance steps in Variant Configuration.

2. **Class maintenance**
   The class maintenance requires the existence of characteristics.

3. **Material master maintenance**
   Like the characteristic maintenance, the material master maintenance doesn't require model maintenance steps in Variant Configuration (at least with regard to the steps described above).

4. **Configuration profile**
   The configuration profile requires the existence of the configurable material master, specific classes (of the variant classes), and the link "material master – variant class" (classification).

5. **Pricing**
   At least in the standard version, the pricing settings require the existence of the material master.

6. **Bill of materials**
   The BOM requires the existence of the material master.

7. **Routing**
   Before you create routings, the material master and BOM are usually already available. This is not mandatory, however.

8. **Object dependencies**

The maintenance of the object dependencies requires that all objects that are addressed in the syntax have already been created. Otherwise you cannot release the object dependencies. These objects include:

▶ Characteristics with their values (mandatory)

▶ Material master

▶ BOM and objects at the item level (material master, class, and document)

These descriptions indicate the necessary sequence of the modeling steps, which also defines the sequence of the sections.

The first items in the list, namely, characteristic maintenance and class maintenance, refer to both master data and tools from the classification system. The modeling process could also be described starting with tools from the classification system or with the material master. The following sections first deal with tools from the classification system and then discuss material, material BOM, and routings as well as configuration profiles and object dependencies.

2.2 Tools from the Classification System

Variant Configuration uses the tools of the classification system to map the variability of the configurable product. Variant Configuration mainly uses variant classes and class nodes but also other class types. This section focuses on the specific aspects that you must consider when using tools from the classification system in Variant Configuration. In this context, basic knowledge is mandatory.

Tools from the classification system are used in Variant Configuration to model the value assignment interface via variant classes to facilitate the maintenance of BOMs using so-called class nodes and to limit the selection conditions for classified BOM items. This section explains these usages in detail.

In general, you can divide the classification system into four basic functions:

▶ Characteristic management

▶ Class management

▶ Classification

▶ Search

These four functions are the basis for the structure of this section. In this context, the description focuses on the special aspects from the point of view of Variant Configuration and the usages in Variant Configuration.

2.2.1 Characteristic Management

Characteristics can map fields in the value assignment screen of the configuration. In this context, you create a characteristic for each property for which variability exists, from which a selection is to be made in the configuration.

A characteristic is consequently a field in the value assignment interface of the configuration.

In the traditional usage of the classification system, characteristics enable you to describe additional properties of the object for a group of objects, that is, the objects of a class. You also use this form in the two usages of the classification system (class node and classified BOM items), which are introduced in Section 2.7, Object Dependencies for BOM and Routing.

The characteristic master is formatted in various tabs. These are as follows (see Figure 2.2):

- Basic data
- Descriptions
- Values
- Additional data
- Restrictions

When maintaining these tabs, you must take some factors into account. Let's first have a look at the header view (see Figure 2.2).

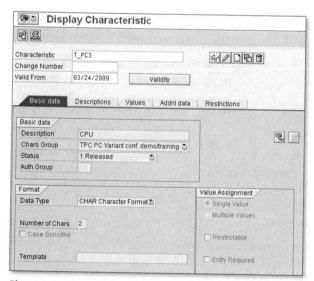

Figure 2.2 Views of the Characteristic Master

The characteristic name must be assigned externally, that is, by the user, and entered into the Characteristic field. The name must start with a letter. In addition to letters, the name can contain numbers and underscores. Although you can also use hyphens, within Variant Configuration you should avoid using them, because this makes the writing of object dependencies more complex. The name of the characteristic is not the internal key of the characteristic. Therefore, you can change the name of the characteristic retroactively. This is still possible even when the characteristic has been assigned to classes or when it has been classified or configured. There is only one exclusion criterion for the modification of characteristic names: the usage of the characteristic name in object dependencies. The same applies to class names.

In general, you can create and change characteristics using Engineering Change Management. For this purpose, enter a change number into the Change Number field in the header. Refer to Section 10.2, Change Management, in Chapter 10 for more information on Engineering Change Management.

Basic Data Tab

For characteristics, you have to maintain the fields in the Basic data tab. All other tabs are optional (see Figure 2.2).

The following list illustrates which fields you can, and sometimes must, maintain:

▶ **Description**
In the Basic data tab, you must enter the description in the logon language. Descriptions are generally language-dependent. You can also maintain long texts (see the first button next to the Description field), which you can call via the F4 help during the configuration.

▶ **Characteristic group**
You can assign a characteristic group. Characteristic groups are created in Customizing. They consist of only a key and a description. Characteristic groups don't have a functional meaning and are solely used for searches and to implement plain, single-level structures.

▶ **Characteristic status**
Like almost all SAP objects, characteristics must have a status, which you maintain in the Status field. The Customizing may already prompt a characteristic status. The characteristic status is not relevant for the classification and configuration. It is exclusively evaluated in the class maintenance.

▶ **Authorization group**
The characteristic can have an authorization group, which you maintain in the Auth.Group field. This authorization group is also only relevant for the characteristic maintenance and not for the classification and configuration.

▶ **Data type**

The characteristic must have a data type.

> ## Data Types in Variant Configuration [+]
>
> In Variant Configuration, you can only use the CHAR character format and the NUM numerical format without any restrictions. *Without any restrictions* refers to the unrestricted usage of these formats in object dependencies.
>
> All other formats are subject to restrictions when they are used in object dependencies. Consequently, they are rarely used in Variant Configuration.
>
> You can query date and time formats in object dependencies, but you cannot use them in bills. All other formats cannot be used in object dependencies at all. This includes the currency format and all user-specific formats.

For the *character format*, you have to define the number of characters for the language-independent key of the characteristic values. This number is limited to 30. You can only use longer keys if you implement modifications. You can use the corresponding checkbox to enable the case sensitivity for keys of the characteristic values. This should be avoided within Variant Configuration if possible (this is not always possible), because it represents a source of error for the writing of object dependencies. Except for this, object dependencies are not case-sensitive. You can also use templates for character formats. However, this option is rarely used, because you mainly use fixed lists of allowed values in Variant Configuration.

For *numerical formats*, the Format section looks different than shown in Figure 2.2. Here, you must specify a number of digits, some of which you can define as decimal places. The decimal point does not count as a character. You can also use units of measurement. However, you have to define them in Customizing first and cannot address them in object dependencies.

In addition, you can allow for negative values and an exponential presentation of numeric values. The numerical format usually also allows for interval values. These interval values cannot be addressed in object dependencies, so you shouldn't use them in Variant Configuration.

Next to the Format section in Figure 2.2 is the Value Assignment section. This section includes settings for the value assignment type and required characteristic.

▶ **Value assignment type**

Characteristics can have a single value (one value at most) or multiple values (any number of values). You can also use single-value and restrictable characteristics. Here, you can easily restrict the list of allowed values using constraints within object dependencies. Moreover, these characteristics also consider each value assignment a restriction. After the value assignment, the list of the allowed values is consequently restricted to the value that was used for the value assignment. Only the configurator of the IPC can use restrictable characteristics with multiple values.

▶ **Required characteristic**
It can be mandatory to enter a value for a characteristic. Such characteristics are called *required characteristics*. If a characteristic is only supposed to be a required characteristic under certain conditions, you mustn't select this checkbox. This is then dynamically controlled via object dependencies.

Descriptions and Values Tabs

All other tabs of the characteristic master can be maintained optionally. This applies to Descriptions tab, for example (see Figure 2.3). The menu of the characteristic maintenance also enables you to change the language to create the description of the characteristic and of the characteristic values.

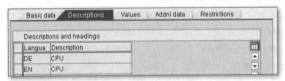

Figure 2.3 Descriptions Tab

The Values tab enables you to define a list of allowed values, a check table, a check function module, or a catalog (see Figure 2.4).

▶ For the CHAR *character format*, you can only specify single values. You can maintain descriptions for the characteristic maintenance, and you can translate the descriptions into other languages using the respective button, Description for Value, or the conversion in the menu.

▶ In addition to single values, you can also enter interval values as the allowed values for the NUM *numerical format* — irrespective of whether the checkbox for interval values is selected in the basic data. You cannot assign descriptions to characteristic values for the numerical format.

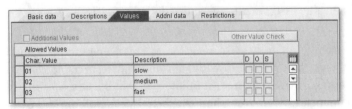

Figure 2.4 Values Tab

► You can assign *documents* and *long texts* to characteristic values. They are available in the configuration as you know it from the classification or search in the classification system.

Additional Data and Restrictions Tabs

In the Additional data tab (see Figure 2.5), you can create so-called reference characteristics using a reference to a table field.

Reference characteristics enable you to elegantly access information in database fields in object dependencies. One problem is that you can usually access characteristics but not database fields in object dependencies. You could solve this problem via function modules in combination with variant functions. This procedure is described later on in this chapter. Another option is the use of reference characteristics, which are links to a field of a database table or structure. As a result, you only address a characteristic in the object dependencies but access the database field. In the standard version of the classification system, you can only access the table whose entries are supposed to be classified — and this usually only with read access. In contrast, you can easily access numerous database tables in Variant Configuration. You have read access and sometimes even write access to all of the tables and structures listed as follows.

Read and Write Access with Reference Characteristics	**[+]**
Read access means the content is evaluated in the context of object dependencies; write access means you can change the content of the fields.	

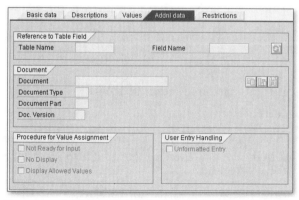

Figure 2.5 Additional Data Tab

Let's take a closer look at this option and start with the configuration in the SD document.

During the configuration in the SD document, you have read and write access to the following tables:

- SDCOM: SD communication structure
- VCSD_UPDATE: Update structure for some VBAP fields
- SCREEN_DEP: Structure for changing the behavior of characteristics for the value assignment

During the configuration in the SD document, you have read access to the following tables only:

- VBAK: Header of the sales and distribution document
- VBAP: Item of the sales and distribution document
- VBKD: Commercial data
- MAEPV & MAAPV: Material master fields
- VEDA: Contract data
- VBPA_AG, VBPA_WE, VBPA_RE, VBPA_RG: Contract partner (sold-to party, ship-to party, bill-to party, and payer)

Let's take a look at the configuration in the purchasing document. During this configuration, you have *read and write access* to the following tables:

- MMCOM: MM communication structure
- SCREEN_DEP: Structure for changing the behavior of characteristics for the value assignment

During the *BOM and routing explosion*, you have *read and write access* to the following tables:

- STPO: Fields of the BOM item
- PLPO: Fields of the operation in the routing
- PLFH: Fields of the assignment of production resources and tools
- PLFL: Fields of the sequence detail screen

In the Additional Data tab, you can assign a *document*. Similar to the characteristic values, this document is provided in the configuration via the input help and can be displayed if necessary.

The settings for the Procedure for Value Assignment section enable you to define that a characteristic is not ready for input or not displayed at all during the configuration.

This can be useful for characteristics to which you only assign values via object dependencies and/or that have a mere technical character. You can change this behavior using procedures via the SCREEN_DEP structure.

It can be helpful to reserve characteristics exclusively for Variant Configuration. This can be done in the Restrictions tab by entering class type 300 or 200.

2.2.2 Class Management

Variant classes and class nodes are class types that have been created especially for Variant Configuration. Class types are configured in Customizing.

In the standard version, class type 300 (for configurable material masters, general maintenance task lists, and networks) and class type 301 (for configurable model service specifications) are defined as variant classes. This can be done using the indicator with the same name in Customizing as the class types. As a result, you can access this class type when creating the configuration profiles. The system then generates the value assignment interface for the configuration. Class types 200 (for material masters) and 201 (for documents) are provided as class nodes in the standard version.

Similar to the variant classes, this is also a checkbox. When you select this option, the class maintenance provides the Additional Data tab.

With the Additional Data tab, you can allow for including such classes as items in BOMs of configurable materials (see Figure 2.6).

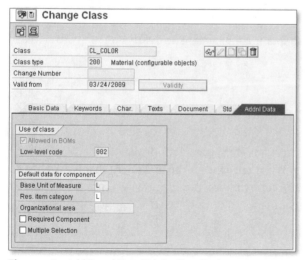

Figure 2.6 Additional Data Tab in Class Nodes

A class that is used as an item prevents you from having to include all materials that are classified in this class in the BOM — including the corresponding selection conditions. A class that is used as an item in the BOM is replaced by exactly one material item in the BOM explosion. However, this requires the configuration of some settings, which are described in detail in Section 2.7, Object Dependencies for BOM and Routing, in the context of object dependencies for BOMs. The Additional Data tab displays a low-level code, which is analogous to the one for material masters.

In addition, you can use the Required Component and Multiple Selection checkboxes to configure these to be generated in scenarios in which you need to be able to manually respond to the respective inconsistency messages. This is important in the following cases:

▶ If no appropriate material can be found in the class for the BOM explosion

▶ If more than one appropriate material is found in the class for the BOM explosion

Variant classes collect characteristics that are supposed to be provided as a value assignment interface for the configuration. In this context, the maintenance of such classes is identical to the maintenance of common material classes, that is, to the maintenance of classes of class type 001. As already mentioned, the difference between the class maintenance of class nodes (class types 200 and 201) and common classes (class types 001 and 017) is that additional data can be set for class nodes.

In the Variant Configuration classes, you can use the following functions or elements:

▶ Class-specific characteristic adaptation (characteristic overwrite)

▶ Class hierarchies

▶ Multiple classification

▶ Organizational areas

▶ Documents

The Always display document checkbox (for permanent display in the search of the classification system in this case) is not relevant for the document that is assigned to the class in the Document tab. The configuration profile, however, provides a similar option including start logo function.

2.2.3　Classification

The classification of class nodes is similar to the classification of common classes. You assign the material masters or documents to the class nodes using the transactions for classification (Transactions CL20N and CL24N) or master data maintenance (Transactions MM01, MM02, and similar; Transactions CV01N and CV02N). Usually, you cannot configure these material masters. Moreover, class nodes have characteristics to which values are assigned during the classification. Because this classification is supposed to replace a class node with exactly one material master, you shouldn't assign objects with the same value to class nodes. You can use the settings in the basic data to avoid this.

You assign variant classes to the configurable material via the classification. In addition to transactions (Transactions CL20N, CL24N, MM02, MM02, etc.), you can also implement the assignment using the maintenance transaction of the configuration profile (Transaction CU41). In retrospect, it doesn't matter which transaction you use – it's always an assignment of material to class, which holds true even for the maintenance transaction of the configuration profile (not profile-specific).

You assign the configurable material to the variant class via the classification. Usually, no values are assigned during this process. Instead of the material, the item in the sales order, for example, is supposed to be assigned a value or configured. Nevertheless, the system provides a value assignment interface when you assign the configurable material. This only serves to restrict the allowed values for each characteristic. The functions are the same as the functions for the class hierarchies or batch classification. Therefore, you speak of object hierarchies for configurable materials or materials that are subject to management in batches.

As already mentioned, you can also use the standard classification of material masters (class type 001) in Variant Configuration. If such traditionally classified material masters are included as items in configurable BOMs, you don't necessarily have to maintain selection conditions. For further details, refer to Section 2.7, Object Dependencies for BOM and Routing.

2.2.4 Search

The classification provides several options for searching for objects in the classification system. You can use the following transactions:

▶ CL30N for object searches in classes or class hierarchies

▶ CL31 for searches in class types

▶ CT12 for the determination of the where-used list for characteristics or characteristic values

You can also use these search functions in Variant Configuration. For example, the object search in classes (Transaction CL30N) displays the configured objects in addition to the configurable material masters that are classified in variant classes. As you can see in Figure 2.7, configured objects can be material variants or items of SD documents such as sales orders. Customizing settings for the class type of the variant classes enable you to navigate directly to the classified material masters and configured objects from the search result. You can also implement similar settings directly from the search result using the ENVIRONMENT menu item.

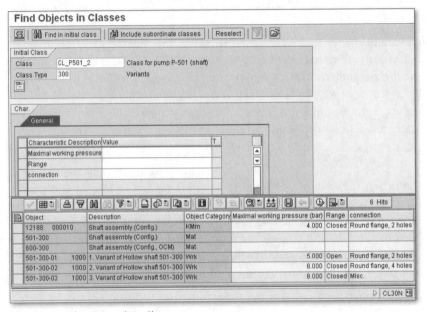

Figure 2.7 Object Search in Classes

In addition to using the Where-Used List for Characteristics/Characteristic Values (Transaction CT12) for the initial usage, you can also use it for the object search. If you start the where-used list for a specific characteristic value of a characteristic in the classification and configuration, the system will display a search result including the objects to which this value is assigned as the classification and configuration (see Figure 2.8).

This kind of object search only works for one characteristic value. However, there are no restrictions regarding class, class type, or object category. In Variant Configuration, this where-used list transaction is not only relevant for the object search. All usage types of characteristics or individual characteristic values are evaluated in Variant Configuration. This plays a significant role if you want to change the variant model by adding or deleting characteristic values.

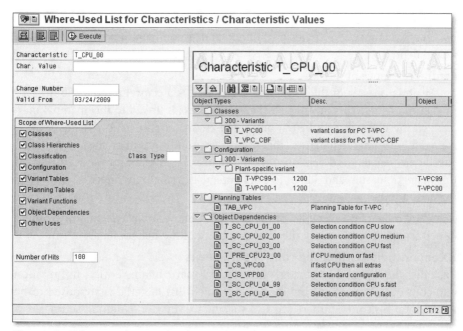

Figure 2.8 Where-Used List in the Characteristic/Characteristic Value Environment in Variant Configuration

2.3 Material Master, BOM, and Routing

Variant Configuration also uses the traditional material master, the material BOM, to map the configurable product. Besides the BOM, the production process also requires information from the routing. You have to take into account some factors when maintaining this master data if you want to do this for configurable products. These factors are introduced in the following sections but without referring to object dependencies. Basic knowledge is required here.

In addition to the tools from the classification system, the material master, BOM, and routing of the configurable material must be mentioned as the master data of the variant model.

2.3.1 Material Master of the Configurable Material

Let's first have a look at what exactly a configurable material master is. It is characterized by the following features:

▶ **One material number for numerous variants**

   The configurable material master is a material master, that is, a material number under which all variants of a product are stored. The concrete variant is defined by a combination of a material number and a configuration, that is, a value assignment.

▶ **Central reference object**
The material master of the configurable material is a central reference object for further modeling. This means that the variant class is linked to this material master, that the BOM and configuration profile are material-specific, that the routing is linked to the material master, that the variant conditions in pricing are material-specific in the standard version, that the material and planning variants refer to this material master, and that object dependencies sometimes directly address this material number.

These are only a few examples of the central reference objects. Sometimes, these dependencies directly query the configurability of the material master.

▶ **It contains information and control parameters**
Basically, the material master is a data record that contains critical information and control parameters of the usages for which this material is supposed to be applied. In this context, there are also specific Variant Configuration settings.

The following sections discuss some characteristics of the control parameters in the material master maintenance. Remember the *view concept* of the material master. It indicates that the views for the material master have to be arranged as tabs whose corresponding usages are supposed to be able to use the material (see Figure 2.9).

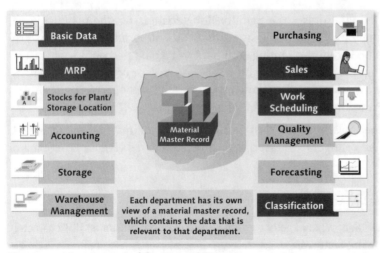

Figure 2.9 View Concept of the Material Master

Bear in mind that each department has its own view of the material master record, which maps the data that is relevant for that department.

Basic Data

In the basic data of the material master, the material master must be set to configurable.

Numerous settings in the Customizing of the material master control to what extent a material master can or has to be configurable. You can activate or deactivate Variant Configuration as a whole via the global settings in the Customizing of the material master.

In addition, you can define for each material type in the corresponding Customizing whether a material master is configurable.

For more information, refer to Section 4.3.1, Configurable Material Master, in Chapter 4.

The following list introduces the possible options for controlling whether a material master is configurable via the material type:

► Material masters of the KMAT material type are generally configurable, because the respective checkbox is preset in the Customizing of this material type and can therefore not be deleted in the material master.

► For material masters of different material types, you can select the checkbox for each material master in the standard version if you want. This is possible because the indicator is not preset in the Customizing of the material type and is provided as ready for input via the field and screen control.

► You can create your own enterprise-specific new material types for generally configurable material masters (similar to KMAT).

► You can create your own enterprise-specific new material types for optionally configurable material masters (similar to FERT and others).

► You can hide the screen module with the indicator for configurable material masters for material types via the screen control and thus create material masters that are generally not configurable.

► If the respective screen module is displayed, you can also use the field control to only hide the corresponding checkbox or set it as not ready for input.

Configurable Material Checkbox

You can also select the checkbox that ensures that the material is configurable retroactively, but you cannot deselect it retroactively in the material master maintenance.

It is critical that not only the header material of a configurable product has a configurable material master. The header materials of all BOMs of assemblies in which object dependencies are supposed to be assigned must also be configurable. You can say that a configurable material must be provided everywhere the configuration is supposed to have effect.

Another important aspect of the Basic Data tabs is the Weights in the corresponding screen module. Sales processes require you to make specifications here. These specifications can differ depending on the configuration. If the specification of weights depends on the configuration, you can determine weight specifications via object dependencies using reference characteristics with reference to the VCSD_UPDATE structure and enter the specifications into the corresponding fields of the SD document. For the determination, the system can read the figures from the material master via MAAPV.

Classification

In the Classification view, you must assign the variant class (or several variant classes). The value assignment screen enables you to specifically restrict the list of the allowed values for the individual characteristics for each material.

Sales and Distribution

In the Sales views, you must consider the following factors in addition to the required entry fields:

▶ A specific item category group is required.

▶ In Variant Configuration, item category group 0002 (configuration) is used in the standard version. The configurable material with a sales order configuration scenario *(SET)* is an exception here. In this case, you use 0004 (make-to-order production/assembly).

The item category group is used to find the item category for the SD document, which in turn, controls the transfer of requirements, distribution, and pricing (see also the explanations of the configuration profile in Section 2.5, Overview of Object Dependencies, and of customizing in Chapter 4, Customizing SAP ERP for Variant Configuration).

Furthermore, you need to run the availability check against individual customer requirements. Here, you cannot combine requirements. Otherwise the link to the configuration in the item of the SD document will be lost. However, this availability check always has a negative result, which means you have to check against the replenishment lead time.

MRP

In Material Requirement Planning (MRP) views, you can use MRP groups.

▶ **MRP type**
In general, MRP should be set as the *MRP type*.

▶ **MRP lot size**
The *MRP lot size* is usually calculated exactly.

▶ **Procurement type**
In-house production or external procurement can be used as the *procurement type*. In in-house production, the procurement elements (planned order, production order, and so on) reference the configuration from the SD document and explode the BOM and routing accordingly. For external procurement, the configuration is copied to the purchase order in addition to the material number.

▶ **Special procurement key**
In MRP views, you can still use *special procurement keys*, such as 50 = phantom assembly. However, particularly for the phantom assembly, you must bear in mind that object dependencies don't take this into account. That means object dependencies must be defined as if this special procurement key was not set.

▶ **Replenishment lead times**
For configurable materials, you can also specify *replenishment lead times* for scheduling in the material master. The in-house production time can be maintained depending on the lot size but not depending on the configuration.

▶ **Strategies**
The standard version provides some *strategies* (and thus strategy groups), which you can use for configurable materials. If you want to use MRP but not planning, you should adhere to strategy 25. For planning, you can use strategies 54 and 56. You can also use the transactions for assembly processing (80–89) here. For more details on the strategies, refer to Section 3.3, Planning and Variant Configuration, in Chapter 3 and Chapter 4, Customizing SAP ERP for Variant Configuration.

In the MRP views, the setting for the availability check is also available; it is set to the individual customer requirement. The BOM is also supposed to be exploded with regard to the individual customer requirement.

Work Scheduling

The Work Scheduling view is required for in-house production. In this context, no special aspects are to be mentioned that are relevant for Variant Configuration. If the configurable material is supposed to be used for order change management, an overall profile must be specified here. Initially, it is solely used as an activation indicator.

For the other views that are listed in Figure 2.9 but not highlighted, you don't have to consider special aspects with regard to Variant Configuration.

2.3.2 Super BOM of the Configurable Material

For the configurable material, you require BOM information for MRP, production, costing, and other areas, for example, in sales and distribution, where you have to explode the BOM for product costing or if you want to implement a multilevel configuration.

In this case, you require a super BOM for the configurable material. As indicated in Figure 2.10, this BOM structure can have several levels. The header materials of the assemblies in whose BOMs the configuration is supposed to be evaluated via object dependencies must also be configurable. Every material in whose BOM configurable assemblies are supposed to be integrated must be configurable. That means all materials above a configurable assembly must be configurable in a multilevel BOM structure.

[+]

Technical Category of BOMs and Variant Configuration

There isn't a specific technical category for these configurable BOMs. In general, simple material BOMs but also variant BOMs are used in real life. Let's have a closer look at these two categories:

▶ **Simple BOM**
A simple BOM means that a BOM, strictly speaking, a BOM number, contains exactly one BOM for exactly one material.

▶ **Variant and multiple BOMs**
A variant BOM is a BOM number under which one BOM is stored for each material master of several similar material masters.

Bear in mind that a configurable variant BOM is a BOM (group) for several configurable material masters.

With regard to content, configurable BOMs are super BOMs. Accordingly, they contain all possible material or document components that can be required for any configuration.

The components can be assigned directly as material or document items or indirectly via class items. The function for assigning class items in BOMs is coupled to the indicator for configurable materials in the material master of the BOM header.

Figure 2.10 shows the configurable BOM, that is, the super BOM (❶). You require a BOM explosion for the individual configurations (❷). This BOM explosion for a specific sales order, that is, for a specific configuration, can be implemented in two ways:

▶ **Dynamic BOM**

In general, you can use a dynamic BOM as the exploded BOM (see Figure 2.10). This is useful if the variant model is complete. That means the entire variability of the configurable product is integrated with the variant model as a possible configuration and the BOM explosion is ensured for each of these configurations via object dependencies. In this case, you only require the super BOM structure including object dependencies as the BOM master data. No separate order BOMs are required. If a usage, for example, the planned order, production order, or sales order, requires BOM information, this information is dynamically determined from the super BOM by means of the object dependencies and configuration and is made available only temporarily. The system can copy the corresponding component lists from here but doesn't generate explicit order BOMs.

▶ **Order BOM**

If it is not sufficient to explode the BOM dynamically, you can use the settings in the configuration profile to allow for the creation of order BOMs. Order BOMs are material BOMs that are specific for an item of an SD document, in most cases a sales order. This is necessary if the variability for the configurable product is supposed to be higher in sales and distribution than in the variant model — particularly in the super BOM. This procedure enables you to implement order-specific product development using the order BOM.

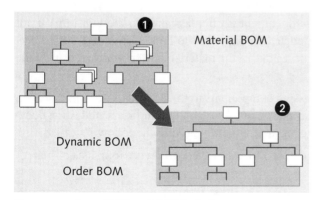

Figure 2.10 Super BOM and BOM Explosion

Each item in a BOM requires an *item category*. In the maintenance of these super BOMs, you can use all item categories for material BOMs. Some of them are listed as follows:

- L: Stock item
- N: Non-stock item
- R: Variable-size item
- D: Document item
- Class item

You usually use the L (stock item) item category for the direct assignment of material components.

You can also use the N (non-stock item) and R (variable-size item) item categories for the direct assignment of material components. As the name implies, non-stock items are not procured from stock; they are directly procured for the order. A specific aspect of non-stock items is that material components with the KMAT material type can only be included in BOMs with this item category and that no direct procurement process — as actually provided for this item category — is started.

Variable-size items are also of particular interest for Variant Configuration. You can, for example, change the sizes via object dependencies with the reference characteristics. Note that you generally have to determine the variable-size item quantity (ROMEN field) via object dependencies as well when you change the sizes (ROMS1–ROMS3 fields). You cannot determine it automatically or use the variable-size item formula.

The D (document item) item category is used if documents are supposed to be directly included in BOMs. The document is often assigned to the BOM in Variant Configuration if object dependencies are supposed to be used to control the selection. This is not possible for object links to the material master, which is usually the common method.

Class items serve to indirectly assign components to the BOM. For this purpose, you can use class nodes for material masters (class type 200) and class nodes for documents (class type 201). Class nodes are frequently used in the following cases:

- If only one component is supposed to be selected from very long lists of materials or documents
- If the same lists of materials or documents are often used in various BOMs, which can then be replaced by class nodes
- If the super BOM is supposed to be clear and to have a structure similar to the exploded BOM

These cases can occur individually or together. Section 2.7, Object Dependencies for BOM and Routing, describes the explosion control of the class nodes and super BOMs in detail.

2.3.3 Super Task List for the Configurable Material

For in-house production scenarios, the configurable material requires information for the production process. This information is also included in scheduling, capacity planning, and costing. For this purpose, you require a super task list for the configurable material. Production can take place at several levels. In this case, several of such routings are necessary. If the header material of the routing is configurable, you can explode the routing using the configuration via the object dependencies.

Task List Types and Variant Configuration [+]

There isn't a specific task list type for these configurable routings. You can use all task list types for Variant Configuration. The only exceptions are inspection plans (task list type Q), which don't allow for an assignment of object dependencies. Usually, standard routings are used.

With regard to content, configurable routings are super task lists. Accordingly, they contain all possible operations and operating facilities that may be required for any configuration. In contrast to BOMs, you can only assign the operations and operating facilities directly. An indirect assignment (as for class items) is not possible here.

Similar to BOMs in Figure 2.10, Figure 2.11 shows the configurable routing, that is, the super task list (❶). You require a routing explosion for the individual configurations (❷). This routing explosion for a specific sales order, that is, to a specific configuration, can be implemented in two ways, just like the BOM:

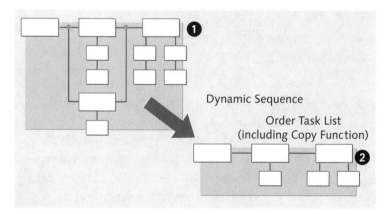

Figure 2.11 Super Task List and Routing Explosion

▶ **Dynamic sequence**
You can use a dynamic sequence, that is, a dynamic routing (see Figure 2.11). If a usage, for example, the planned order, production order, or sales order (for costing, for example), requires routing information, this information is dynamically determined from the super task list by means of the object dependencies and configuration and is made available only temporarily.

▶ **Order routings**
If a dynamic sequence isn't sufficient, you can create order routings. To create such order-specific routings, you can use a simulatively exploded routing as the template. This doesn't require any specific settings, such as in the configuration profile.

The structure of a standard routing consists of a header, sequences, operations, suboperations, and assignments to operations and suboperations (see Figure 2.12).

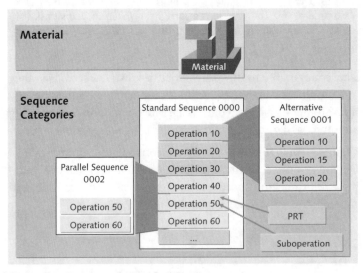

Figure 2.12 Structure of a Standard Routing

▶ **Header**
The configurable material is assigned to the header, the highest-level structure element of the routing. You cannot assign object dependencies to the header. Consequently, you cannot control the task list selection via object dependencies.

▶ **Sequences**
Below the header, you can define several sequences. There is always exactly one standard sequence, which you cannot link to object dependencies. If you

want to process operations in parallel, you must collect them in so-called parallel sequences. There can be several of these sequences. Similarly, you can create one or several alternative sequences for operations that are supposed to be processed alternatively to operations of the standard sequence. You can provide object dependencies for these two sequence types to control their selection and assignment in the routing explosion.

▶ **Operations**

Operations are listed within sequences. Regarding the content, these are super sequences. You can control the selection of the individual operations via object dependencies. In Variant Configuration, suboperations are managed the same way as operations. This holds particularly true for object dependencies.

▶ **Assignment of production resources and tools**

You can assign tools or other operating facilities to operations as production resources and tools. Regarding the content, these are once again super lists from which you can select the required production resources and tools via object dependencies in accordance with the corresponding configuration. You can assign additional objects, such as work centers, components, inspection characteristics, and trigger points, to operations. However, you cannot control these assignments via object dependencies.

In addition to making a selection, you can also implement changes via the reference characteristics at all three points in the routing where you can assign object dependencies. For further details, refer to Section 2.7, Object Dependencies for BOM and Routing.

2.4   Configuration Profile and Configuration Scenarios

You cannot implement a configuration without a configuration profile. Therefore, you require a configuration profile for the configurable material before you can start with the configuration. In this case, it doesn't matter if the configuration is already implemented in the SD document or is initially supposed to be implemented in a simulation transaction. This section introduces the setting options for the configuration profile and the configuration scenarios. It provides an overview of the configuration scenario that should be used in the different enterprise scenarios.

2.4.1   Overview of the Configuration Profile

If you try to call the value assignment interface for a configurable material without a configuration profile to implement a configuration, the call will fail. In the simulation, the system outputs the corresponding error message. In the SD document, analogous information is displayed.

The configuration profile contains essential control settings for the configuration process and object dependencies for the sales configuration (high-level configuration). You don't necessarily require a configuration profile to explode BOMs or routings for a configurable material.

You create configuration profiles for configurable material masters specifically for each material. Give the configuration profile a name and assign a variant class type to it. In most cases, only one configuration profile is created for a configurable material. In this case, the name of the configuration profile doesn't matter. You can change the name of the configuration profile retroactively.

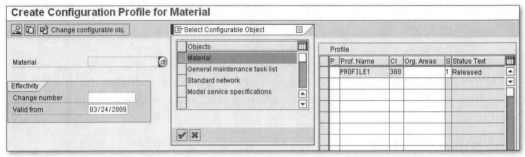

Figure 2.13 Creating a Configuration Profile

Exceptions are the rare cases where you assign several active configuration profiles to a configurable material. These cases in which use several active configuration profiles are listed below:

- ▶ **At the header level of the configurable product**
  Here, you use several active configuration profiles in exceptional cases. In these exceptional cases, the sales and distribution department is supposed to decide which configuration scenario is to be used for a specific order. The sequence can be sorted here. This configuration scenario is discussed more later in this section.

- ▶ **At the level of configurable assemblies**
  If a configurable assembly is used in various configurable products, it may be possible and necessary to have different configuration profiles of this assembly for the different configurable products. You cannot use any configuration profile of an assembly for all configuration profiles of a header material. The system automatically selects the (hopefully only) possible profile.

The assignment of a *variant class type* is not very interesting but is necessary. It is not interesting, because the standard version provides only one variant class type for material masters — class type 300. By assigning this variant class type to the configuration profile, you check whether variant classes of this class type have already been assigned to the configurable material master. If this hasn't yet been done, you can implement this retroactively when creating the configuration profile.

The system automatically locks configuration profiles of materials that don't have a variant class assignment. After assigning a variant class, you must release this lock manually, because the system does not automatically release it.

[!]

It doesn't matter whether you assign the variant classes to the material master via the transactions for classification, material master maintenance, or configuration profile maintenance. You cannot assign variant classes specifically for configuration profiles.

2.4.2 Configuration Profile in Detail

The detailed view of the configuration profile contains the following options:

- Assignment of object dependencies
- Setting of the configuration scenario
- Settings for the BOM explosion (levels of explosion, BOM application, filter, and so on)
- Settings for languages, values, configurators, pricing, and variant determination
- Allowed screens
- Scenario-specific settings

These options are provided in the Basic data and Configuration initial screen tabs (see Figures 2.14 and subsequent figures). The Configuration initial screen tab is subdivided into several tabs.

Figure 2.14 Basic Data in the Detail Screen of the Configuration Profile

Basic Data Tab

In addition to the options you already have to configure in the initial screen, the Basic data tab enables you to assign object dependencies and organizational areas and provides the activation indicator for the start logo (see Figure 2.14).

The Organizational Area field is a tool from the classification system, which enables you to filter characteristics within classes. You can therefore define in Variant Configuration that the value assignment interface is not supposed to display all characteristics of the variant classes. In this case, you activate organizational areas in the variant class and then assign the characteristics to these activated organizational areas. In addition, you specify one or several organizational areas in the configuration profile, which enables you to select the characteristics that the value assignment interface is supposed to display for this configuration profile. An authorization object is also available for organizational areas.

If you select the Start Logo checkbox, the system will check for every configuration start whether a document is directly (not via the document link) assigned to the variant class and display the result as the start information.

Configuration Initial Screen Tab

The Configuration initial screen tab contains the Configuration Parameters and User Interface tabs. For the configuration scenarios with Order BOM or Sales Order (SET), a tab with scenario-specific settings is added (see Figure 2.15).

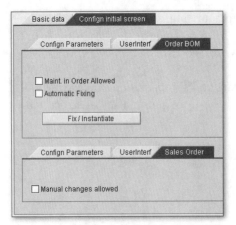

Figure 2.15 Tabs for Scenario-Specific Settings

Configuration Parameters Tab

The configuration scenario is mainly defined by the settings in the Process and BOM Explosion sections (see Figure 2.16).

These options are described in detail in Section 2.4.3, Overview of Configuration Scenarios. At this point, only the other options are briefly discussed here. If you use a (single-level or multilevel) BOM explosion, the system will display the Application and Filter fields. The BOM application that is defined here is used for the BOM explosion in the configuration simulation and in all applications of the supply chain for which no specific BOM application is found. The filter is active in the high-level configuration and enables you to filter the BOM for the following:

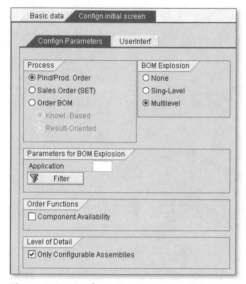

Figure 2.16 Configuration Parameters Tab

- Object category (material, class, document, and text)
- Item category
- Item status
- Sort key

You can also activate the check of the Component Availability. However, selecting this checkbox doesn't have an effect on the availability check in SD documents such as sales orders. Here, only the Customizing in the SD environment is relevant. This applies to the availability check at both the header level and component level.

This checkbox in the Configuration parameters tab in the configuration profile only enables you to run such an availability check at the component level in the configuration simulation. This check is also only simulative, which means the result doesn't affect the requirement and stock situation.

If you select the multilevel BOM explosion, the system will display an additional checkbox, which enables you to restrict the BOM explosion to Only Configurable Assemblies. However, this method is not advisable owing to performance reasons. Configurable assemblies are configurable material masters with BOMs. Only BOMs of nonconfigurable components wouldn't be exploded. This information, however, is not essential, because it is independent of Variant Configuration.

User Interface Tab

The User Interface tab always provides the Interface Design and the Settings button (see Figure 2.17).

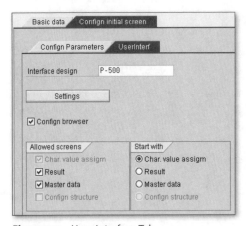

Figure 2.17 User Interface Tab

The system only displays further options in this tab if you want to use a BOM explosion. In the Interface Design field, you define a name under which you can maintain an interface design. If the field is blank, you cannot create an interface design for this configuration profile. Although you can use one interface design for multiple configuration profiles, in real life this is hardly possible, because it requires that the two configuration profiles use exactly the same variant classes. This usually only works if you create multiple configuration profiles for a material master. In this case, you maintain the interface design in the value assignment interface, for example, of the configuration simulation, for which the respective authorization is necessary.

The Settings option provides settings for the following elements:

- Language
- Presentation of the allowed characteristic values
- Scope of the characteristics
- Pricing
- Default values
- Configurator

These settings are specific for configuration profiles. If you are in the value assignment screen during a configuration (for example, in the sales order or the configuration simulation), you can find and change these settings from the configuration profile by going to VIEW • SETTINGS. If you don't save the changes during the configuration, they will be referred to as current settings, which are deleted when the corresponding transaction is completed. If you save the changes, these settings will be user-specific, used across configuration profiles, and have the highest priority for the respective user. Accordingly, in extreme cases, you can have three competing settings for the options mentioned above, which can be evaluated and compared at any time:

- Current settings
- User settings
- Profile settings

If you use a BOM explosion, the User Interface tab additionally displays the Allowed screens and Start with options. In addition to selecting the Characteristic Value Assignment checkbox (see Figure 2.17), which is always active and should be used as the initial screen, you can also allow for the navigation to the Result checkbox (for BOM and routing) and Master data checkbox (for super BOM and super task list).

In addition, you can activate a screen using the Configuration Structure (see Figure 2.18) or activate a configuration browser (see Figure 2.19). You require one of these settings if you want to implement a multilevel configuration.

Additional setting options are available for configuration scenarios with order BOM or sales order (SET). This is discussed further when we introduce the scenarios in this chapter (see Sections 2.4.5, Order BOM Scenario, and 2.4.6, Sales Order (SET) Scenario).

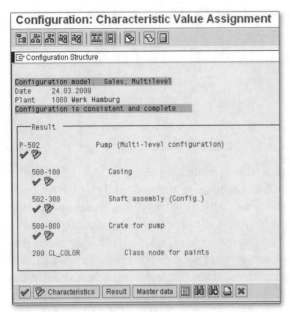

Figure 2.18 Configuration Structure

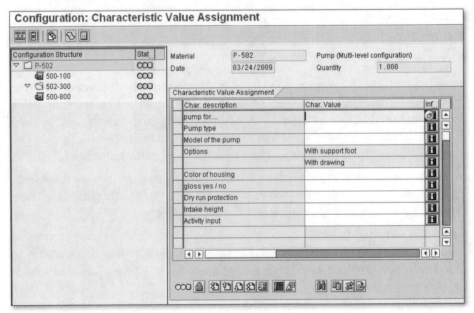

Figure 2.19 Configuration Browser

2.4.3 Overview of Configuration Scenarios

The most important aspect of the configuration profile is the definition of the configuration scenario. The configuration scenario answers the following three questions:

1. Does the SD document require a BOM explosion?

2. How many levels are supposed to be used for the configuration?

3. Does the BOM explosion need to be adapted manually? If so, where is the manual adaptation supposed to be implemented?

These questions are discussed further when the configuration scenarios are described. You can basically distinguish between the following four configuration scenarios:

- Planned/production order without BOM explosion
- Order BOM
- Sales order (SET)
- Planned/production order with BOM explosion

The following sections introduce you to these scenarios in more detail.

2.4.4 Planned/Production Order Without BOM Explosion Scenario

This scenario assumes that the variant model is complete. All possible value assignment combinations are fixed and mapped via the value assignment interface. Additionally, the BOM and routing are also completely modeled, that is, the variant model contains the corresponding BOM and routing explosion for each allowed value assignment combination. Consequently, you don't have to manually adapt the BOM and routing explosion. For this scenario, this is not allowed for the BOM anyway.

This scenario is characterized by its high performance and the fact that it is easy to handle for sales and distribution. A single-level configuration is usually implemented for this scenario. You only require a configuration profile for the header material, and the system queries all configuration information via the respective variant class. Because the configuration has only one level, no BOM explosion is required in sales and distribution from the Variant Configuration perspective.

Planned/Production Order Without BOM Explosion Scenario	[Ex]

Buying a car is a good example of such a configuration scenario. Here, you select the respective model — the configurable header material. Based on this header material, you then select all extras via the configuration's value assignment. These are only properties of the header material.

For pricing, solely this header material and this value assignment are used, too. No BOM information is required, and the model is complete. When selecting the extras, you have to adhere to the catalog; additional extras are not allowed.

The super BOM and super task list are also completely modeled so that the system can automatically explode the BOM and routing for any allowed value assignment combination. This applies particularly to planned and production orders. (This may be the reason for the name of this scenario.)

The answers to the three questions that characterize a scenario are as follows:

1. There is no BOM explosion in sales and distribution.

2. The configuration has only one level.

3. You aren't allowed to manually adapt the BOM explosion.

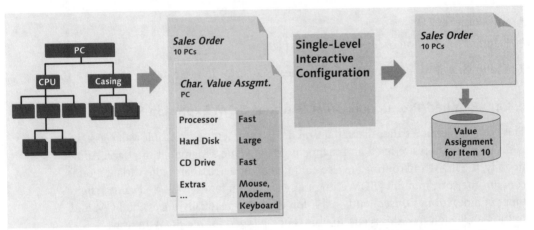

Figure 2.20 Overview of the Planned/Production Order Without BOM Explosion Scenario

The following box summarizes Figure 2.20.

[+]

Overview of the Planned/Production Order Without BOM Explosion Scenario

Enterprise scenario:
- Complete model
- Single-level and easy configuration with high performance in sales and distribution

Settings in the configuration profile:
- Process: planned/production order
- BOM explosion: none
- (No additional scenario-specific settings)

Configuration process:

▶ Sales order is opened for the header material.

▶ Values can only be assigned to the header material.

▶ No BOM/routing explosion possible.

▶ The main item in the sales order is the only result.

▶ The value assignment is stored with reference to the main item.

2.4.5 Order BOM Scenario

This scenario is mainly characterized by the fact that the variant model is not complete. Consequently, you cannot predefine all allowed value assignment options, and they are not modeled in the value assignment interface. The super BOM and super task list are therefore also incomplete. There can be various reasons for this, for example:

▶ **Full flexibility with regard to customer requirements**
You want to map all possible customer requirements. Even the most exotic requirement is supposed to be allowed at first. For this purpose, it must be possible to specify the customer requirement. Furthermore, you must be able to adapt the exploded BOM and routing specifically for each order — up to engineering to order.

▶ **Reduction of the maintenance work**
The variant model is not supposed to map the full variability of the product, because this would increase the maintenance work considerably and go beyond the scope of the corresponding development project. For example, the variant model covers the product only up to 95% or 98%, and the remaining percentage is covered via manual adaptations.

▶ **Automation of routine work**
In areas such as special machine production or plant construction, you want to automate routine work regarding BOM and routing maintenance with Variant Configuration. These aspects are included in the variant model. This enables you to lay the foundation for the order-specific development in the form of manual adaptation via the configuration and via a respectively automated explosion of BOM and routing.

This scenario focuses on the maintenance option of order-specific BOMs. For the other two questions that characterize the scenario (BOM explosion and configuration levels), two variants are possible.

▶ **Standard variant without BOM explosion**
In the standard variant, the procedure is similar to the planned/production order without BOM explosion scenario. The BOM is not exploded in sales and distribution — because the incomplete modeling makes a correct explosion impossible. Without BOM explosion, only the single-level configuration is possible.

▶ **BOM explosion in the SD document**
You activate the second variant using the Maintenance in Order Allowed checkbox in the Order BOM tab in the configuration profile. In this variant, you can explode the BOM in the SD document. However, this option is somewhat hidden. You call this BOM explosion via the Engineering button in the value assignment screen. The exploded BOM cannot be modified here, but you can find additional configurable materials in the product structure and implement a multilevel configuration in the SD document if necessary.

Figure 2.21 shows the two steps of the order BOM scenario.

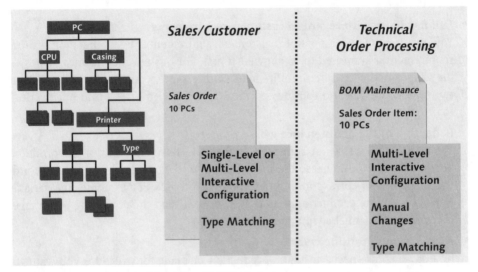

Figure 2.21 The Two Steps of the Order BOM Scenario

As you can see in Figure 2.21, the order BOM scenario consists of two steps.

Step 1: Configuration in Sales and Distribution

The first step is the configuration in sales and distribution. Depending on whether the Maintenance in Order Allowed checkbox is selected in the configuration profile, this step is analogous to the following scenarios:

▶ Planned/production order without BOM explosion scenario (if the indicator is not set)

▶ Planned/production order with BOM explosion scenario (if the indicator is set)

Step 2: Technical Postprocessing

In addition to the first step, this scenario also allows for technical postprocessing, which is started in the BOM menu for order BOMs. Here, you can generally implement a multilevel configuration. The only exception is the header material, because the configuration of the header material is restricted to the SD document. However, a multilevel configuration is only possible in the following cases:

▶ If the BOM contains configurable assemblies

▶ If these assemblies exist in the exploded BOM (that is, no decelerated BOM explosion or object dependencies cause them to not be displayed)

▶ If a released configuration profile, also with the order BOM scenario and the same form of storage for the order BOM, exists for these configurable assemblies

In addition to implementing the multilevel configuration, the technical postprocessing also enables you to manually change the exploded BOM. You can make any modification when changing the BOM manually. This includes the following modifications:

▶ Deleting items

▶ Modifying items in the item detail screen

▶ Adding items (also as copies)

▶ Manually replacing class nodes (see also Section 2.7, Object Dependencies for BOM and Routing)

If an added item is configurable and if the corresponding configuration profile is available, you can also configure this item. In general, you can create a single-level order BOM. In some cases, you can also use multilevel order BOMs. If the following requirements are met, you can create multilevel order BOMs:

▶ If the configuration profile indicates that a multilevel BOM explosion is used

▶ If the BOM explosion displays the BOM

▶ If a material BOM is available for this material

▶ If the configuration profile indicates that a result-oriented order BOM is used

▶ If the configuration profile indicates that a knowledge-based order BOM is used

> The usage of a knowledge-based order BOM requires that an active configuration profile with exactly this configuration scenario is available for the configurable assembly for which an order BOM is supposed to be created.   **[+]**

Similar to the order-specific material BOMs, you can also use order-specific routings. For this purpose, the routings exploded in the simulation according to the configuration and are then used as templates for such order routings. This method is also used for the creation of routings for so-called material variants and is described in Section 2.10, Material Variants.

For the order BOM scenario, numerous additional setting options are available in the configuration profile. You can configure the settings in the Configuration Parameters tab and in the Order BOM tab, which is only provided for this scenario.

When selecting the order BOM, the configuration parameters enable you to select either knowledge-based or a result-oriented storage. Here, you select the type of storage for the order BOMs. The storage in a model must be uniform; that is, all BOMs must either be knowledge-based or result-oriented.

- **Knowledge-based**
  To create an order BOM, the super BOM is copied and all changes are made and stored in the super BOM.

- **Result-oriented**
  In contrast to the knowledge-based BOM, the template is the already exploded BOM in this case. The changes are implemented here.

The result-oriented order BOM has a higher performance level, is considerably leaner, and has practically no disadvantages compared to the knowledge-based order BOM. Theoretically, you could skip updates when changing the configuration for knowledge-based order BOMs. Because the order BOM is a copy of the super BOM, that is, it also contains the object dependencies, changes that are made to the configuration directly affect the next BOM explosion. This does not apply to result-oriented order BOMs. However, because you usually call the technical postprocessing of the order BOM again when you make changes to the configuration, this is only a theoretical disadvantage.

If you call a result-oriented order BOM in the technical postprocessing after the changes have been made to the configuration, the BOM will be exploded again. Manually created, changed, or deleted items are not included in this new explosion.

For the order BOM scenario, you can select either a single-level or a multilevel BOM explosion. You cannot define no BOM explosion or a two-level BOM explosion. This BOM explosion setting only affects the SD document if you select the Maintenance in Order Allowed checkbox. In this case, the BOM is exploded accordingly. In general, however, you cannot create order BOMs in the SD document; consequently, you also cannot manually change the BOM. You can only create order BOMs in the transaction for technical postprocessing (for example, Transaction CU51). The following options are available here:

▶ If you use a single-level BOM explosion, you can create one order BOM only.

▶ If you use a multilevel BOM explosion and result-oriented storage, you can optionally create one order BOM for each BOM in the exploded BOM structure.

▶ If you use a multilevel BOM explosion and knowledge-based storage, you can optionally create one order BOM for each BOM in the exploded BOM structure, provided that the corresponding BOM header material is configurable and has a configuration profile with a knowledge-based order BOM scenario.

For assemblies with configuration profiles with the order BOM scenario, it is not relevant whether the single-level or multilevel BOM explosion is defined. However, there is one exception to that.

Exception	[!]
There is one critical exception. Scenarios with a single-level BOM explosion cannot be used for the IPC.

You can find additional settings for the order BOM scenario in the Order BOM tab. This includes the following options:

▶ **Maintenance in Order Allowed**
As already described, the Maintenance in Order Allowed checkbox enables you to implement a BOM explosion in the SD document with a multilevel configuration — if it exists.

▶ **Automatic Fixing**
The Automatic Fixing checkbox generally creates order BOMs in the technical postprocessing. However, this function only has an effect if no manual changes are made to the exploded BOM in the technical postprocessing. Accordingly, the system generally doesn't create order BOMs in the SD document, irrespective of this checkbox. The system also creates an order BOM if manual changes are implemented in the technical postprocessing, again irrespective of this checkbox.

▶ **Fix button**
In addition to the automatic fixing function, the technical postprocessing generally also provides the option of manual fixing. That means you manually create order BOMs — with regard to BOMs — that have not been changed manually. If you want to have the system create not only a single order BOM when you use this manual fixing option but a complete multilevel structure, you can define this using the Fix button in the Order BOM tab.

► **Instantiate button**
For result-oriented storage of the order BOM, the Fix button can also be used for instantiation; therefore, the button is called Fix/Instantiate (see Figure 2.15). You only require this function if the BOM contains the same material as an assembly several times. Without instantiation, these items with the same material number cannot be differentiated. You can create a uniform order BOM only for these assemblies. If you don't want this, you can generate several different instances using this button.

The answers to the three questions that characterize a scenario are somewhat complicated and are as follows:

1. There is no BOM explosion in sales and distribution in the standard version. The nonstandard version (Maintenance in Order Allowed) enables you to implement a BOM explosion.

2. The standard version uses a single-level configuration in sales and distribution (in the nonstandard version, the configuration can also have multiple levels in sales and distribution). The technical postprocessing generally enables you to implement multilevel configurations, provided that the appropriate configuration profiles are available.

3. You can manually adapt the BOM explosion, but only in the transactions for the technical postprocessing and not in sales and distribution.

[+] **Overview of the Order BOM Scenario**

Enterprise scenario:
► Incomplete model
► Single-level configuration in sales and distribution in the standard version
► Multilevel configuration in the nonstandard version allowed (Maintenance in Order Allowed)

Settings in the configuration profile:
► Process: Order BOM
► Knowledge-based or result-oriented storage
► BOM explosion: Single-level or multilevel
► Additional scenario-specific settings:
  — Indicator: Maintenance in Order Allowed
  — Indicator: Automatic Fixing
  — Manual Fixing/Instantiating

Configuration process:

▸ Sales order is opened for the header material.

▸ In the standard version, the values can only be assigned to the header material.

No BOM/routing explosion possible:

▸ In the nonstandard version, multilevel configuration with BOM/routing explosion via the engineering button.

▸ Technical postprocessing possible but not mandatory.

▸ In this context, you can create and change order BOMs (and order routings) as required.

2.4.6 Sales Order (SET) Scenario

You use the sales order (SET) customer scenario if you want to sell several configurable products. If you include the configurable products in an SD document as individual main items, you can configure these configurable products. These configurations, however, are completely independent of each other, and you cannot define dependencies between the configurations of the individual main items. The trick is now to define a so-called package material, set material, or phantom material. (See Figure 2.22)

You use this material to create a short BOM for sales and distribution, which collects the configurable products between whose configurations you want to map dependencies.

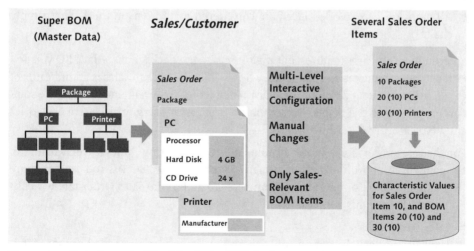

Figure 2.22 Basic Concept of the Sales Order (SET) Configuration Structure

This package material is also a configurable material. You can, for example, create a configuration profile with the sales order (SET) scenario for this material. You can also assign object dependencies to the BOM for this package material, and you can use class nodes. For the configuration profile, you need to assign a variant class to the package material. This variant class usually contains characteristics that define the scope of the "package" and possible central properties. According to the value assignment with regard to this variant class, you can then explode the BOM of the package material. Furthermore, you assign all object dependencies to the configuration profile mentioned here. These object dependencies represent the dependencies between the individual configurable products.

A critical aspect of this scenario is that the header material is only a phantom, as already mentioned. This main item must be configurable in the SD document but isn't allowed to trigger requirements. In the standard version, you can do this by assigning item category group 0004 instead of the common item category group, 0002, to the material master in the SD views. This way, an item category that exactly meets these requirements is found for the sales order. For more information on this, refer to Chapter 4, Customizing SAP ERP for Variant Configuration.

If the header material doesn't trigger requirements in the main item, this material is no longer relevant in the further supply chain, for example, for MRP, purchasing, production, and so on. You must therefore ensure that the following requirements are met for all items of the exploded BOM:

▶ All items of the exploded BOM are included in the sales order as subitems.

▶ All items of the exploded BOM are considered main items in the further supply chain.

You can meet the first requirement if all items of the package material's BOM is relevant to sales, which is also mandatory here. The relevance to sales is an indicator of the status in the item detail of the BOM. The second requirement is met in the standard version with the respective settings in the Customizing for the item categories and item category determination.

For the sales order (SET) scenario, some additional setting options are available in the configuration profile, but not as many as for the order BOMs. You can configure the settings in the Configuration Parameters tab and in the Sales Order tab, which is only provided for this scenario.

▶ **Sales Order tab**
The Sales Order tab contains only the Manual changes allowed checkbox. Similar to the maintenance of order BOMs, this checkbox enables you to implement manual changes to the exploded BOM in the sales order. Here, basically the same

changes as in the technical postprocessing are allowed. The following lists the differences from the technical postprocessing and thus the specific aspects:

▶ The manual changes are already implemented in the sales order or in similar SD documents and nowhere else.

▶ No order BOM is created. This is impossible at this level anyway.

▶ All manual changes become relevant for the further supply chain, because they are included in the sales order as subitems that trigger requirements.

▶ **Configuration Parameters tab**
In the Configuration Parameters tab, you select the Sales Order (SET) checkbox. You can also define a single-level or multilevel BOM explosion here. To not use a BOM explosion here doesn't make sense and therefore this option cannot be selected.

For the settings in the Configuration Parameters tab, note the following:

A Single-Level BOM Explosion is not IPC-Compatible	**[!]**
The single-level BOM explosion is also not IPC-compatible here. The system would ignore such a configuration profile. This applies to both the header level and the assembly level.	

What does a configuration process look like in this configuration scenario?

1. The sales order for the configurable package material is opened, and the configuration of this header material is implemented.

2. The BOM is exploded in accordance with the configuration. The system only displays items that are relevant to sales, that is, items that are included as subitems in the sales order. You can manually change this exploded BOM — provided that this has been allowed in the configuration profile.

3. The system then determines all configurable items including those that were added manually. A released configuration profile must have been created for these configurable items. All configuration scenarios are possible here, except for the planned/production order with BOM explosion scenario. A complete value assignment must be implemented for this second configuration level. This can partly be done via object dependencies. It is critical that each of these configurable products has its own complete configuration within the package material. Only this configuration is available for further MRP and production processes.

No requirements are triggered at the package material level. Consequently, the configuration is not available at this level outside the sales order and can therefore not be evaluated.

[+]
If the BOM explosion is multilevel and if additional configurable components are found at lower levels, you can also implement a two-level configuration.

Besides the main item, the sales order has also subitems. Only subitems trigger requirements. Pricing can be implemented either at the header level or at the subitem level but is usually implemented at the subitem level (that is, at the requirement level here). The configuration is also always stored with reference to the corresponding main item or subitem of the sales order.

The component level of the package BOM is completely flexible, which means the products can be configurable or nonconfigurable. Configurable products include:

▶ **Completely modeled products**
These are products with the planned/production order without BOM explosion configuration scenario.

▶ **Incompletely modeled products**
This refers to products with the order BOM configuration scenario including all setting options.

▶ **Package materials**
Products with the sales order (SET) configuration scenario. You can also use "package in package" models.

The answers to the three questions that characterize a scenario are as follows:

1. The BOM is exploded in sales and distribution.

2. The configuration is generally multilevel.

3. You can allow for a manual adaptation of the BOM explosion via the settings in the configuration profile. This adaptation is already implemented in sales and distribution and doesn't lead to an order BOM.

[+] **Overview of the Sales Order (SET) Scenario**

Enterprise scenario:

▶ Packages of several configurable products whose configurations are supposed to be adapted to each other.

▶ Generally multilevel configuration in sales and distribution.

▶ The models of the configurable products can be complete or incomplete, which leads to an almost full flexibility regarding the configuration scenarios.

▶ You can change the scope of the package in the sales order.

Settings in the configuration profile:

▶ Process: Sales order (SET)

▶ BOM explosion: Single-level or multilevel

▶ Additional scenario-specific settings: Manual changes allowed

▶ Additional settings:

  ▶ In the material master: *Item category group* 0004

  ▶ In the BOM: All items are relevant to sales

Configuration process:

▶ Sales order is opened for the package material.

▶ Configuration of the package material.

▶ BOM explosion of the package material.

▶ Possible manual changes to this BOM.

▶ Configuration of this BOM's configurable products.

▶ Additional configuration levels if the BOM explosion is multilevel.

▶ One main item (which doesn't trigger requirements) and the corresponding subitems in the sales order.

▶ Generally no order BOM allowed for main items.

▶ Order BOMs possible for subitems if allowed in the configuration profile.

▶ All configuration scenarios except for planned/production order with BOM explosion allowed for subitems.

2.4.7 Planned/Production Order with BOM Explosion Scenario

Similar to the scenario without BOM explosion, this scenario assumes that the variant model is complete. All possible value assignment combinations are fixed and mapped via the value assignment interface. Moreover, the BOM and routing are also completely modeled; that is, the variant model contains the corresponding BOM and routing explosion for each allowed value assignment combination. Consequently, you don't have to manually adapt the BOM and routing explosion. For this scenario, this is not allowed for the BOM anyway. (See Figure 2.23)

In contrast to the scenario without BOM explosion, sales and distribution uses the BOM explosion here and the multilevel configuration in most cases. It is thus more similar to the sales order (SET) scenario.

The following lists the differences from the sales order (SET) scenario:

▶ **Multilevel configuration**

You do not necessarily have to use a multilevel configuration. If the system already queries all required information for the BOM and routing explosion of a configurable assembly at the header level, this doesn't have to be done again.

If you want to implement a multilevel configuration, the configurable assemblies must also use the planned/production order with BOM explosion scenario. No flexibility is given here.

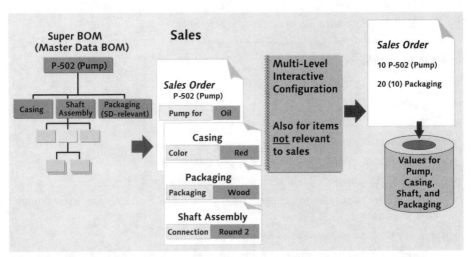

Figure 2.23 Overview of the Planned/Production Order with BOM Explosion Scenario

▶ **Triggering of requirements**
The requirement must be triggered at the level of the header material (i.e., main item in the sales order).

▶ **Relevance to sales of BOM items**
The items in the BOM don't have to be relevant to sales. Therefore, subitems are not required but can be used. In this scenario, subitems are not relevant for Variant Configuration, particularly not for the storage of the configuration.

Enterprise scenarios for the planned/production order with BOM explosion are sales processes for which BOM explosions are requested despite complete modeling.

[Ex] **Complete Modeling and BOM Explosion**

In particular, this applies if the configurable assemblies in the BOM of the header material are supposed to be sold separately, that is, if they are individual configurable products. In this case, you can create specific variant classes and configuration profiles for these configurable assemblies and write the object dependencies for the value assignment interface, the BOM explosion, and the routing explosion and assign them to the model.

You don't want to have to do this work twice. Therefore, you should try to implement a multilevel configuration and use these already created configurations of the assemblies also for the configuration of the header material.

The only available settings in the configuration profile here are the Planned/Production Order checkbox and the option to select either a single-level or a multilevel BOM explosion. For a multilevel configuration, the configurable assemblies also require these settings. At the assembly level, it doesn't matter if you use a single-level or multilevel BOM explosion. Only the setting for the levels of the BOM explosion at the header level is critical here. Its significance is explained in the following:

▶ **Single-level BOM explosion at the header level**
Only a single-level and thus, at the most, a two-level configuration is possible.

A single-level BOM explosion is not IPC-compatible; the system would ignore such a configuration profile. This applies to both header level and assembly level.

▶ **Multilevel BOM explosion at the header level**
A multilevel BOM structure and thus a configuration with more than two levels is possible.

The answers to the three questions that characterize a scenario are as follows:

1. The BOM is exploded in sales and distribution.

2. A multilevel configuration is possible.

3. You aren't allowed to manually adapt the BOM explosion. This applies both to the header BOM and to the BOMs of the configurable assemblies.

Overview of the Planned/Production Order with BOM Explosion Scenario   [+]

Enterprise scenario:

▶ Complete model

▶ Configurable assemblies with their own configuration profiles

▶ (Planned/production order with BOM explosion as well)

▶ Multilevel configuration possible in sales and distribution

Settings in the configuration profile:

▶ Process: Planned/production order

▶ BOM explosion: Single-level or multilevel

▶ (No additional scenario-specific settings)

Configuration process:

- Sales order is opened for the header material.
- Value assignment for the header material and BOM explosion.
- Multilevel configuration possible in accordance with the BOM explosion.
- Routing explosion possible in addition to BOM explosion.
- The result is a main item that triggers requirements in the sales order; the multilevel configuration is stored with reference to the main item.
- If required, subitems, which are not relevant for Variant Configuration.

These descriptions complete the overview of the configuration scenarios.

2.5 Overview of Object Dependencies

In this section, we'll discuss the object dependencies in detail. This context deals with the high-level configuration (sales configuration, in the dialog, for the value assignment interface) and with the low-level configuration (BOM and routing explosion, also without dialog).

First, this section provides you with a general overview and explains the basic principles and the general rules of object dependencies.

2.5.1 Types of Object Dependencies and Assignment

The purposes of object dependencies are the following two tasks:

- **Support the configuration process of the value assignment interface**
  Object dependencies support the configuration process in the value assignment interface, for example, in the sales order. Here, the configuration result must be complete and consistent. This can be achieved in different elegant ways. The aim is to not provide not allowed value assignment combinations, to determine or implement value assignments (where required) via object dependencies, and to provide default values.

- **Support BOM and routing explosions**
  Object dependencies generate BOM and routing explosions in accordance with the configuration in the sales order, for example. During this process, the elements that are not required are deleted from the super BOM and super task list. You can then change the elements that were not deleted via object dependencies.

Four dependency types are available: preconditions, selection conditions, procedures, and constraints. Every type has its specific area of use. However, you can also map different tasks of object dependencies in different ways, using different types of object dependencies. The following sections introduce the various types of object dependencies and describe how you can assign them.

Preconditions

You can use preconditions for the configuration process in the value assignment interface. Without preconditions, you can assign any value from the list of the allowed values to any characteristic — irrespective of the value assignment of other characteristics. You can map dependencies via preconditions by only allowing specific values or even complete characteristics under certain conditions. For this purpose, you must assign the corresponding preconditions to the characteristic or characteristic value.

Selection Conditions

Selection conditions explode BOMs and routings in accordance with a configuration. All super BOM and super task list elements for which no selection conditions are provided are generally included in the exploded BOM and routing. These elements are called *non-variable parts*, that is, generally required parts. In contrast, elements to which selection conditions are assigned are called *variant parts*. If none of the selection conditions are met for an element, this item will not be included. Elements in the BOM and routing to which you can assign selection conditions are the following:

▸ BOM items

▸ Sequence assignments in the routing

▸ Operations in the routing

▸ Production resources and tools in the routing

Selection conditions are also used for the dynamic conversion of characteristics into required characteristics. If a characteristic is only supposed to be a required characteristic under certain conditions, you must define the characteristic as an optional characteristic (do not select the Entry Required option in the characteristic definition). The system then assigns the corresponding selection condition that contains exactly this condition to the characteristic.

Procedures

Procedures can be used to assign values. You can also assign several procedures to an object. Procedures are processed exactly once after the start of the configuration; you can define a sequence for this. The assignment of values can be done successively, that is, a field can be overwritten several times, even if the previous value is reused.

In the BOM or routing explosion, you can make changes to the details of the elements that were included in the exploded BOM and routing. As described in the context of the tools from the classification system, you use reference characteristics with reference to the STPO, PLPO, PLFH, and PLFL tables for this purpose. (For operations, you can only change the fields that are also contained in the PLPO_CFMOD structure.)

You can also read the current assigned values in the procedures and use this knowledge, for example, to increase the component set in the BOM item by one or reduce the setup time in the routing operation by five.

[!]   **The Assignment of Values for Reference Characteristics**

For assignments of values in the BOM and routing via procedures, the system simply overwrites the individual fields of the exploded structures.

You must therefore ensure that no function that you know from the standard BOM and routing maintenance is active. All dependencies must be remodeled via object dependencies. That means, for example, that if you change a size in the item, you must use the procedure to ensure that the variable-size item quantity is adapted. Consequently, you shouldn't change fields with strong dependencies, such as the material number in the BOM item or the work center in the operation.

You must assign procedures to the element whose detail is supposed to be changed. The elements to which such procedures can be assigned are the same elements that were already listed for selection conditions, namely, BOM items, sequence assignments in the routing, operations in the routing, and production resources and tools in the routing.

You can also use procedures to assign values to characteristics in the value assignment interface. In this context, the following types of assignments can be distinguished:

▶ **Fixed assignment of values**
The set value cannot be deleted or overwritten by the user. Constraints cannot delete or overwrite this set value either. The only general exception is that procedures can overwrite each other. The following rule applies: The last one wins.

▶ **Dynamic assignment of values**

These values are also referred to as dynamic default values. If a default value is supposed to depend on the current configuration, you don't define it in the characteristic definition. For example, if the default value is supposed to depend on the already implemented value assignment of other characteristics, the value is set via procedures.

The dynamic assignment of values must be specifically indicated in the procedure. The default setting is the fixed assignment of values.

You assign procedures for the value assignment interface to the configuration profile. You can also assign such procedures to characteristics or characteristic values.

However, you should use the configuration profile here, because this ensures full control over the processing sequence via the respective settings. Moreover, procedures for characteristics or characteristic values are not supported by the IPC.

Constraints

You can use constraints to assign and check values. They therefore include functions of both procedures and conditions. You can use constraints to design the configuration in the value assignment interface; there are no restrictions with regard to the configuration scenarios.

In the multilevel configuration, constraints play a particular role, because only they can be used to map dependencies between all objects of a multilevel configuration.

In addition to this, constraints can evaluate equations, variant tables, and variant functions to a greater extent than any other type of object dependencies. This also applies to the elegant restriction of allowed values for the value assignment of characteristics. The performance and the fact that no settings have to be configured for the processing sequence are further advantages. Constraints are collected in so-called dependency or constraint nets. These nets can be assigned to the configuration profile only. Accordingly, they are not available for the BOM and routing explosion.

| Excursus: Actions as a Type of Object Dependencies | [+] |

In this context, we must also mention actions. Actions are an additional type of object dependency. However, you shouldn't use them for new models. Actions enable you to set values, but procedures and constraints provide many more functions and, with regard to performance, they are not as critical as actions. Therefore, actions are not supported by the IPC. You can assign actions to all objects to which you can assign procedures.

Tables 2.1 and 2.2 summarize the aspects just described. Table 2.1 lists key words for which the individual types of object dependencies can be used in the two areas of use.

Area of Use: Value Assignment Interface	Area of Use: BOM/Routing Explosion
Preconditions	
Disallow/allow	–
Selection conditions	
Dynamic required characteristics	Variant parts
Procedures	
Fixed or dynamic assignment of values	Changes via reference characteristics
Dynamic default value	Class node value assignment
Constraints	
Assignment of values	–
Checking values	
Single-level/multilevel configuration	
Additional functions, high performance	

Table 2.1 Overview: Areas of Use of Object Dependencies

Table 2.2 shows where you can assign object dependencies — depending on which of the two areas of use you want to use.

Assignment for: Value Assignment Interface	Assignment for: BOM/Routing Explosion
Preconditions	
Characteristic and value	–
Selection conditions	
Characteristic	BOM item
	Operation, sequence, production resources and tools of the routing
Procedures	
Configuration profile	BOM item
(characteristic and value)	Operation, sequence, production resources and tools of the routing
Constraints	
Configuration profile (in net)	–

Table 2.2 Overview: Assignment of Object Dependencies

2.5.2 The Procedural and Declarative Character of Object Dependencies

These two types of object dependencies support both a procedural and a declarative approach to the configurator. You can assign these two approaches to the types of object dependencies.

▶ **Declarative object dependencies**
Constraints (and actions) are declarative object dependencies. For these object dependencies, the result only depends on the initial condition. The point in time when the object dependencies are processed and the sequence in which the object dependencies are processed are not relevant. The syntax has an analogous structure.

▶ **Semi-declarative object dependencies**
Preconditions and selection conditions are semi-declarative object dependencies. The syntax is merely declarative. Here, only the condition is described — without any reference to the processing. The evaluation, however, is procedural, as described in Section 2.5.6 in the context of the processing sequence of object dependencies. Preconditions and selection conditions are processed at defined points in time in the configuration process.

▶ **Procedural object dependencies**
Procedures are merely procedural object dependencies. Like preconditions and selection conditions, they are also processed at a defined point in time in the configuration process. Additionally, you can define the exact processing sequence in the assignment of the procedures. Procedures also enable you to use the corresponding procedural language elements, such as $x = x + 1$ for successive calculations or assignments of default values as described in the explanations of the syntax.

Now that we've described the characters of the individual types of object dependencies, the following section introduces the two storage types: local and global object dependencies.

2.5.3 Global and Local Object Dependencies

You can distinguish between local and global object dependencies. Local object dependencies have purely numerical names, for example, procedure 4711. You can only create names by assigning numbers internally. Local object dependencies are created, changed, and deleted via the assignment. Local object dependencies can also only exist in the assignment. If the connection is deleted, these object dependencies will also be deleted.

Global object dependencies have non-numerical names, for example, procedure PROC_CHAR1. Here, you usually use an external number assignment. You must create a short text and manually set the status to Released. Specific transactions are available for the maintenance of global object dependencies. However, you can also create global object dependencies (except for constraint nets) via the assignment and change them for one-time usage. You can also use global object dependencies multiple times by assigning them multiple times.

[+]   Because of this multiple usability and the resulting increased performance and simplified evaluation, you should use global object dependencies in Variant Configuration.

2.5.4   Status of Object Dependencies

Object dependencies, both local and global, have a statuses. You cannot release object dependencies until the syntax is free of errors. Having no syntax is also an error. The system releases local object dependencies automatically when the syntax that is created before saving is free of errors.

You cannot save syntax that contains errors. If you do, the system automatically sets the status to Locked, for both local and global object dependencies.

[!]   **Releasing Locked Object Dependencies**

After a correction has been made, you must always manually set the status to Released. In contrast to local object dependencies, global object dependencies must be released manually. Note that the status of global object dependencies doesn't affect the assignment. You can assign global object dependencies with any of the following statuses:

▶ In Preparation

▶ Released

▶ Locked

You can also delete the assignment without considering the status. Similarly, you can also change the status for object dependencies that have already been assigned.

You can define the status of object dependencies in Customizing.

2.5.5   Object Dependencies in Classification and in Variant Configuration

You can use object dependencies in classification and in Variant Configuration. In the classification system, object dependencies are only active in the classification but not in the search or other functions. For the classification, you assign object dependencies to characteristics, characteristic values, or classes. Note the following:

▸ **Object dependencies for characteristics and values**
If characteristics with assigned object dependencies are used in the classification (classes that aren't variant classes) and in Variant Configuration (variant classes), the object dependencies are also active. This may be intended. For example, if the same characteristics are used for the classification in class nodes and in the configuration, it is often preferred that the same rules apply everywhere. Apart from that, you should avoid this by restricting the usage of characteristics in the characteristic definition to specific class types or by only assigning object dependencies to characteristics in the class-specific overwrite process. This includes prerequisites, selection conditions, and procedures.

▸ **Object dependencies for classes**
Object dependencies that are assigned to classes are only active in the classification. You can also assign object dependencies to variant classes; however, this doesn't have any effect in Variant Configuration, so you should avoid doing this.

Similar to the forms described for Variant Configuration, you can assign preconditions, selection conditions, and procedures to characteristics and characteristic values in the classification.

Procedures for Classes	[+]
Procedures are the only object dependencies that can be assigned to classes. In the classification, the procedures that are assigned to classes perform the tasks that procedures perform for the configuration profile in Variant Configuration.	

2.5.6 Execution Sequence of Object Dependencies

With the execution sequence of object dependencies, you have to distinguish between two tasks:

▸ Configuration in the value assignment interface

▸ BOM and routing explosion

The following sections explain these tasks in detail.

Configuration in the Value Assignment Interface

You use all types of object dependencies for the configuration in the value assignment interface. You can use settings, for example, those that you can define in the configuration profile, to control when object dependencies are processed for the value assignment interface. In the standard version, these object dependencies are processed every time you change to another screen or every time data is released (via the ⏎ key).

[+] As an alternative to this default setting, you can also have the system process object dependencies only when requested or when data is released (via the [↵] key).

All constraints of the whole model are active during the entire processing of object dependencies. Only then can you ensure the declarative character of the constraints. As a result, each change that is made to object dependencies with regard to constraints requires the system to process the corresponding constraint.

The following steps are carried out every time you start processing object dependencies:

1. **Resetting the values**
   The system resets all values that were set by procedures. As already mentioned, this may lead to changes to the conditions of the constraints so that the corresponding constraint needs to be processed.

2. **Processing the procedures**
   In this step, the system processes the procedures. During processing, the procedures are evaluated exactly once. The following sequence applies:

   ▶ Configuration profile

   ▶ Characteristics (that were used for the value assignment)

   ▶ Characteristic values
   According to SAP's recommendation, procedures are only assigned to the configuration profile. You can arrange these procedures for the configuration profile using the sorting function. This enables you to define the exact processing sequence. If procedures are also assigned to characteristics, the system considers only the characteristics to which values have already been assigned.

   You can define a sequence for each characteristic via the sorting function. You cannot specify the sequence in which the characteristics are evaluated. The same applies to characteristics values. That means if procedures are also assigned to characteristic values, the system considers only the characteristic values that were used for the value assignment.

   As already mentioned, constraints are active during the entire processing of procedures. The procedures may lead to changes to the conditions of the constraints so that the corresponding constraint needs to be processed.

3. **Processing the preconditions**
   The second-to-last step is the processing of the preconditions. Processing sequences are not relevant here. This processing has no effect on procedures and constraints.

4. **Processing the selection conditions**

The last step is the processing of the selection conditions. Processing sequences are not relevant here either, and it has no effect on procedures and constraints.

| More Complex Processing Sequence due to Actions | **[+]** |

This list deliberately omitted actions, because they make the processing sequence much more complex and shouldn't be used.

BOM and Routing Explosion

For the BOM and routing explosion, you can only use selection conditions and procedures that were directly assigned to these objects. The following sequence applies:

1. **Evaluating the selection conditions**

First, the system evaluates the selection conditions. It copies elements if no selection conditions have been assigned to them or if at least one of the assigned selection conditions has been met.

2. **Evaluating the procedures**

The system then evaluates the procedures for the selected elements, that is, for the elements to which no selection condition has been assigned or for which at least one of the assigned selection conditions has been met. For each element only the procedures that were assigned to exactly this element are evaluated — and this in the sequence or sorting defined.

For this process, we deliberately omitted actions, because they shouldn't be used and would considerably increase the effort required for the processing of object dependencies.

2.5.7 Basic Syntax Rules

Object dependencies follow their own syntax rules. The syntax isn't programmed with ABAP and doesn't correspond to any other common programming language. So-called variant functions enable you to integrate function modules with all types of object dependencies — and consequently also with the functions of the ABAP world. In the object dependency syntax, you can address variant functions, which then call ABAP function modules.

The result of the processing of object dependencies can be as follows:

▸ Preconditions and selection conditions are Boolean expressions. That is, they return a binary result.

- Procedures return any values as the result.

- Constraints return the corresponding values for the assignment of values and a binary result for the consistency check.

Names of Classes and Characteristics

The syntax is language-independent and therefore uses language-independent names of classes, characteristics, characteristic values, material masters, and documents. These names are directly included in the syntax. The only exceptions are characteristic values in the character format (CHAR). These are enclosed in single quotation marks. The SKEY syntax element and a name that is enclosed in single quotation marks are always required when the name contains a hyphen.

[+]

Variables in Object Dependencies

You can also solely use constraints with variables for objects and characteristics. These variables are defined specifically for each constraint.

For classes and documents, you must bear in mind that the object key consists of multiple parts. You must specify the object category for material masters and documents. An example of class, material master, and document in the syntax of object dependencies is provided in Listing 2.1.

```
(300)class1
(MATERIAL)()(NR='material2')
(document)()(TYPE = 'DRW', VERSION = '00', PART = '000',
           NR = 'D4545').
```

Listing 2.1 Class, Material Master, and Documents in Object Dependencies

Case Sensitivity

You don't have to distinguish between upper- and lowercase. However, there is one exception: characteristic values for the CHAR character format for which the Case Sensitive checkbox was selected. So if you don't need this checkbox, you shouldn't use it. (Unfortunately, you must select this checkbox for reference characteristics for variant conditions. This is discussed further in Section 2.8, Pricing for Configurable Materials.)

You can also include language-independent comment lines in the syntax (for example, an asterisk (*) in the first column) and use language-specific comments in object dependencies. However, this information is included not in the syntax but in the basic data of the object dependencies under EXTRAS • DOCUMENTATION.

$self, $parent, $root, and Other Syntax Rules

You can include blanks and line breaks anywhere. Object dependencies also allow for lists of several statements, which you have to separate with commas.

Regarding the objects that can be addressed in object dependencies, you have to distinguish between constraints and all other types of object dependencies, including simple object dependencies. In constraints, you can address all objects of a product structure. In all other types of object dependencies, you can address a maximum of three objects. This is done solely with reference to the storage location of the object dependencies, that is, no structure as illustrated in Figure 2.24, for example, is given. This has the following two consequences:

▸ A simple dependency (precondition, selection condition, or procedure) is assigned to the component at the lowest level of the BOM structure.

▸ All higher-level assemblies and the header material can be configured. (The cross is supposed to map the respective indicator in the material master.) All of these materials should have the corresponding configuration profiles in this environment so that they can be configured, too.

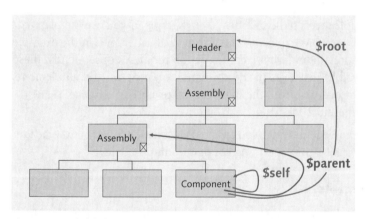

Figure 2.24 $self, $parent, and $root in Object Dependencies

As a result, the following three objects can be addressed in this dependency:

▸ **$SELF**
If $SELF. precedes a characteristic name, you can address the characteristic value assignment at the object level itself. Procedures can generally only set values at this level.

▸ **$ROOT**
If $ROOT. precedes a characteristic name, you can address the characteristic value assignment at the header object level. This is the highest instance with respect to

the configuration structure. It represents the highest level where the characteristic value assignment is stored.

▶ **$PARENT**

If $PARENT. precedes a characteristic name, you can address the characteristic value assignment at the next instance above the $SELF level. If the header material of the single-level BOM in which the dependency is assigned at the component level also has a specific value assignment with regard to the configuration, this is the $PARENT instance. Otherwise this level moves further upward.

[+]

> **Syntax**
>
> You can use uppercase (as in the list) or lowercase for these syntax elements as desired. The syntax is not case sensitive. An example of a procedure with lowercase could be:
>
> `$self.char3 = $parent.char2 + $root.char1`

In the example in Figure 2.24, you can address the first assembly under the header material, because the levels between $PARENT and $ROOT are not addressed. The same would apply to possible levels below the $SELF level. This means no $grandparent, $children, or similar related dependencies exist.

It is possible for multiple levels to be identical. For example, if you use the planned/production order without BOM explosion scenario at the header level in the example above, the system ignores all underlying configuration profiles. Consequently, these levels are not displayed as instances in the configuration process, and all elements can address the header material from the BOM structure as $ROOT. and as $PARENT. These two levels are therefore identical.

Even all three levels can be identical if you assign object dependencies at the header material level, that is, at the corresponding configuration profile level.

2.5.8 Syntax Elements

In addition to the already introduced names of characteristics, characteristic values, material masters, and documents as well as $self, $parent, and $root, you can also use other syntax elements in object dependencies. The following logical operators are available:

▶ Negation: NOT

▶ Conditions: IF

▶ Links: AND, OR

▶ Relational operators:

　▶ Less than: <, LT

　▶ Less than or equal to: >=, LE

- ► Equal to: =, EQ
- ► Greater than or equal to: >=, =>, GE
- ► Greater than: >, GT

The following arithmetic operators are available:

- ► Basic arithmetic operations: +, -, /, *
- ► Standard functions:
  - ► Trigonometric functions: sin, cos, tan
  - ► Exponential function (base e): exp
  - ► Logarithm functions (base e and 10): ln, log10
  - ► Inverse trigonometric functions: arcsin, arccos, arctan
  - ► Square root: sqrt
  - ► Absolute value: abs
  - ► Sign function: sign
  - ► Decimal fraction: frac
  - ► Integral fraction: trunc
  - ► Rounded up: ceil
  - ► Rounded down: floor

The following character string operators are available:

- ► The link: ||
- ► The conversion to lowercase characters: LC
- ► The conversion to uppercase characters: UC

You can optionally use in to query lists. For numeric characteristics, these lists can also contain general intervals. Examples are illustrated in Listing 2.2.

```
color in ('r', 'b', 'g')
width in (100, 200 - 400, 600).
```

Listing 2.2 Examples of a List Query with "in"

specified enables you to query whether a characteristic has an assigned value. Negative queries are also possible. Examples are illustrated in Listing 2.3.

```
color specified
specified color
not specified color.
```

Listing 2.3 Examples of a List Query with "specified"

You can use `type_of` to query the object at the corresponding level in object dependencies (except for constraints). In this case, the variant class at the header level is generally queried as shown in Listing 2.4.

```
type_of($root, (300)class1)
type_of($root, (material(300)(nr='material1')))
```

Listing 2.4 Examples of a List Query with "type of"

No Negative Queries for Constraints

As the only exception to negative queries with `not in`, `not type_of`, and `not spec-ified`, constraints do not allow for these negations. These are syntax errors in constraints.

Additional syntax elements are listed as follows:

▶ **mdata**
This element queries master data.

▶ **$set_default, ?=**
This element is provided for setting default values (only allowed in procedures). This can also be done via `?=` instead of `=` for fixed assignments of values. You can also use the `?=` syntax element when calling variant tables and variant functions.

▶ **$del_default**
This element enables you to delete default values that were set with `$set_default` (only allowed in procedures).

▶ **$sum_part**
This element adds up a numeric characteristic in an exploded BOM (only allowed in procedures).

▶ **$count_part**
This element is provided for adding up the component set in an exploded BOM (only allowed in procedures).

▶ **$part_of**
This element can be used to query objects in an exploded BOM (only allowed in constraints).

▶ **$subpart_of**
This syntax element queries objects in an exploded BOM of any depth (only allowed in constraints).

▶ **$set_pricing_factor**
This syntax element enables you to retroactively assign a factor to a condition record for pricing.

This is only a brief introduction to these syntax elements. They are detailed in the following sections, particularly in Section 2.6, Object Dependencies for the Value Assignment Interface or the Sales View, and Section 2.7, Object Dependencies for BOM and Routing.

2.5.9 Variant Tables and Functions

You can use so-called variant tables in all types of object dependencies. This enables you to easily build table-like structures and address them in object dependencies. The columns are defined by characteristics. You can query tables in conditions, for example, to allow only for value assignment combinations that are listed in the table, or you can use tables to set values for individual characteristics. The assignment of values with tables can be implemented in the same way as any assignment of values with procedures or constraints. An assignment of default values with variant tables and procedures is also possible.

In object dependencies, a table is called via the `table` keyword followed by the table name and the characteristics from the table and the characteristics from the variant class of the value assignment interface in brackets. An example is illustrated in Listing 2.5.

```
table  tab_pc ( char_type_package  =  $self.char_type_package,
                char_casing        =  $self.char_casing,
                char_cpu           =  $self.char_cpu       )
```

Listing 2.5 Sample Call of a Variant Table

You can use this table call in conditions to check whether the implemented value assignment is allowed according to the table. Alternatively, you can use a procedure to have the system infer the third characteristic from the first two characteristics, for example.

You can also use variant tables to evaluate database tables. In this case, the variant table is linked to the database table in its definition (appropriate maintenance transaction). Afterward, this link is activated.

Variant functions enable you to use function modules in all types of object dependencies. This allows you to use more complex functions than allowed by the rather restricted syntax of object dependencies. The maintenance of variant functions is analogous to the maintenance of variant tables. Input and output parameters of such variant functions are also implemented via characteristics.

Table TAB_PC		
PACKAGE	CASING	CPU
LUXURY	TOWER	Fast
LUXURY	MINITOWER	Medium
ECONOMY	MINITOWER	Medium
ECONOMY	DESKTOP	Standard
ECONOMY	TOWER	Fast

Figure 2.25 Example of a Variant Table

Such a variant function is called in the same way as a variant table. In object dependencies, you use the `function` keyword followed by the name of the function module (which is also the name of the variant function) and the characteristics from the variant function as well as the characteristics from the variant class of the value assignment interface in brackets. Listing 2.6 shows an example of a variant function call.

```
function   z_function1
     ( char_type_package    =  $self.char_type_package,
            char_casing      =  $self.char_casing,
            char_cpu         =  $self.char_cpu )
```

Listing 2.6 Sample Call of a Variant Function

2.5.10 Evaluation Function for Object Dependencies

Various functions are available for the evaluation of object dependencies, including common functions such as dependency lists and where-used lists of object dependencies. As already mentioned for the tools from the classification system, you can also evaluate the usage of objects in object dependencies. In this context, you should pay particular attention to the where-used list of characteristics, characteristic values, and classes.

Variant Model Browser

Another option to display and evaluate master data with regard to the modeling of the variant model including object dependencies is the variant model browser. You can call it using Transaction CUMODEL (see Figure 2.26). In the initial screen of this transaction, you can enter the configurable material and the plant for which the BOM is supposed to be exploded. BOM usage, quantity, key date, and change number are optional entries in the initial screen. The browser lists all significant objects that describe a variant model. This allows for an evaluation of the variant model's structure. You can navigate to the details of the following objects:

► Characteristics

► Assignment of object dependencies (including syntax)

► Configuration profile

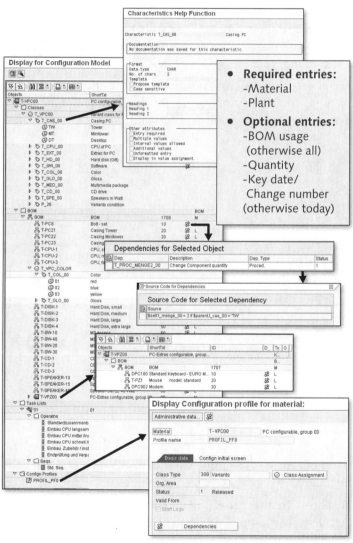

Figure 2.26 Transaction CUMODEL Variant Model Browser

In addition to these evaluation functions, which focus on the master data of the variant model, you're provided with further evaluation options for the configuration process.

Trace

You can have the system log the processing of object dependencies. For performance reasons, the system doesn't permanently document the processing of object dependencies in detail. You can, however, configure and activate such a documentation in the form of a trace at any time. You can find the trace from the user interface of the configuration via the menu path EXTRAS • TRACE • SETTINGS. The system then displays the setting options as shown in Figure 2.27.

You can define the level of detail for the documentation. You can also use additional filters. This function identifies possible errors during the development of the object dependencies. In day-to-day business, you will deactivate this trace for performance reasons.

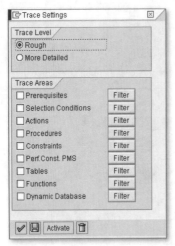

Figure 2.27 A Trace and Its Setting Options

Analysis Tool

In addition, you can carry out an analysis during the configuration process using the analysis tool (see Figure 2.28).

You can find the analysis tool from the user interface of the configuration via the menu path EXTRAS • ANALYSIS. The result of this analysis is a description of the actual status of the current configuration.

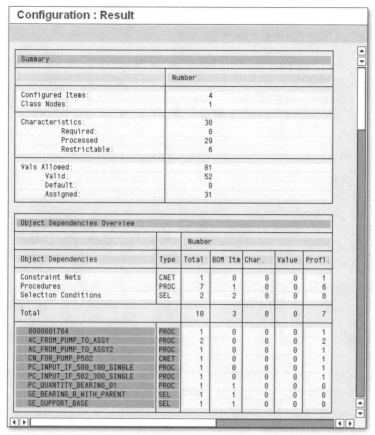

Figure 2.28 The Analysis Tool

The description includes the following:

▶ A summary of the configured instances, class nodes, characteristics, and characteristic values

▶ A summary of the object dependencies with the option to navigate to the syntax (If you double-click on Object Dependencies, the system will display the name, status, short text, and the entire syntax of the corresponding object dependencies in a new window.)

▶ A list of all BOM explosions including assigned object dependencies

▶ An evaluation of the class nodes

▶ A detailed list of the characteristic value assignment including assigned object dependencies for characteristics and values

▶ An evaluation of the configuration profiles, including object dependencies

2.6 Object Dependencies for the Value Assignment Interface or the Sales View

As already described, object dependencies are required for two usages: the high-level configuration (sales configuration, in the dialog, for the value assignment interface) and the low-level configuration (BOM and routing explosion, also without dialog). The following section discusses the first usage in more detail.

2.6.1 Product Modeling Environment PMEVC

Various transactions or methods are provided for the maintenance of object dependencies for the value assignment interface. The most important maintenance environment for object dependencies for the value assignment interface, both local and global, is the PMEVC product modeling environment (see Figure 2.29). This section therefore focuses on it. Section 2.7, Object Dependencies for BOM and Routing, introduces additional maintenance options.

The PMEVC product modeling environment has been available as Transaction PMEVC since SAP ERP Release 5.0. A similar function is also part of the IPC. PMEVC is short for Product Modeling Environment Variant Configuration. The concept behind this transaction is to create an environment in which you can maintain the entire variant model via the model structure from the high-level configuration perspective. Similar to the CUMODEL variant model browser, you can first obtain an overview of the existing model structure and then navigate to details.

You can also create and change numerous components of the configuration model from this product modeling environment. This enables you to create and change all types of object dependencies, both global and local. The same applies to configuration profiles, variant tables, and IPC data.

The product modeling environment uses an additional editor for the maintenance of object dependencies. In contrast to the traditional object dependency editor, you can use the following elements:

- Context-sensitive input help
- Drag-and-drop
- Object dependency wizard for preconditions, selection conditions, and table-based constraints

You can call the context-sensitive input help via the function key F4 or the second button in the editor (see Figure 2.29). All types of syntax elements as well as characteristics and characteristic values are provided. You can insert them at the respective position in the object dependency syntax.

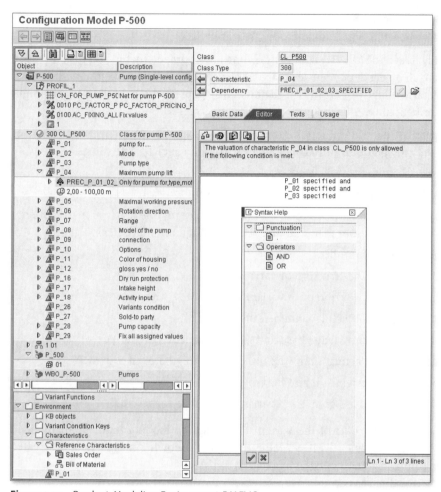

Figure 2.29 Product Modeling Environment PMEVC

The drag-and-drop function enables you to copy individual characteristic values from the lists on the left (see Figure 2.29) to the editor in PMEVC. As the example in Listing 2.7 shows, the characteristic and the characteristic value are transferred in the form of an equation.

```
char1 = 'value1'
```

Listing 2.7 Example of a Syntax Generated via Drag-and-Drop

The object dependency wizard enables you, as already mentioned, to create preconditions and selection conditions as well as table-based constraints without having to write the syntax yourself. The wizard queries all necessary information, and the system creates the syntax and all other required data.

Not only does the PMEVC product modeling environment enable you to completely maintain all object dependencies that are relevant for the value assignment interface; PMEVC also allows for nearly all maintenance steps of the modeling for the high-level configuration. This includes the following aspects:

▶ Maintenance (creation, modification, display, assignment) of object dependencies for the configuration profile, characteristics, and characteristic values

▶ Class-specific characteristic adaptation

▶ Simple classification (no multiple classification)

▶ Creation of a configuration profile (several profiles for one material master are not possible, but you can maintain all existing configuration profiles)

▶ Usage of change numbers

▶ Maintenance of the structure of variant tables

▶ Maintenance of the content of variant tables

▶ Modification of object dependencies for the BOM

▶ Creation of the knowledge base and runtime version for the IPC

▶ Material-specific activation of the IPC as the configurator in SAP ERP

▶ User interface design — maintenance

▶ Maintenance and assignment of variant conditions for pricing

Most of the master data of the Variant Configuration model cannot be maintained via PMEVC. Consequently, it is required that the following master data be created via the common transactions in advance.

▶ Characteristics

▶ Classes

▶ Material masters

▶ BOMs including all objects at the item level

▶ Routings including all objects at the operation level

▶ Variant functions

▶ Change numbers

▶ Objects that couldn't be created in PMEVC before Release ERP 6.0

After this introduction of PMEVC, the following section describes an example of its usage.

2.6.2 Example

After these rather theoretical explanations, it may be helpful to provide you with a practical example of how you can use PMEVC to create object dependencies for the value assignment interface. For this purpose, you use the object dependency wizard in PMEVC.

To map dependencies between the individual characteristics of the value assignment interface, you should use tables or, if they don't become too long, variant tables. The advantage of tables is that you can read the dependency type from them more easily than when evaluating the syntax of the object dependencies directly. Furthermore, the usage of variant tables has a major advantage if the model is "alive." If the dependencies in the model change, you only have to change the content of the variant table, without having to modify the syntax of the object dependencies.

The most elegant way to evaluate the table is to write object dependencies that query the table and only allow for value assignments that comply with the table. This is supposed to be implemented in such a way that no disallowed value assignments are possible. The list of the allowed values for each characteristic is supposed to be dynamically restricted in such a way that only allowed value assignments are possible. For the selection of such "elegant" object dependencies, you must use a type for which the user doesn't have to specify a point in time when the object dependencies are supposed to be processed. You should therefore use constraints.

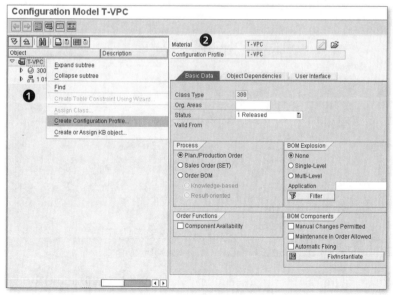

Figure 2.30 Getting Started in PMEVC: Class Assignment and Creation of a Configuration Profile

In Figure 2.30, PMEVC was called with material T-VPC. The system has found the BOM for the material. These are initially the two only entries in structure (❶). You now require at least the variant class and the configuration profile. As already mentioned, the variant class, including its characteristics, needs to be created outside PMEVC. In Figure 2.30 the existing variant class is already assigned.

This variant class (or a complete group of variant classes) was previously included in the PMEVC environment (see the bottom left of Figure 2.29, here under ENVIRON-MENT • CLASSES and context menu). Now, assign the variant class from the environment at the bottom to the material master at the top via drag-and-drop.

In contrast to the variant class, you can create the configuration profile directly from PMEVC. As you can see in Figure 2.30 (❷), this function is available in the context menu at the material master level. Similar to the common transaction for the creation of configuration profiles, you can create a configuration profile with the default values. In contrast to the common transaction, there are also default values for the name and variant class type.

In addition to the two previously mentioned steps, the model structure consists of four objects: material master, configuration profile, variant class, and BOM.

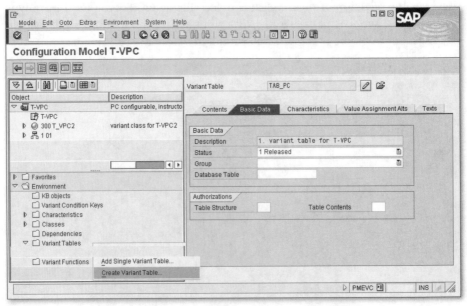

Figure 2.31 Variant Tables: Creation and Content Maintenance with PMEVC

A variant table (see Figure 2.31) is supposed to map the allowed combinations for the value assignment of numerous characteristics. In the figure's example, this includes the three characteristics, Special wish, Casing, and CPU. For this purpose, you must perform the following steps:

1. **Creating a variant table**

    In the first step, you create the variant table via the context menu in the environment, because it isn't provided here yet.

2. **Name of the table**

    Then a window, Create Variant Table, opens in which you enter a name for the table. Engineering Change Management is optional and not supposed to be used in this example.

3. **Description of the table**

    The system then displays the detail screen (Figure 2.31) with five tabs. In the Basic Data tab, enter a description (language-dependent), and then release the variant table.

4. **Assigning characteristics**

    In the Characteristics tab, specify the three mentioned characteristics in any sequence.

5. **Entering the table content**

    Finally, you can enter the allowed combinations of the value assignment with regard to the three characteristics as rows in the first tab, Contents.

After the variant tables have been created and their content has been maintained, you require object dependencies. Object dependencies are supposed to read the table and dynamically restrict the value lists of the corresponding characteristics in such a way that only value assignments from the table are possible. For this purpose, start the table constraint wizard via the context menu for the configuration profile, as shown in Figure 2.32.

The wizard guides you through the individual steps of the creation of constraints and queries you for all required information. It is possible for the wizard to dynamically adapt the steps to the already specified information. The wizard processes the following steps:

1. **Start**

    The first step of the wizard provides information on the procedure for the creation of a constraint with reference to a variant table.

2. **Mode of Action**

    The second step queries about the mode of action. Here, you're supposed to restrict the value lists for characteristics. This is the first selection option in this step. The difference from other selection options is described in the following.

It is possible that the wizard won't provide the Value Restriction option. In this case, the required prerequisite that the characteristics whose value lists are supposed to be restricted are indicated as restrictable in the characteristic definition is not met.

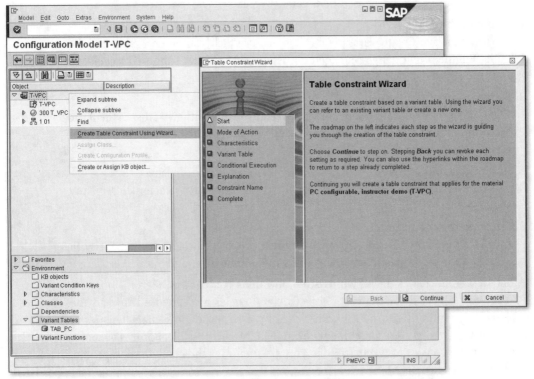

Figure 2.32 Creating Object Dependencies Using the Table Constraint Wizard

3. **Configuration Object**

   Next, the wizard queries you for the configuration object. In addition to characteristics from classes, you can also use and evaluate other reference objects in constraints. Because this example is supposed to restrict characteristics of a variant class, the corresponding variant class is also the configuration or reference object.

4. **Variant Table**

   After the configuration object step, the variant table selection is implemented. Because this example contains only one variant table in the PMEVC environment, no comprehensive selection can be made.

5. **Explanation**

   In the Explanation step, you can assign a language-dependent long text to the object dependencies, that is, to the table constraint. It also lists the characteristics of the variant table to exclude characteristics for object dependencies if required.

6. **Constraint Name**

   For constraints, the name must be assigned externally, that is, by the user. This request including a short text is made in the Constraint Name step.

7. **Complete**

   All steps are completed. A description of these steps are provided once again in detail before the user can complete the table constraint.

The system then displays the completed table constraint (see Figure 2.33). After saving, you can directly test it by calling the configuration simulation from PMEVC via the Test button.

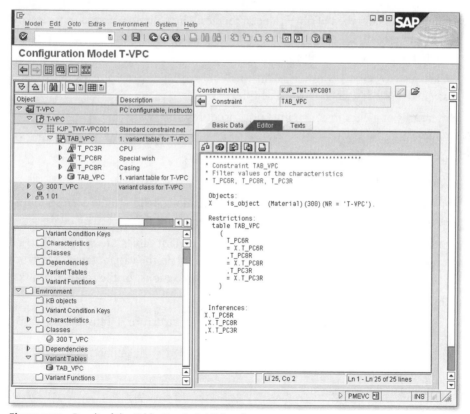

Figure 2.33 Result of the Table Constraint Wizard

The value assignment interface displays the three characteristics, Special wish, Casing, and CPU. Initially, the corresponding input help ‹F4› provides all values from the variant table for each characteristic. However, if you start assigning values to any of the three characteristics, the system only provides the values for the other two characteristics that lead to an allowed value assignment according to the variant table. The same applies to two characteristics to which values have been assigned.

A special feature of the traditional configurator is that if the allowed value range for a characteristic is restricted to a value, the system automatically sets this value. This is only done for required characteristics in the IPC. Note that the characteristics need to be restrictable according to the characteristic definition. As a result, the lists of the allowed values for characteristics to which values have been assigned are restricted to the value assignment, as already described.

2.6.3 Variant Tables in Detail

The first usage of variant tables was introduced in the example above. You can address variant tables in all types of object dependencies. The columns of variant tables are always characteristics. The rows represent value assignment combinations. You can use variant tables for different purposes:

- Value restrictions in constraints (in this context, values can already have been assigned — as described in the example)
- Inferences of values in constraints or procedures
- Conditions as preconditions, selection conditions, if conditions in procedures or constraints, and as a condition part in constraints
- Consistency checks via constraints

These purposes are discussed in more detail in the introduction to the corresponding types of object dependencies in Sections 2.6.4 through 2.6.7.

At this point, the structure and maintenance of variant tables are supposed to be introduced (see Figures 2.31 and 2.34). You can maintain the variant table's structure and content via PMEVC. There are also specific transactions for the following tasks:

- Content maintenance (Transaction CU60)
- Creating, changing, and displaying the table structure (Transactions CU61 through 63)

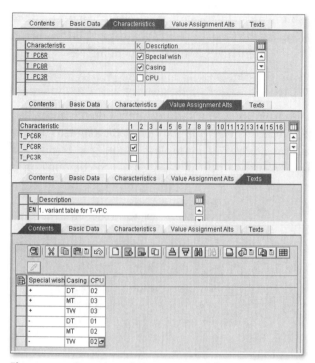

Figure 2.34 Structure and Maintenance of the Variant Table (PMEVC)

The Basic Data tab of variant tables (see Figure 2.31) contains names, the description (language-dependent short text), the status, and the group. The status can be adapted via Customizing and also includes the content maintenance and usage in object dependencies as well as distribution locks for the content and structure. The same aspects apply to groups as to characteristic and class groups; it is also a separate list in Customizing. Furthermore, you can couple the variant table with a database table in the basic data. This is described in detail later in this section. The basic data is complemented by the authorization groups for the content and structure maintenance.

The Characteristics tab, which is assigned to a variant table, provides the columns of the variant table. For the maintenance of the table content and for the usage in object dependencies, the settings in the characteristic definition, such as single-level/multi-level, restrictable, required characteristic, default values, or object dependencies, are irrelevant. In the characteristic view, you can define a first key. This key is merely a prerequisite and is only significant if you want to infer values from the variant table. This can be done via constraints or procedures. For value restrictions, conditions, or mere consistency checks, the system ignores the key information.

For inference of values, constraints with an inference part can evaluate more than the Value Assignment Alternative (key in the Characteristics tab) that is specified in the tab. You can create these additional alternatives in the corresponding view.

In PMEVC, you can also maintain the elements of the Content directly in the variant table.

Besides maintaining content from PMEVC, you can implement the content via a common transaction, namely, Transaction CU60. In addition to the standard display (❶ in Figure 2.35), this transaction also allows for displays as a matrix (❷) and as a list (❸). These last two displays enable you to easily decide which combination is supposed to be used (decision table).

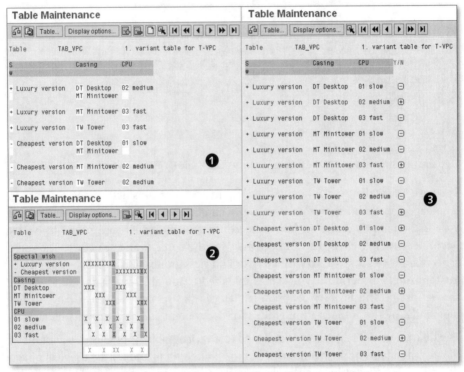

Figure 2.35 Variant Tables: Content Maintenance with the Common Transaction

[!] **IPC-Compatible Table Content**

You can also select multiple values for each field in the standard display. This display isn't IPC-compatible, but you can transfer it to such a display using the respective menu entry (Add up, Untag).

As already mentioned, you can use variant tables in all types of object dependencies. Basically, the syntax always looks the same (see Listing 2.8).

```
table <table name>
(<column characteristic>=<interface characteristic>,...)
```

Listing 2.8 Syntax Concept for Calling Variant Tables

You can break down this visual similarity as follows:

- The call starts with the keyword `table`.
- Then the name of the table is specified.
- Next, all columns of the variant table that are supposed to be evaluated are listed in brackets. You don't always have to evaluate the entire table. Because the characteristics of the table don't have to correspond to the characteristics of the user interface, the system assigns the corresponding characteristic to the value assignment interface after each column characteristic.

Figure 2.33 shows an example of such a table call in constraints. Note that X. precedes the characteristics from the value assignment interface. This refers to material T-VPC, as you can see in the constraint directly above the table call. As mentioned in Chapter 1, you cannot address objects (the material master in this case) as flexibly as in constraints. In this case, you can only use the `$self.`, `$root.`, or `$parent.` levels. The table access from Figure 2.33 would then be as shown in Listing 2.9.

```
table TAB_VPC (   T_PC6R = $self.T_PC6R,
                  T_PC6R = $self.T_PC6R,
                  T_PC6R = $self.T_PC6R)
```

Listing 2.9 Table Access in Preconditions, Selection Conditions, and Procedures

Characteristics of the value assignment interface for which the values are supposed to be inferred must be linked with `$self.`; `$parent.` and `$root.` are also allowed.

Figure 2.36 Coupling of Variant and Database Tables

Coupling with database tables is also supposed to be discussed (see Figure 2.36) in the context of the maintenance of the variant tables. The background for this function is the requirement to also address database tables in object dependencies. However, this is not directly possible in object dependencies. There are two options for addressing database tables in object dependencies.

- **Directly addressing a database table (variant function)**
  You create a variant function and address the database table in the function mod-

ule that corresponds to the variant function. The variant function is used in object dependencies.

▶ **Indirectly addressing a database table (coupled variant table)**
You create an appropriate variant table and couple it with the database table. If the coupling is active, you can directly include the variant table in any type of object dependencies. Then the database table is also addressed.

The second option uses an assignment of the database table in the maintenance of the basic data of the variant table structure. You can also distinguish between two scenarios here.

Scenario 1: Starting Point Database Table

An existing database table is supposed to be addressed using this approach. You must create the appropriate characteristics, that is, characteristics with the appropriate format, in this scenario. You only have to create characteristics for the fields, that is, the columns of the database table, that are supposed to be addressed in the object dependencies. Then create a variant table that contains exactly these characteristics as columns. You must include the name of the database table in the basic data of the variant table. This allows for a field assignment. All characteristics of the variant table are linked to the columns of the database table. Finally, you must activate the coupling. Afterward, you can evaluate the database table in all types of object dependencies by addressing the coupled variant table in the syntax.

Scenario 2: Starting Point Variant Table

An existing variant table that may already be used in object dependencies is supposed to be converted into a database table.

[Ex]   This scenario is used, for example, when variant tables reach a size that may lead to performance problems. However, there may also be other reasons.

In this case, you need to create an appropriate database table. With regard to the format, the fields of the table must be created in such a way that they can include at least the values of the corresponding characteristics. The key fields of the database table are not relevant for Variant Configuration. Only the key combination of the variant table is critical for the evaluation of object dependencies. Of course, you must select the database table key in such a way that uniqueness is ensured for later content.

Analogous to the first scenario, you must now couple the variant table with the still empty database table, activate it, and assign the corresponding fields. Finally, the content of the variant table is transferred to the database table using Transaction

CU59. In this second scenario, you don't have to adapt all object dependencies that use this variant table. They have exactly the same function as before.

> Bear in mind that the content of the variant table is inactive when the coupling has been activated. **[+]**

In the standard version, you maintain the content directly in the database table. You can lock the content maintenance of the variant table via the status. You can also delete the content of the variant table to prevent misunderstandings.

Additional maintenance in the variant table with a delta transfer of this data to the database table is also possible, but this isn't very useful, because you cannot delete rows when implementing a delta transfer, and problems may occur due to the key of the database table.

2.6.4 Constraints in Detail

As already mentioned, you can use all types of object dependencies for the configuration in the value assignment interface. Constraints are the most important type, or the default type, of object dependencies, because they can be used for nearly all tasks. You can therefore perform the following tasks using constraints:

- Setting values
- Checking values (in consistency checks)
- Addressing any objects that are also in the multilevel configuration (not only $self, $parent, or $root)
- Working with a high performance level
- Working in a declarative way (This means you don't have to consider a processing sequence or similar factor in the modeling process.)

Constraints are collected in dependency nets (also referred to as constraint nets). Dependency nets are generally assigned to the configuration profile. It doesn't matter whether a large number of constraints are assigned to a dependency net. However, the number of dependency nets in a configuration model should be kept to a minimum, but you cannot always ensure that only one dependency net is assigned to the configuration profile of the header material of the configuration profile.

> **Need for Multiple Constraint Nets** **[Ex]**
>
> If, for example, individual configurable assemblies of the overall model are sold separately, all constraints that are related to this assembly must also be assigned to the configuration profile of this assembly via a dependency net.

Within the dependency net, the constraints are local object dependencies; the dependency net itself is global. Note that both have a status. Constraints are consequently only active if the following conditions are met:

- The constraint is released.
- The dependency net is released.
- The dependency net is assigned to a configuration profile that is considered during configuration.

A constraint consists of at least two and at the most four sections. Objects and restriction are the mandatory sections here. The following explains the four sections according to their sequence:

1. **Objects**
   In this section, you enter all used classes and objects and define variables.

2. **Condition**
   This section is only used to (optionally) specify a central condition under which the constraint is supposed to be evaluated.

3. **Restriction**
   In this section, you define equations, tables, and functions for inferences of values and/or value checks.

4. **Inferences**
   This section enables you to (optionally) enhance the inference and restrict characteristic values.

The objects section is mandatory and must contain the objects that are addressed in the constraint. Objects can be classes, material masters, and documents. The declaration has the following syntax:

- **Class**
  (<class type>)class name, for example, `(300)T_VPC`

- **Material master**
  (Material)(<class type>)(Nr = '<material number>'),
  for example, `(Material)(300)(Nr = 'T-VPC')`

- **Document**
  (Document)(<class type>)(Type = '<document type>', Version = '<document version>', Part = '<document part>', Nr = '<document number>'), for example, `(Document)(017)(Type='DRM',Version='01',Part='000',Nr='T-VPC')`

In the objects section, you can also locally define variables for the constraint; the constraint in Figure 2.33 uses such an object variable. Object variables for classes are declared by `is_a`, and for all other objects by `is_object`, as shown in Figure 2.33. The constraint in Figure 2.33 would be as shown is Listing 2.10 without variables.

```
objects:       (300)t_vpc
restrictions:  table tab_vpc (  t_pc6r = (300) t_vpc. t_pc6r,
                                 t_pc8r = (300) t_vpc. t_pc8r,
                                 t_pc3r = (300) t_vpc. t_pc3r).
```

Listing 2.10 Sample Constraint: Table Call Without Using Variables

Note that in this context you can also declare the variant class instead of the material master in the objects sections, as done here. After the equals sign, the syntax is as follows: `<object>.<characteristic>`.

In the objects section, you can also define characteristic variables, which are class-specific. You therefore don't have to use object variables here. In the syntax, this is implemented with `where` and a list of the variables separated by a semicolon. The constraint in Figure 2.33 would be as follows with characteristic variables (Listing 2.11).

```
objects:       (300)t_vpc where  ?spe = t_pc6r ;
                                 ?cas = t_pc8r ;
                                 ?cpu = t_pc3r.
restrictions:  table tab_vpc (  t_pc6r = ?spe,
                                 t_pc8r = ?cas,
                                 t_pc3r = ?cpu).
```

Listing 2.11 Sample Constraint: Table Call Using Characteristic Variables

---

Specific Variable Names　　　　　　　　　　　　　　　　　　　　　　　　**[+]**

As shown here, question marks are often used as variables so that they can be easily identified. You cannot use variables for characteristics from tables, that is, for columns.

If the objects section lists multiple objects, you use a comma to separate them.

The restriction section is the second section that is mandatory in constraints. Here, you can carry out consistency checks. The constraint reports an inconsistency when the restriction section is not met. You can also use constraints to infer values. For this purpose, the equations must be solved for the value that is supposed to be inferred, or you must use the inference section. This is further detailed in the context of the inferences section below. The restriction section enables you to restrict values (see Figure 2.33). You can use variant tables and variant functions.

You can also work with conditions in the restriction section. The `if` syntax element is provided for this purpose. Additional syntax elements, such as `then` or `else`, are not available. As shown in Listing 2.12, first the statement and then the if condition is provided:

```
restrictions:
(300)t_vpc.m1 = 'a'     if (300)t_vpc.m2 = 'x1',
        false           if (300)t_vpc.m2 = (300)t_vpc.m3.
```

Listing 2.12 Restrictions with if Conditions

This restriction section contains two statements that are listed with a comma. The first statement consists of the assignment of the a value for the m1 characteristic. The second statement only leads to an inconsistency message if the condition after false is met. The false syntax element basically generates inconsistency messages and can only be used in constraints.

In constraints, you can use a condition section. The condition section must always be inserted between the objects and the restriction section. It contains exactly one logical expression. The constraint is not processed until the condition of the condition section is met, which is very good regarding the performance. You can also work with variables and variant tables in the condition section, as Listings 2.13 and 2.14 show.

```
objects:    (300)t_vpc where    ?os = t_pc01;
                                 ?hd = t_pc04,
            (300)t_vpr where    ?tr = t_pr02.
condition : specified ?hd.
restrictions : ?tr = ?os
```

Listing 2.13 Sample Constraint with Condition Section

```
objects:    (300)t_vpc where    ?spe = t_pc6r;
                                 ?cas = t_pc8r;
                                 ?cpu = t_pc3r.
condition:    table tab_vpc (   t_pc6r = ?spe,
                                t_pc8r = ?cas,
                                t_pc3r = ?cpu).
restrictions: false.
```

Listing 2.14 Sample Constraint with Table Call in the Condition Section

In Listing 2.13, the restriction section is evaluated under the condition that a value has been assigned to the t_pc04 characteristic. In this case, the system checks whether the same values have been assigned to the t_pc01 and t_pr02 characteristics. If no values have been assigned to t_pc01, the system copies the value assignment of t_pr02 to this characteristic.

In Listing 2.14, an inconsistency message is output if the (300)t_vpc variant class in the value assignment interface corresponds to a row of the tab_vpc variant table in the configuration. This is used if inconsistent value assignment combinations are collected in variant tables. You can also create such constraints with the table constraint wizard. In this case, you must select the Checking Inconsistent Combinations entry as the mode of action. In this context, have a look at Figure 2.32 in Section 2.6.2, Example.

The inferences section, which is optional just like the condition section, is the fourth section of a constraint. This section is always the last section of the constraint and enhances the evaluation of the restriction section. This "enhanced evaluation" can refer to equations, variant tables, variant functions, and restrictable characteristics. The syntax of the inferences section is merely a list of characteristics.

Consequently, an equation, $V = L * W * H$, for example, in a restriction section without a subsequent inferences section is only evaluated for calculation in such a way that V is the product of L, W, and H. However, if the constraint has the following structure, the fourth value is inferred from any three values (see Listing 2.15).

```
objects:      (300)t_vpc where
                    v = t_pc91 ;    w = t_pc92;
                    b = t_pc93 ;    h = t_pc94 .
restrictions:       v = l * w * h .
inferences:         v, l, w, h .
```

Listing 2.15 Sample Constraint with Equation and Inferences Section

Variant tables and functions for which more than one value assignment alternative is defined are additional examples (see Figure 2.34 in Section 2.6.3, Variant Tables in Detail). If you don't use the inferences section here, the system can only evaluate the first key, that is, the first value assignment alternative. However, if the constraint has the following structure and if two additional value assignment alternatives exist for the first and third characteristics and for the second and third characteristics, the system can infer the remaining third characteristic from the table from any two characteristics to which values have been assigned (see Listing 2.16).

```
objects:      (300)t_vpc where    ?spe = t_pc6r ;
                                  ?cas = t_pc8r ;
                                  ?cpu = t_pc3r.
restrictions: table tab_vpc ( t_pc6r = ?spe,
                              t_pc8r = ?cas,
                              t_pc3r = ?cpu).
inferences: ?spe, ?cas, ?cpu.
```

Listing 2.16 Sample Constraint with Variant Table and Inferences Section

For usage with restrictable characteristics, have a look at the constraint in Figure 2.33 in Section 2.6.2, Example.

Now that we've discussed constraints in detail, the following sections deal with the other types of object dependencies. As already mentioned, you can use all types of object dependencies to design the value assignment interface in sales and distribution.

2.6.5 Preconditions

You can use preconditions to disallow individual characteristic values or entire characteristics for the value assignment interface. If you don't use preconditions, values can be assigned to any characteristic of the value assignment interface in any sequence. You can select any value from the list of the allowed values — irrespective of the value assignment of other characteristics. In this context, you have to find answers to two questions:

1. What is supposed to be disallowed? That is, which characteristic or which characteristic value is supposed to be dynamically disallowed?

2. When is it supposed to be allowed? That is, when is the corresponding characteristic or characteristic value supposed to be allowed within the scope of the value assignment?

The storage location of the respective precondition answers the first question. The syntax answers the second question. For example, if the XYZ engine is only supposed to be offered for the sport version of a car configuration, a precondition must be assigned to the XYZ characteristic value of the characteristic for the engine selection (What is supposed to be disallowed?). The syntax contains the prerequisite that the sport version was selected for which the XYZ+ engine is allowed (When it is supposed to be allowed?).

What effect does it have when the `$self.version = 'sport'` precondition is assigned to the XYZ characteristic value?

▸ **Assigning the Sport value to the Version characteristic**
If the Sport value is assigned to the Version characteristic, the list of allowed values in the characteristic for the engine will provide all values as if no precondition exists.

▸ **Assigning a different value than Sport to the Version characteristic**
If a different value than Sport is assigned to the Version characteristic, the XYZ value will be missing in the list of allowed values in the characteristic for the engine.

▶ **No value assigned to the Version characteristic**
Note that the precondition is considered to be met if no value is assigned to the Version characteristic. In this case, the list of allowed values of the engine characteristic would include all engine values. In the standard version all allowed values are initially available. During the value assignment, the system hides the values that are no longer allowed. If you don't want to use this standard logic, you must implement this by adding `$self.version = 'sport'` and `$self.version specified`. The system then first provides all values that can be generally selected, and the list of allowed values is gradually extended.

▶ **Sequence of the value assignment**
The precondition mentioned in the previous item is elegant in one direction only: if values are assigned first to the version and then to the engine. If you start by assigning values to the engine, you can select any engine and any version. Only then is the precondition evaluated. It retroactively disallows the XYZ engine. This leads to an inconsistency message if the XYZ engine and a version other than sport is selected. You can avoid this by assigning a precondition to the sport version. Another option is to force a processing sequence. A value is not supposed to be assigned to the Engine characteristic until the version is known. In this case, you use a precondition for the characteristics. You assign a precondition of the `$self.version specified` (When it is supposed to be allowed?) to the engine characteristic (What is supposed to be disallowed?).

▶ **Multiple preconditions**
You can also assign multiple preconditions to characteristic values or characteristics. In this case, the value assignment is only allowed if all preconditions are met. It can be considered an And link. You can only implement an Or link between preconditions when you include the conditions in a precondition. A precondition can be any complex condition using any brackets, negations, and concatenation with `and` and `or`.

Values or characteristics that are excluded via preconditions are not displayed in the value assignment interface by default. However, you can use the settings in the configuration profile (see Figure 2.37), for example, to define that disallowed characteristic values or characteristics are displayed but not used for the value assignment.

In Figure 2.37, the settings were called via the menu during the configuration. The With excluded characteristics checkbox was selected here. As a result, the disallowed value, XYZ, is displayed but cannot be selected. Similarly, disallowed characteristics would be displayed, but you couldn't assign values to them.

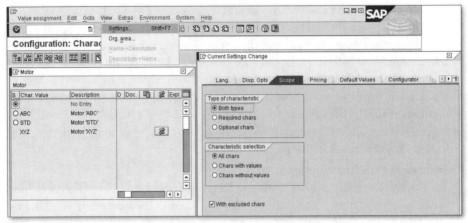

Figure 2.37 Displaying Excluded Characteristic Values (Preconditions)

You can use variant tables in all types of object dependencies, that is, also in preconditions. The syntax of a precondition that uses the table from Figures 2.31 and 2.34 could be, as shown in Listing 2.17.

```
table TAB_VPC (    T_PC6R = $self.T_PC6R,
                   T_PC8R = $self.T_PC8R,
                   T_PC3R = $self.T_PC3R)
```

Listing 2.17 Example: Variant Table in a Precondition

This example has the same syntax as a procedure. Instead of $self, you can also use $parent or $root everywhere — provided that this is correct with regard to content. You could assign such a precondition with exactly this syntax to one of the three mentioned characteristics; however, this wouldn't be an elegant solution.

Considering the increasing elegance of the solution, the following three options can be used to disallow values.

▶ Precondition for the characteristic (as described)

▶ Precondition for the characteristic value (as described later)

▶ The most elegant solution: A constraint with this variant table and restrictable characteristics (see Figure 2.33)

If you selected the variant with the precondition for the characteristic, the precondition wouldn't affect the value assignment of the three characteristics. Not until values have been assigned to all three characteristics that are addressed in the variant table does the precondition become active and output an inconsistency if the table doesn't contain this value assignment combination.

However, preconditions for characteristic values are more elegant than this variant with a precondition for the characteristic but still not as elegant as the variant with a constraint. For example, if you assume that values are assigned to the characteristics in a fixed sequence and the T_PC3R characteristic is the last characteristic to which a value is assigned, you can provide preconditions for the characteristic values of this characteristic. The precondition for the value '03' would have the following syntax (see Listing 2.18):

```
table TAB_VPC (   T_PC6R = $self.T_PC6R,
                  T_PC8R = $self.T_PC8R,
                  T_PC3R = '03')
```

Listing 2.18 Sample Variant Table in a Precondition for Characteristic Value '03'

Analogous preconditions also apply to the other values of this characteristic. Compared to the first variant, that is, the precondition for the characteristic, this has the advantage that only the values that lead to a consistent value assignment are provided for the third characteristic. Preconditions for characteristics or characteristic values don't require restrictable characteristics.

2.6.6   Selection Conditions

Selection conditions can dynamically convert optional characteristics into required characteristics.

Optional Characteristics   **[+]**

This technology requires that the Entry Required option in the characteristic is not set, that is, the characteristic is actually a so-called optional characteristic.

If you assign a selection condition to such a characteristic, the characteristic dynamically becomes a required characteristic if the condition is met. Let's illustrate this with an example.

Sport Version Requires a GPS   **[Ex]**

For the car configuration, the selection of a GPS is supposed to become a prerequisite for the sport version. In this context, the rules and the syntax must be analogous to the preconditions. The selection conditions must be assigned to the GPS characteristic (What is required?); the content of the syntax is the condition of the sport version (When is it required?). The syntax is analogous to the precondition mentioned previously: $self. version = 'sport'. The selection condition with this syntax is assigned to the GPS characteristic. This way the selection of a GPS is required for the sport version.

Compared to the rules of preconditions, there are two differences, which must be considered:

▶ **Multiple selection conditions for a characteristic**
If multiple selection conditions are assigned to a characteristic, it is sufficient for the characteristic to become a required characteristic when one selection condition is met.

▶ **No value assigned to a characteristic in a selection condition**
If a selection condition addresses characteristics to which no value has been assigned, the selection condition is not met. In contrast to a precondition with the same syntax, the selection condition above outputs "false," that is, not met if no version is selected.

2.6.7 Procedures

Procedures are object dependencies that set values. In contrast to other types of object dependencies, they can affect processing sequences. For this purpose, the configuration profile is assigned with procedures, and a sorting is transferred. There is also the option of assigning procedures to characteristics and characteristic values. However, this method is not supported for the configurator of the IPC and is consequently not discussed here. Instead, we'll discuss in detail the assignment to the configuration profile. Procedures allow for *fixed* or *dynamic* assignments of values. The following list describes in detail what this means.

▶ **Fixed assignment of values**
A fixed assignment of values refers to an assignment of values that cannot be overwritten by users or constraints. Attempts would result in inconsistencies. Procedures can also not overwrite *external* assignments of values, that is, values set by users or constraints. The same applies to users and constraints. The following rule is the only exception: Procedures can overwrite each other. If there are several such competing assignments of values, the last one wins.

▶ **Dynamic assignment of values**
Only procedures allow for a dynamic assignment of values. Such dynamic assignments of values are assignments of values that can be overwritten by other object dependencies or users. Every assignment of values that is not dynamic has more priority, whether it is implemented before or after a procedure with a dynamic assignment of values. If there are several competing dynamic assignments of values, the last one wins.

Listing 2.19 illustrates an example of a fixed assignment of values with a procedure.

```
$self.alarm = 'green' if    $self.temperature > 80,
$self.alarm = 'yellow' if   $self.temperature > 100,
$self.alarm = 'red' if      $self.temperature > 150.
```

Listing 2.19 Procedure: Example of a Fixed Assignment of Values and Overwriting

Although the `alarm` characteristic has only one value, no inconsistencies occur. Because procedures can overwrite each other, temperatures above 100 would overwrite the value `green` to `yellow`, and temperatures above 150 would overwrite the values `green` and `yellow` to `red`. If each line was one procedure, the sorting sequence would define the processing sequence. A different processing sequence would lead to a different result but not to inconsistencies.

A dynamic assignment of values for the `alarm` characteristic can have the following structure:

```
$self.alarm ?= 'green' if $self.temperature > 80.
```

The following syntax has exactly the same functionality:

```
$set_default($self, alarm, 'green') if $self.temperature > 80.
```

For such dynamic assignments of values, you also speak of dynamic default values. If default values are used via procedures and not via settings in the characteristic definition, they can also be deleted via procedures of the following structure:

```
$del_default($self, alarm, 'green') if ...
```

Another possible variant is the following:

```
$del_default($self, alarm, $self.alarm) if ...        .
```

The syntax of both the `$set_default` statement and the `$del_default` statement consists of the corresponding keyword followed by the following arguments in brackets, again separated by commas:

1. Argument: `$self`

2. Argument: `characteristic`

3. Argument: `value`

To delete the default values, you can also keep the third argument neutral by selecting `$self.characteristic`.

As already mentioned in Section 2.5, Overview of Object Dependencies, some syntax elements that can be used only in procedures. In addition to `$set_default` and `$del_default`, mentioned previously, these include negations, such as `not specified` or `not type_of`, but also `$sum_parts`, `$count_parts`, `$set_pricing_factor`, and successive calculations of the `$self.length = $self.length + 10` syntax. In all other types of object dependencies, this syntax would lead to errors in the syntax check.

Please also refer to Section 2.6.9, Variant Functions. As already mentioned, you can also use variant tables to infer values via procedures. A prerequisite here is that a key is defined in the definition of the variant table. If there are multiple value assignment alternatives in the variant table, procedures only allow for evaluating the first alternative. As already mentioned, the variant table shown in Figure 2.34 can be queried with the following syntax shown in Listing 2.20 in procedures.

```
table TAB_VPC (    T_PC6R = $self.T_PC6R,
                   T_PC8R = $self.T_PC8R,
                   T_PC3R = $self.T_PC3R)
```

Listing 2.20 Procedure with Variant Table: Fixed Assignment of Values

Because the first two characteristics in Figure 2.34 are selected as key fields, you can also use $parent or $root instead of $self when calling the table — provided that this is correct with regard to content. A procedure with this syntax would wait for the value assignment of the T_PC6R and T_PC8R characteristics. Afterward, the T_PC3R characteristic is assigned with a fixed value. The syntax in Listing 2.21 allows for a dynamic assignment of values (question mark for assignments of values).

```
table TAB_VPC (    T_PC6R = $self.T_PC6R,
                   T_PC8R = $self.T_PC8R,
                   T_PC3R ?= $self.T_PC3R)
```

Listing 2.21 Procedure with Variant Table: Dynamic Assignment of Values

2.6.8   Reference Characteristics

With regard to object dependencies for the value assignment interface, you can read and sometimes even change information from the environment of the configuration profile using reference characteristics. Section 2.2, Tools from the Classification System, listed the tables and structures to which read or write access is granted with reference to the tools from the classification system. You can, for example, map customer-specific features. The customer as the sold-to party can be queried, for example, via the VBAK table and here via the Sold-to-Party field. From the user interface, you can often query the names of tables and fields by following the F1 HELP • TECHNICAL INFORMATION menu path of the corresponding field.

The VCSD_UPDATE structure enables you to change individual fields of the items of the SD document, such as the net and gross weight. For this purpose, you would create reference characteristics for the NTGEW and BRGEW fields of this structure. Furthermore, you can use the specifications that are defined in the material master to determine the weight. However, you cannot directly access the material master, that is, table MARA. Instead, you have to do this via table MAAPV.

Access via Table MAAPV [Ex]

As a sample scenario, let's say you first read the net weight from the material master and then change it successively depending on the value assignment from the configuration. Finally, you should multiply the weight by the order quantity and add 10% as the gross weight. If the linked fields were selected as the names for the reference characteristics, the syntax could be as shown in Listing 2.22.

```
$self.vcsd_update_ntgew = mdata $self.maapv_ntgew + 10 if ...,
$self.vcsd_update_ntgew = $self.vcsd_update_ntgew + 20 if ...,
...
$self.vcsd_update_ntgew = $self.vcsd_update_ntgew *
                          $self.vbap_kwmeng     ,
$self.vcsd_update_brgew = $self.vcsd_update_ntgew * 1.1     .
```

Listing 2.22 Structure VCSD_UPDATE and Procedures

Reference characteristics with reference to the SCREEN_DEP structure enable you to change the display of characteristics dynamically. You can also define a characteristic in the characteristic definition in such a way that the characteristic is displayed in the value assignment interface as follows:

- Ready for input
- Only displayed
- Hidden

You can dynamically override this via the SCREEN_DEP structure by assigning the names of the characteristics to an reference characteristic with reference to the field:

- INPUT for characteristics that are ready for input
- NO_INPUT for characteristics that are not ready for input
- INVISIBLE for hidden characteristics

If, for example, P_INVIS is an reference characteristic with reference to SCEEN_DEP-INVISIBLE, the following procedure hides the P_01 characteristic: $self.p_invis = 'p_01'.

Particularly for hiding characteristics dynamically, the following syntax, which has the same result, is also possible: $self.p_01 is invisble

Such a simplified syntax is only available for hiding. Similar to this simplified syntax, you can also hide characteristics using variant tables. A variant table controls when the system is supposed to hide characteristics. The table includes the characteristics as key fields on which it depends whether the characteristics are displayed or hidden. For example, if the value assignment of the char1 characteristic is supposed to control whether the char2 and char3 characteristics are displayed or hidden, you require two additional characteristics (for instance, char4 and char5) that can adopt the values T (for true) and F (for false). The char1, char4, and char5 characteristics must be included in a variant table. The variant table is supposed to have, for example, the content shown in Table 2.3.

char1	char4	char5
AAA	F	F
BBB	T	F
CCC	F	T
DDD	T	T

Table 2.3 Example of a Variant Table for Controlling the Visibility of Characteristics

In addition, a procedure is supposed to ensure that the following characteristics are displayed in a specific way according to the table.

▶ The char2 and char3 characteristics are supposed to be hidden if char1 has the value DDD.

▶ The char3 characteristic is supposed to be hidden if char1 has the value CCC.

▶ The char2 characteristic is supposed to be hidden if char1 has the value BBB.

▶ No characteristic is supposed to be hidden if char1 has the value AAA.

The following procedure (see Listing 2.23) would allow for hiding the characteristics according to Table 2.3.

```
table tab001 (char1      =       $self.char1,
              char4      = inv $self.char2,
              char5      = inv $self.char3)
```

Listing 2.23 Example of a Procedure with Variant Table for Controlling the Visibility of Characteristics

It is critical here that a characteristic be hidden when it has the value T before the = inv syntax element.

A RESET field is additionally provided for the SCREEN_DEP structure so that you can reset the settings to the settings in the characteristic definition.

[+]

Only the Display of the Characteristics Changes

Note that this only changes the display of the characteristics. It doesn't affect the value assignment of the characteristic. Properties, such as being a required characteristic, also remain unchanged. The functionality of SCEEN_DEP must thus be generally differentiated from the functionality of preconditions.

2.6.9 Variant Functions

Variant functions enable you to integrate function modules with all types of object dependencies. In the narrower sense, the variant function is a wrapping around the actual function module. In the syntax of object dependencies, only the variant function can be addressed. It controls when the variant function is called and controls input and output parameters.

Variant functions are used everywhere where the standard syntax is not sufficient to map the desired functionality or where it may improve the performance. This may be possible in the following cases:

▶ During the complex processing of strings

▶ During the evaluation of database tables that cannot be addressed via reference characteristics

▶ During the generation of individually designed messages

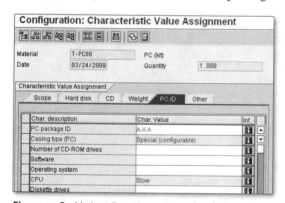

Figure 2.38 Variant Functions: Example of Use

In Figure 2.38, an example of the use of variant functions uses a type of configuration ID. Users like to use the option of generating a string from the entire relevant configuration to store a compressed configuration in addition to the single values. You can also generate this string without Variant Configuration, that is, with a traditional syntax for string processing. This example, however, is also supposed to allow for implementing this in the opposite direction. Such a configuration ID (for simplicity, it only consists of three properties here) is specified via the PC Package ID characteristic. The variant function splits the string into the individual value assignments, and the value assignments are assigned to the corresponding characteristics.

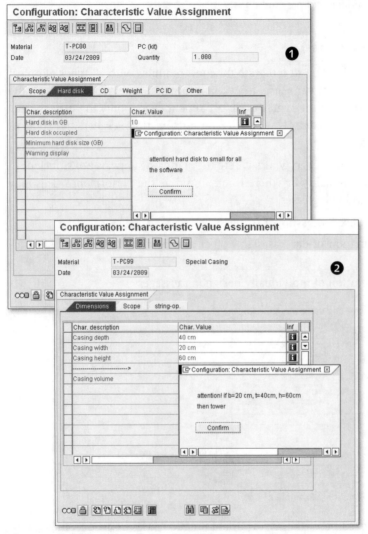

Figure 2.39 Additional Examples of Uses of the Variant Function

Two other examples use the option of displaying customer-specific messages here (see Figure 2.39). This way you can check if the selected software components correspond to the required hard disk size. Apart from that, the system is supposed to output the corresponding message (❶) (see Figure 2.39). The second example refers to the selected casing (❷). Numerous standard casings are provided for selection. However, you can also select your own casing and define the dimensions as required. The system displays a message when the dimensions of a standard casing have been selected.

To be able to use the variant function as described, you must create a function module that includes the required functionality. Therefore, you must provide the following interface parameters for the function module:

- Import: Parameter name = GLOBALS
          reference type = CUOV_00
- Export: No entries
- Changing: No entries
- Tables: Parameter names = QUERY and MATCH
          reference type = CUOV_01 (for both)
- Exceptions: Exception = FAIL and INTERNAL_ERROR

The variant function is created with the same name as the function module and has the same structure as variant tables. You must include characteristics, which are divided into input and output parameters. Like variant tables, you can also use multiple value assignment alternatives here; this would require constraints with inferences sections, however.

Variant functions are called in object dependencies in the same way as variant tables. An example in procedures could have the following syntax (Listing 2.24):

```
function   Z_PCSET_ID
      ( tpc_09  =    $self.tpc_09,
        tpc_10  =    $self.tpc_10,
        tpc_14  =    $self.tpc_14,
        tpc_31  =    $self.tpc_31) if ...         .
```

Listing 2.24 Variant Function in a Procedure

In addition to the `function` syntax element, `pfunction` is also available for procedures. This enables you to use the complete value assignment of the `self`, `$parent`, and `$root` levels in the function module. Here, you can use the function modules of the `CUPR` function group.

> In this context, you can also refer to the appendix, which describes the mentioned examples of variant functions (see Figures 2.38 and 2.39) in detail.   **[+]**

2.6.10 User Interface Design

Another aspect that has nothing to do with object dependencies but belongs to the value assignment interface topic is the user interface design. You can structure multiple characteristics for the value assignment interface, for example, as tabs in which the characteristics are arranged. You can map different groupings of characteristics by using buttons or by aggregating them (see Figure 2.40).

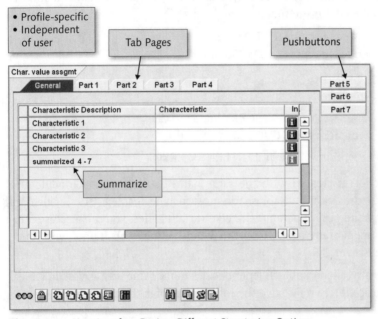

Figure 2.40 User Interface Design: Different Structuring Options

Basically, the user interface design is specifically created for a configuration profile. It is not user-specific. There are two prerequisites for designing the user interface:

▶ Entry of the user interface design's name in the configuration profile as described in Section 2.4, Configuration Profile and Configuration Scenarios.

▶ The corresponding authorization for the user

If these prerequisites are met, you can create new characteristic groups from the value assignment interface via the menu path VALUE ASSIGNMENT • INTERFACE DESIGN • CHARACTERISTIC GROUPING, irrespective of the transaction. You can change existing groups via the menu path VALUE ASSIGNMENT • INTERFACE DESIGN • OVERVIEW. A characteristic group contains the following information regarding its design:

► **Name of the characteristic group**
The name is used for display if no language-dependent description is maintained. In addition, the name also lexicographically defines the sequence of the characteristic groups.

► **Language-dependent descriptions**
These are used as titles of the groups; for buttons, you can also define labels.

► **Display settings**
The following variants are available:

  ► Tab

  ► Aggregate

  ► Pushbutton (Optionally, this can also be used together with Aggregate.)

  ► Sequence

The Tab, Pushbutton, and Aggregate settings are shown in Figure 2.40. Here, it is necessary to consider the following:

► You can set sequences for sales and distribution and technical postprocessing (engineering, order BOM). This option is used if you want to have different sequences for the two use cases.

► The system displays the General tab even if the user interface design is not maintained. This tab is automatically not displayed if it doesn't contain any characteristics. Characteristics are no longer included in the general tab if you assign them to other tabs or to the characteristic groups of the Aggregate category.

► Characteristics of buttons are no longer included in the general tab if Aggregate is selected in addition to the Pushbutton checkbox. The name of the description of the Aggregate characteristic group is generally displayed as a characteristic in the General tab.

► You can use characteristic groups from the user interface design as filters for printing the characteristics in individual applications. For this purpose, you have to select one of the three checkboxes, Sales Printouts, Purchasing Printouts, or Engineering Printouts, in the settings of the characteristic groups in addition to the display checkboxes Tab, Aggregate, Pushbutton, and Sequence. For each Sales, Purchasing, or Engineering application, you can define at most one characteristic group.

► You can assign any number of characteristics to each characteristic group. Characteristics can also be assigned to several groups but not for IPC. You can change the sequence of the characteristics in the characteristic group as required. Without user interface design, the sequence of the characteristics from the class definition applies.

2.7   Object Dependencies for BOM and Routing

You can explode BOMs and routings using object dependencies. If you already use BOM and routing explosions in the configuration, the explosion of BOMs and routings is also critical for the high-level configuration (sales configuration, in the dialog, for the value assignment interface). This is generally implemented in low-level configuration (BOM and routing explosion, also without dialog).

Before going into details on the maintenance of object dependencies for BOMs and routings, we should mention some aspects that apply to maintenance of object dependencies in general. Specific transactions or methods are also available for the maintenance of object dependencies.

2.7.1   Local and Global Object Dependencies

Local object dependencies can only be maintained – created, changed, or deleted – via their assignment to objects. For this purpose, you need to navigate to the maintenance transactions of the objects to which the local object dependencies are assigned. This includes the following transactions:

▶ BOM and routing maintenance transactions (described in this section)

▶ characteristic and configuration profile maintenance transactions (described in Section 2.6, Object Dependencies for the Value Assignment Interface or the Sales View)

Another maintenance environment for local object dependencies is the product modeling environment (Product Modeling Environment for Variant Configuration, PMEVC). However, the PMEVC cannot be used to maintain object dependencies for routings, and only to a limited extent for BOMs.

The process for global object dependencies theoretically consists of two steps:

1. **Maintenance of global object dependencies**
   Use Variant Configuration Transactions CU01, CU02, CU21, and CU22.to maintain global object dependencies, irrespective of their assignment to objects.

2. **Assignment of global object dependencies**
   To assign a global object dependency to an object (BOM, routing, characteristic, configuration profile), use the object's maintenance transaction.

You can carry out these two steps separately or as one step. If you want to perform them in one step, start with the second step, the assignment. If you try to assign non-existing global object dependencies, you can create the object dependencies directly via the assignment. Except for constraint nets and thus for constraints, you can even change global object dependencies via their assignment. This can be done only for one-time use, however. If global object dependencies are assigned several times, object dependencies cannot be changed via the assignment. You can also use the PMEVC to completely maintain global object dependencies, but the restrictions for the BOM and routing apply in this case too.

As already mentioned, local object dependencies are created and directly assigned from the maintenance transaction of the corresponding object. In Figure 2.41, BOM maintenance was called for this example (❶). The menu path EXTRAS • OBJECT DEPENDENCIES • EDITOR provides a general method to create local object dependencies. The system then displays a small window (❷) where you can select the object dependency type. This takes you to the editor.

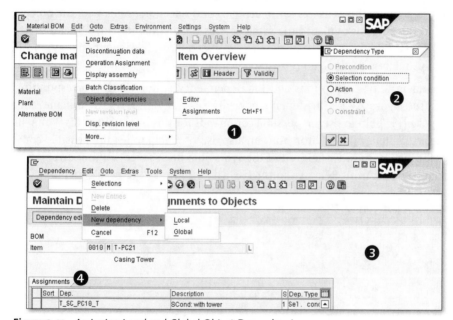

Figure 2.41 Assigning Local and Global Object Dependencies

The menu path EXTRAS • OBJECT DEPENDENCIES • ASSIGNMENTS provides a general method to create global object dependencies. The 🗟 icon (if available) has exactly the same function as this menu path. The system then opens a screen (❸). The list of assignments (❹) indicates if object dependencies have already been assigned. Local object dependencies are indicated by their purely numeric names, and global object dependencies by their non-numeric names. From this screen (as shown in Figure 2.41), you can create additional local object dependencies via the menu path EDIT • NEW DEPENDENCY • LOCAL. You assign global object dependencies irrespective of whether they already exist or are supposed to be created via the assignment by including the name in the list of assignments. If the global object dependencies already exist, the system displays a row as shown in Figure 2.41.

If no global object dependencies exist yet, the system displays a window in which you can create the new object dependency with or without Engineering Change Management and with or without a template. You can use both local and global object dependencies as templates for global object dependencies.

The description is maintained in the Basic Data screen of the object dependencies (see Figure 2.42). In general, the status is initially set to In Preparation and cannot be changed to Released at this point. To release it, the syntax must be free of errors. An empty object dependency is also an error. The dependency can be assigned to a dependency group; this is useful for structuring and selection as it is with the objects in the classification system, and the same applies for the maintenance of authorizations. The last option in the basic data is the selection of the dependency type. All data in this tab can be changed retroactively.

You can navigate to descriptions and the dependency editor via the corresponding buttons (see Figure 2.42). The dependency editor initially displays a command line. You can use F1 help to list the possible commands for this line. Below this, you will find the actual lines of the editor. You can also use F1 help to list the possible commands for the line number field. After the syntax has been saved without errors, you can change the status to Released in the basic data and save the new dependency.

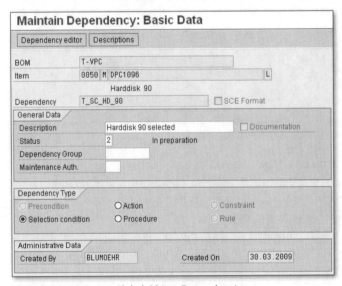

Figure 2.42 Creating Global Object Dependencies

2.7.2 Selection Conditions for BOM and Routing

To explode BOMs and routing, you can use object dependencies that are directly assigned to this master data. For this purpose, selection conditions and procedures are available.

Selection conditions basically control which items are so-called variant parts. All items from the super BOM without selection conditions are transferred to every exploded BOM. You could also speak of non-variable parts here, even though a general option of changing them via procedures is provided. The syntax of such selection conditions contains a logical expression. This expression can consist of any number of separate logical expressions linked with `and` or `or`. In the syntax, characteristic value assignments are queried at the `$parent` or `$root` level. Just as with the preconditions in Section 2.6.5, you can also use variant tables here.

You can assign selection conditions for BOMs and routings to the following elements:

- BOM items
- Sequences (except for the standard sequence) of the routing
- Operations and suboperations of the routing
- Production resource and tool assignments in the routing

You cannot assign selection conditions to BOM and routing headers, which means you cannot control selection of the BOM and routing via object dependencies. You can implement something similar for routings if you don't use parallel or alternative operations in the standard version. In this case, instead of creating alternative routings, you can map them as alternative sequences in a routing. You can then provide selection conditions for them.

The rules for selection conditions, as described previously, also apply here: A comparison test of a characteristic to which values haven't been assigned is considered as not met. When multiple selection conditions are assigned, it is sufficient that one condition is met to transfer the corresponding element to the explosion (or link).

2.7.3 Class Nodes in BOMs

Class nodes are BOM items containing material classes or document classes for which the usage in BOMs has been allowed via the class definition. In a super BOM, a class node is used as an indirect assignment of all materials or documents that have been classified in the corresponding class. In the exploded BOM, however, the class node must be replaced by an object of the corresponding class.

Class nodes are used for the following reasons:

▶ **Simplified BOM maintenance**
If a group of objects is required in the same form in several BOMs, it is easier to collect these objects in a class and include only this class in the BOMs than to directly include the entire group of objects in the BOMs.

▶ **More elegant BOM maintenance**
A class node in the BOM simplifies a BOM much more than a group of objects. It is possible for the super BOM to have the same number of elements (class nodes in this case) as the exploded BOM (material or document items in this case).

▶ **Increased performance for the BOM explosion**
Class nodes with numerous assignments can be replaced more quickly than a large number of selection conditions.

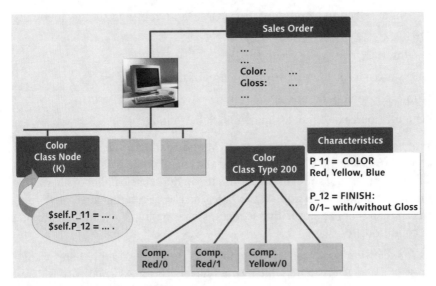

Figure 2.43 Class Nodes in BOMs

To explode class nodes, you require information regarding the characteristics through which the objects have been classified in the class node. In the example in Figure 2.43, these are the P_11 and P_12 characteristics, which contain the value assignment regarding color and finish. To explode class nodes, you can use four different methods:

▶ Value assignment using a procedure

▶ Automatic search of value assignment and replacement

▶ Value assignment using a constraint

▶ Manual replacement of the class node

Let's take a closer look at these methods.

Value Assignment Using a Procedure

The method of a value assignment using a procedure can be universally used and should therefore be preferred. You can apply this method to all configuration scenarios (see Section 2.4) and together with the configurator of the ERP system and the IPC. Here, a procedure is assigned to the BOM item of the class node. In the syntax, values are assigned to the characteristics of the class node. If values have already been assigned to these characteristics in the value assignment interface, the syntax is as follows (Listing 2.25):

```
$self.P_11 = $parent.P_11,
$self.P_12 = $parent.P_12.
```

Listing 2.25 Procedure for the Class Node — Forwarding the Customer Request

Instead of the $parent level, you can query the $root level. However, if values haven't been assigned to these characteristics in the value assignment interface, they have to be derived from other value assignments. In this case, the syntax can be as follows (Listing 2.26):

```
$self.P_11 = 'red' if $parent.char1 = 'value1',
$self.P_12 = '1'   if $parent.char2 = 'value2'.
```

Listing 2.26 Procedure for the Class Node — Value Assignment Without Customer Request

For this purpose, you can also use a variant table, which would look as follows (Listing 2.27):

```
table tab_cl_n1 ( char1 = $parent.char1,
                  char2 = $parent.char2,
                  p_11    = $self.P_11,
                  P_12    = $self.P_12).
```

Listing 2.27 Procedure for the Class Node — Value Assignment Without Customer Request but with Table

Automatic Search of Value Assignment and Replacement

The method of automatic search of value assignment and replacement can be used for configuration scenarios with the configurator of the ERP system but is not compatible with the configurator of the IPC. (This means that modeling for the IPC always requires a procedure for the class node.) The ERP configurator searches the levels above the $self level for value assignments of the characteristics of the class node if it cannot find a value assignment for the class node. This may be the case if no value assignment has been implemented for the class node by means of a procedure. It is therefore sufficient to include the characteristics of the class node in the variant class of the header material in order to obtain a value assignment in the sales order as shown in Figure 2.43.

Value Assignment Using a Constraint

The method of value assignment using a constraint requires that BOM information is saved while the constraint is active. Consequently, this method requires the order BOM, or the sales order (SET) configuration scenario. Here, the condition (of saving of BOM information)is met because the system creates an order BOM or generates items in the sales order that are relevant to sales and to requirements while the constraint is processed, that is, during the interactive configuration.

Manual Replacement of the Class Node

The method of manual replacement of the class node also requires the *order* BOM or the sales *order (SET)* configuration scenario. A class node that has been manually replaced by a material or document item must also be saved during interactive configuration. This can be done by creating an order BOM or by generating an item that is relevant to sales or to requirements in the sales order.

For smooth replacement of class nodes, it is required that the characteristic value assignment is complete, and that the search result is unambiguous. This results from the following methods of the configurator:

- **No values have been assigned to the characteristics of the class node**
  (in Figure 2.43 neither P_11 nor P_12 have assigned values)
  In this case, the system doesn't search for possible objects. The class node is retained in the simulated explosion of the BOM. This doesn't lead to error messages in the standard version. Exceptions are the order BOM and sales order (SET) configuration scenarios and class nodes with the corresponding indicator in the definition. A class node that hasn't been replaced is interpreted as a text item in the downstream supply chain; that is, it is ignored. In this case, the example would not include a color. If this is what you want, you should use the corresponding selection condition for the BOM item of the class node instead.

- **Values are not assigned to all of the necessary characteristics**
  In this case, the configurator starts an object search with this partial value assignment. It uses any object of the search result for the replacement. You cannot affect the result in the standard version. An appropriate user exit is available, however.

▶ **Values are assigned to all characteristics of the class node: unambiguous result**
Values are assigned to all characteristics of the class node, and an unambiguous result is found in the subsequent object search of the configurator. In this case, the unambiguous result is used for the replacement. Strictly speaking, this is the only clear explosion of the class node.

▶ **Values are assigned to all characteristics: ambiguous result**
Similar to the incomplete value assignment, any object is used for the replacement here. The same problems as mentioned for the other methods occur.

▶ **Values are assigned to all characteristics: no result**
No replacement is implemented in this case. The class nodes remain in the simultatively exploded BOM. The same problems as mentioned for the other methods occur here as well.

The listed reactions of the configurator with regard to the class node have the following consequences:

1. You should aim for a complete value assignment of all characteristics of the class node.

2. You should avoid identical value assignments of objects in the class node.

3. There is exactly one object for each value assignment combination with regard to the class node — or you only allow for value assignments for which an object is available in the class node.

You can meet the first requirement using required characteristics.

The second requirement can be fulfilled by making the corresponding setting in the basic data of the class node's class definition (Same Classification section).

The third requirement is somewhat more difficult to fulfill. A completeness check in the form mentioned is not provided as a standard function. If you use restrictable characteristics, you can restrict their value lists in such a way that only value assignments of existing objects are permitted. To do so, you can again use variant tables or coupled database tables. The tools of the classification system enable you to generate and automatically fill such a database table (Set up Tables for Search, Transaction CLGT).

2.7.4 Classified Materials in BOMs

In addition to the two described options of BOM items with selection conditions and the class node as the BOM item, another method is available for the configurator of the SAP ERP system. This method selects components for the exploded BOM, which can be used for *classified materials* (see Figure 2.44).

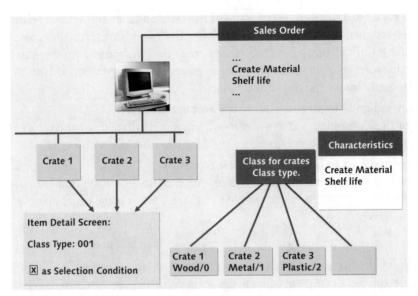

Figure 2.44 Classified Materials as BOM Items

However, this method is not IPC-compatible. As already mentioned, this method provides classified materials. The class type is not relevant here. In most cases, this is material class type 001. These materials are required in the super BOM and are only supposed to be selected if materials with exactly these classification properties are required.

You can use this classification in Variant Configuration. You don't need selection conditions for the items, so you don't have to write them. You cannot include this material class in the super BOM, because it is not a class node. For performance reasons and other reasons, you shouldn't enable this via the Customizing of the classification system for class type 001.

You can therefore only directly include these material masters in the BOM. In contrast to the option that was first introduced (in Section 2.7.2), you don't require selection conditions here. Setting the corresponding indicator is sufficient. You can select the As Selection Condition checkbox in the details of the respective BOM items in the Basic Data tab, which defines that classification is to be interpreted as a selection condition (see Figure 2.45). No selection condition is generated at this point. The system dynamically generates a selection condition from the complete value assignment of the classification during the BOM explosion. This selection condition only allows for selecting this item if values have been correspondingly assigned to all characteristics of the classification in the configuration.

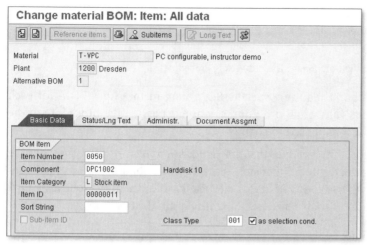

Figure 2.45 Classifying a Class Type as the Selection Condition

2.7.5 Procedures in BOM and Routing

In addition to the previously described usage for the explosion of class nodes, procedures are mainly used with reference characteristics in BOMs and routings to implement changes in the exploded structures. As explained in Section 2.2, Tools from the Classification System, you can make changes to the following details via reference characteristics with reference to the fields of the corresponding database tables:

- BOM items (STPO)
- Operations of routings (PLPO)
- Sequence assignments of routings (PLFL)
- Assignment of production resources and tools (PLFH)

For elements without selection conditions and for elements with selection conditions that are met, the system evaluates the procedures after the selection conditions have been evaluated for the elements that were transferred to the exploded BOMs. (Procedures don't enable you to change the elements that aren't transferred according to the selection condition.)

In this context, changes of the item quantity in the BOM and changes to the size and quantity for variable-size items are good examples. If the QUANTITY characteristic is an reference characteristic with reference to the STPO database table of the BOM items and the MENGE (quantity) field, you can use the following procedure to overwrite the required number of screws for particular casings (see Listing 2.28).

```
$self.quantity = 12   if $root.casing = 'tower',
$self.quantity = 10   if $root.casing = 'minitower',
$self.quantity = 8    if $root.casing = 'desktop'.
```

Listing 2.28 Changing the Component Quantity Using Procedures

Consequently, you don't require a specific BOM item in the super BOM for each number. As a general rule for all procedures, you must assign the procedure to the item whose values it is supposed to change.

The example in Figure 2.46 assumes that the value assignment interface provides the option of querying the desired dimensions of the casing. The BOM always includes the same metal for the various variable-size items. Procedures of the following type determine the size and variable-size item quantity again (Listing 2.29).

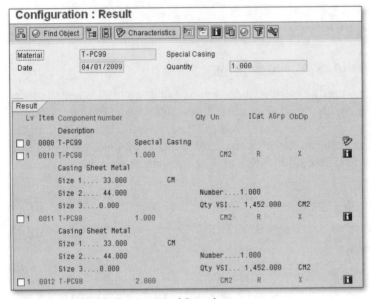

Figure 2.46 Variable-Size Items and Procedures

```
$self.t_stpo_roms1 = $parent.t_20 ,
$self.t_stpo_roms2 = $parent.t_21 ,
$self.t_stpo_romen = $parent.t_20 * $parent.t_p21.
```

Listing 2.29 Changing Variable-Size Items Using Procedures

> **No Checks for Reference Characteristics** [!]
>
> Procedures with reference characteristics for BOMs and routings only change the specified field. Consequently, no checks or updates are carried out as usual if you implement the same changes in the dialog in the corresponding maintenance transactions for the BOM and routing maintenance.
>
> The example in Figure 2.46 requires, for instance, that the system recalculate the variable-size item quantity if changes are made to the size via procedures.
>
> You therefore shouldn't change fields with strong dependencies, such as the material number in BOMs or work centers in task lists, using procedures. The system doesn't read the corresponding master data in this case.

Procedures are normally used in routings for changes to time fields during operations, such as to change standard values. Up to six standard values can be provided. These standard values are stored in the PLPO database table in fields VGW01 to VGW06. So-called *standard value keys* in the work center define the number and meaning of the standard values that are active for the individual operation. For example, if the labor time in an operation is supposed to be increased by five minutes depending on the configuration, and the labor time is the third standard value, an reference characteristic with reference to PLPO and VGW03 (for instance, `plpo_vgw03`) is required. The corresponding procedure could be as follows (Listing 2.30):

```
$self.plpo_vgw03 = mdata $self.plpo_vgw03 + 5 if ...
```

Listing 2.30 Increasing a Standard Value Using Procedures

The `mdata` syntax element ensures that the initial value is read from the super task list. If the time is supposed to be successively increased using several statements, `mdata` must not be included in the subsequent statements. Note that the syntax doesn't ensure that the time is increased by five minutes and not by five hours, for example. To ensure this, the unit of measurement of the standard values needs to be read via an appropriate reference characteristic as well.

2.8   Pricing for Configurable Materials

Of course, the price of a configurable material must depend on its specific configuration. This can be directly mapped, as this chapter introduces. For this purpose, surcharges are directly linked to the values assignments of the configuration. You can also indirectly evaluate the configuration for pricing. For more information, refer to Section 2.9, Product Costing for Configurable Materials.

Within Variant Configuration, product variability leads to price variability. In addition to providing standard pricing tools, Variant Configuration also allows direct value-dependent pricing. In this case, the price consists of a base price plus value-dependent surcharges and reductions.

If you create a sales order as shown below ❶ in Figure 2.47, the value assignment interface provides you with price information. Depending on the setting in the configuration profile, pricing is carried out in the value assignment screen either continuously or only if requested. In the figure, the first case is applied. If pricing is supposed to be implemented only on request, you can request pricing via a pricing icon located directly below the title bar.

You can assign the surcharges and reductions to the model in such a way that they are directly displayed in the list of allowed values (F4 help). The Net Value section (❷ in Figure 2.47) displays the total price. It consists of a base price which is not affected by the configuration, and configuration-dependent surcharges and reductions. Surcharges and reductions can be mapped with the condition technique, and evaluated using the Conditions... button.

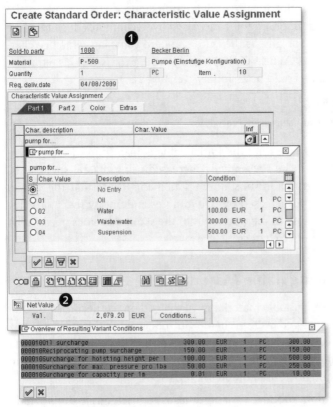

Figure 2.47 Value-Dependent Pricing

If you return from the characteristic value assignment to the sales order and navigate to the conditions of the corresponding order item, the system displays a screen that also specifies value-dependent surcharges.

Pricing in Sales and Distribution

To model such value-dependent pricing in sales and distribution, perform the following steps:

1. Create a reference characteristic with reference to SDCOM-VKOND.
2. Create condition records for the VA00 and VA01 condition types.
3. Check the pricing procedure for these condition types.
4. Link the Variant Configuration records with the model.

We'll now explain these steps in detail. First, you require a reference characteristic with reference to the SDCOM structure and to the VKOND field for the communication between Variant Configuration and the SD document. In Figure 2.48, this is implemented via the P_26 characteristic (❶). The reference characteristic serves to include the variant condition keys. It should have multiple values so that it can include all required keys of the condition records. The reference characteristic is also included in the variant class. For this reason, you should configure the reference characteristic in such a way that it is not displayed.

In the second step, pricing uses the condition technique. The total price consists of condition records of different standard condition types. A base price is required here. The standard version uses condition type PR00. You can also use condition records of the VA00 (for absolute surcharges) and VA01 (for relative surcharges) variant condition types. The example in Figure 2.48 uses VA00 only. You create the keys of the variant condition records and descriptions.

As you can see in Figure 2.48, this can be implemented in the product modeling environment (PMEVC) (❷) or in a sales transaction (Transaction VK30). The actual amounts for the different condition types are entered in Transaction VK11 for sales and distribution.

In the third step, the pricing procedure that is used in the SD document must consider the corresponding condition types. This is already ensured in the standard version; bear this in mind, however, if you create the pricing procedure yourself.

In the last step, the new variant condition records are linked to the model. Define which condition records are supposed to be used when. If the assignment is kept simple, that is, a characteristic value selection leads to a variant condition record in the configuration, you can map this without object dependencies. This direct assignment of condition record to characteristic values can be implemented in the product modeling environment (PMEVC), which is part of the master data maintenance as illustrated in Figure 2.48, beneath ❸. The simulation transaction (Transaction CU50) also allows for such an assignment — but with less elegance. In Transaction CU50, the list of allowed values (F4 help) is called in the value assignment interface. No values are assigned to the characteristic. Select the characteristic value to which a variant condition record is to be assigned, without assigning a value. The list of allowed values contains a Conditions button. Here, you can assign the variant condition without confirming with ↵ . Instead, immediately complete the process with Save. If you implement an assignment via the simulation transaction, you should recall it to check whether all assignments have been saved properly.

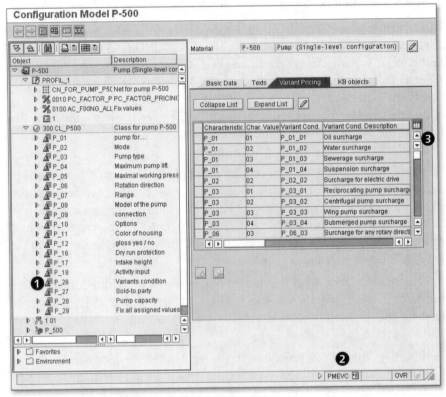

Figure 2.48 Modeling Value-Dependent Pricing

In addition to direct assignments, you can also implement assignments via object dependencies. For this purpose, you assign the name of the variant condition record to the reference characteristic in procedures.

A procedure for item 0010 with the following syntax was assigned to the model in Figure 2.48. [Ex]

```
$SELF.P_26 = 'P_04_1'      if $self.p_04 specified,
$SELF.P_26 = 'P_05_1'      if $self.p_05 specified,
$SELF.P_26 = 'P_28_1'      if $self.p_28 specified,
$SET_PRICING_FACTOR($SELF,P_26,'P_04_1',P_04)
                           if $self.p_04 specified,
$SET_PRICING_FACTOR($SELF,P_26,'P_05_1',P_05)
                           if $self.p_05 specified,
$SET_PRICING_FACTOR($SELF,P_26,'P_28_1',P_28)
                           if $self.p_28 specified.
```

This method is particularly useful if the conditions are complex. You can also use this method if you use factors with the $set_pricing_factor syntax element. This syntax element multiplies a condition record that has already been assigned to the reference characteristic by a factor (which effectively defaults to 1). The corresponding syntax is as follows: $set_pricing_factor($self, <name of the reference characteristic>, <name of the variant condition>, <factor>).

In this case, the factor can even be any complex, arithmetic expression for which numeric characteristics can be used. The factor determination is canceled if the name of the variant condition record hasn't been assigned to the reference characteristic yet, if no values have been assigned to the numeric characteristics in the factor, or if the factor is 0. Note in the example that the procedure's syntax is only correct if the value 0 is not allowed for the numeric characteristics, P_04, P_05, and P_28. Otherwise the assignment of 0 would have the result that the respective variant condition record becomes effective with factor 1.

Consider the Case Sensitivity of Variant Condition [!]

The names of variant conditions are case-sensitive, so the corresponding reference characteristic must also be case-sensitive. This is automatically set. You can also define variant conditions first, and then assign them directly or via object dependencies to the model. This can be implemented in the reverse order (first assign, then define) as well. This variability, however, has the result that no existence check is carried out when variant condition records are used—neither for the assignment to the model nor for the definition of the values of the variant conditions.

The simple value-dependent pricing introduced here is only one of the possible pricing methods. Here, a technique was used that has been especially developed for Variant Configuration. BOM and routing explosion don't play any role in this context.

However, you can also use BOM and routing information for pricing. This is especially helpful if you use order BOMs and order routings. In this context, product costing with quantity structure forms the basis. This is introduced in Section 2.9, Product Costing for Configurable Materials.

Pricing in Purchasing

Similar to the pricing method we described for sales and distribution, you can also model a pricing method in purchasing.

This can be used in various scenarios, such as:

- External procurement of the entire product that is configured in the sales order, for example, in the case of bottlenecks or third-party transactions
- External procurement of individual configurable assemblies of the product that is configured in the sales order
- External procurement of configurable materials without reference to sales

Figure 2.49 illustrates an example of external procurement of a configurable assembly. The sales order was created for a configurable material, P-502, as you can see in the stock/requirements list. The resulting dependent requirement with regard to material 500–800 is supposed to be satisfied through a purchase order.

The configuration of the crate that is to be ordered according to sales order for the P-502 material is copied to the purchase order. You can use the icon to call this configuration from the material data in the purchase order.

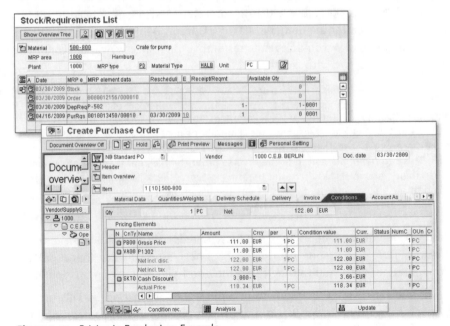

Figure 2.49 Pricing in Purchasing: Example

The characteristic value assignment includes the packaging material and the variant conditions. The latter are not displayed and have been assigned with values using a procedure.

The Conditions tab in the purchase order shows how the total price is determined from a base price (PB00) and a variant condition (VA00), P1302. The modeling process for this, that is, for value-dependent pricing in purchasing, is quite similar to the modeling for sales and distribution described above and roughly consists of the following steps:

1. Create an reference characteristic with reference to MMCOM-VKOND
2. Create condition records with regard to the VA00 and VA01 condition types
3. Check the pricing procedure with regard to these condition types
4. Link the Variant Configuration records with the model

We'll explain these steps in detail.

The communication between Variant Configuration and the purchase order also requires a reference characteristic. It is, however, created for the MMCOM structure, rather than SDCOM. The reference characteristic is also included in the variant class.

Pricing also uses the condition technique here. However, the condition records are maintained in purchasing, rather than sales and distribution. You can create the condition records from the purchasing info record. In this example, the base price €111 of the PB00 condition type and various VA00 variant conditions, such as €11 for P1302, were created in the purchasing info record for material 500–800 and vendor 1000.

The pricing procedure that is used in the purchase order must consider the corresponding condition types. This is already ensured in the standard version; bear this in mind, however, if you create the pricing procedure yourself.

In the last step, the new variant condition records are linked to the model. You cannot use the direct assignment of condition type to characteristic value as described above. This is only implemented for sales and distribution; so here, you must use object dependencies.

Assignment via object dependencies is analogous to the procedure described above. For this purpose, you assign the name of the variant condition to the reference characteristic in procedures. You can also use factors here when using the `$set_pricing_factor` syntax element.

2.9 Product Costing for Configurable Materials

You have various options for implementing product costing for concrete configurations in order to use them as sources for pricing, for example.

As mentioned in Section 2.8, Pricing for Configurable Materials, pricing can be done based on product costing with quantity structure. For this purpose, you carry out sales order costing — that is, product costing with quantity structure — from the sales order considering the corresponding configuration. If order BOM or order routing is available, they are used. Apart from that, the super BOM and super task list are used, that is, they are exploded according to the configuration and object dependencies.

As you can see in Figure 2.50, this sales order costing is called from the SD document via menu path EXTRAS • COSTING. You can control costing directly or indirectly via the settings in the material master. To control it directly, use the settings of the requirements type and requirements class, as described in Chapter 4, Customizing SAP ERP for Variant Configuration.

You can save the sales order costing result. You can also include it in pricing. In the standard version, you can include condition types EK01, EK02, and EK03 in the costing result as a part of pricing.

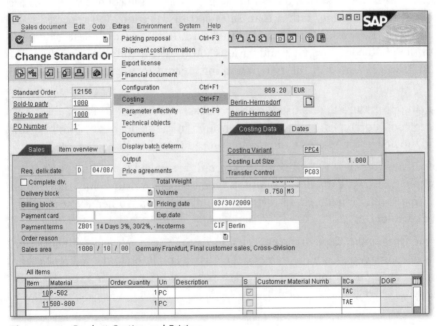

Figure 2.50 Product Costing and Pricing

The EK01 condition type is a base price. You use this condition type if you don't want to carry out value-dependent pricing. In this case, you determine, for example, the price that is based on the EK01 base price and on a certain surcharge for overhead costs and profit. This procedure with the EK01 condition type is particularly helpful if you use incomplete models, order BOMs, or order routings. Changes to and by order BOMs and order routings don't directly affect pricing in the case of mere value-dependent pricing. However, also in the case of pricing including the costing result with the EK01 condition type, you should ensure that another product costing with quantity structure (that is, with BOM and routing information) and pricing is carried out after the order BOM and order routing have been maintained.

The example in Figure 2.50 assumes that value-dependent pricing (as described in the previous section) has already been used. You can therefore not use the EK01 condition type to include the costing result. In simple terms this means you would pay twice; once for requesting something (variant conditions) and once for receiving that something (costing result). For the case explained here, you can use the EK02 and EK03 condition types to calculate the profit margin — EK02 for the total margin or EK03 for a fixed portion. You can define in the Customizing of the order types which condition type is to be used in the sales order pricing. In this case that means the EK02 condition type was assigned to the standard order type.

In the standard version, you can carry out product costing in the following documents:

- Configuration simulation (Transaction CU50)
- SD document (for example, sales order)
- Material costing with quantity structure (Transaction CK11N)

You cannot save the costing result in the simulation transaction but can save it in operational transactions. To do this, you require a configurable object, for example, an SD document. Material costing with quantity structure also requires a configuration to which the costing result refers. For this purpose, a configuration must be stored in the material master, that is, the material must be a material variant. The following section describes material variants in further detail.

2.10   Material Variants

Material variants are the central tool for working independently or anonymously of a configuration in the sales order or other documents in Variant Configuration.

The traditional scenario with configurable materials is make-to-order production, which is triggered by sales orders. Sales orders contain the configuration but also provide pegged requirements. The customer requirement is included in material requirements planning as an individual customer requirement and generates a sales-order-specific planned order, which in turn, is included in a sales-order-specific production order in the case of in-house production. The result of the procurement process is a special stock of the configurable material. This is the only way to assign the requirement of a configuration to a sales order. In this process, the sales order that activates the requirement is the triggering element. You cannot implement procurement or production processes without sales orders.

You can bypass this by means of material variants. In the Variant Configuration environment, you may also want to use make-to-stock production for specific configurations. Material variants enable you to link the traditional make-to-order production in Variant Configuration with make-to-stock production. In Variant Configuration, you can also implement make-to-stock production for specific configurations. In this case, the sales order doesn't contain the configuration. You must find another way to map the configuration for materials that were made to stock. To do so, create specific material masters for materials that are supposed to be made to stock or have already been made to stock. These are generally non-configurable materials masters, namely, material variants.

A material variant is a material master that corresponds to a specific configuration of a configurable material. This can be mapped, because the material master of the material variant contains a link to a configurable material and a configuration for this configurable material. The material master and the configuration clearly describe the corresponding product so that the product — in contrast to the configurable product — can be made to stock and stored anonymously, that is, without reference to a sales order. The link to the configurable material also enables you to manually or automatically determine the corresponding material variant. Consequently, sales and distribution doesn't have to have an overview of the existence of material variants. If required, the system can automatically search for material variants, check the availability, and redirect requirements for the configurable material and the existing configuration (see Figure 2.51).

So far, we've only discussed only the modeling of configurable materials, not yet material variants. Let's recall what has been created for configurable materials:

▶ Material master

▶ BOM, that is, super BOM including object dependencies

▶ Routing, that is, super task list including object dependencies

▶ Pricing in the form of condition records, assignments, and object dependencies

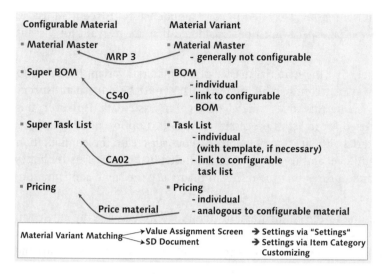

Figure 2.51 Configurable Material and Modeling of the Material Variant

Now you have to model the material variant. For this, you require a specific material master. Because material variants are also supposed to be planned, manufactured, and sold, you also require a BOM, routing, and pricing. In simple terms, you can say that you can either create specific data or use the data of the configurable material. The following sections discuss these options in more detail.

2.10.1 Material Master of the Material Variant

As already mentioned, material variants are specific, non-configurable material masters. However, there are exceptions to this:

▶ Storage of a material variant under the material number

▶ Definition of configurable material variants

You can store a material variant under its configurable material's material number. For this purpose, you must link the material master to itself (as described later) in order to configure it. The second exception is that you can also define configurable material variants. These configurable material variants are only partly configured. With regard to this partial configuration, you can also implement anonymous sub-assemblies for stock. In this case, the final configuration is carried out in the sales order based on the configurable material variant so that the final assembly is also implemented as a make-to-order production for the specific sales order.

In general, the material master is created in correspondence to its required usage. In this context, no special aspects must be considered that are relevant for Variant Configuration.

It is critical, however, that the material master of the material variant be linked to the material master of the configurable material. This must be done plant-specifically. You can also create this link at the client level in basic data. However, this method cannot be used for logistics processes. Why is this option provided, then? If you use this material variant in various plants, you can create the configuration centrally at the client level and then copy it to the individual plants. This facilitates data maintenance and ensures that the different plants use the same configuration, if preferred.

In Figure 2.52, the link to the configurable material, P-500, was already created in the Basic Data 2 tab (❶) and thus at client level. The Configure Variant button enables you to open the value assignment interface for the P-500 material and configure the material variant.

If you have configured at client level (as was done in Figure 2.52), the MRP 3 tab (❷) displays an additional, third button (❸), Copy X-plant configuration, at the plant level. This button enables you to plant-specifically copy the link including the value assignment. Otherwise you have to implement the configuration here.

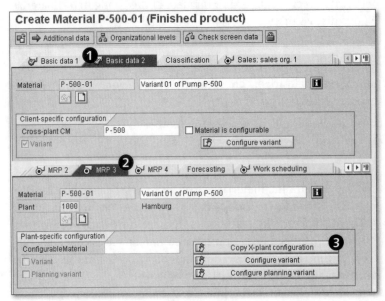

Figure 2.52 Material Master: Linking the Material Variant to the Configurable Material Master

2.10.2 BOM and Material Variant

The following two options are provided for additional master data for the material variant, that is, for the BOM, routing, and pricing:

▸ Using specific data

▸ Creating a link to the corresponding data of the configurable material

This enables you to create the BOM of the material variant via the traditional BOM transactions or the Engineering Workbench. In this case, you're completely independent of the BOM of the configurable material. Support for the creation of such BOMs is not provided in the same way as for the corresponding function for routings, which is described later in this section.

If you don't have to consider any special aspects for the BOM of the material variant compared to the BOM of the configurable material, no specific BOM is necessary for the material variant. This is the case if you want to use a BOM for the material variant that exactly corresponds to the result of the BOM explosion for the configurable material, provided that you use a configuration that corresponds to the material variant.

Here, it is much more elegant to create a link to the BOM of the configurable material. **[+]**
This has the advantage that only one BOM exists and all changes to the super BOM automatically also applies to the material variants.

You can decide for each material variant if you want to use a specific BOM or only a link.

A specific transaction, Transaction CS40 (Create Link to Configurable Material), is provided for the creation of such links. You access this transaction with the material number, the material variant, and the plant. If several BOMs are available for the configurable material, you can use BOM Usage and Alternative BOM to select the BOM to which an assignment is supposed to be made. These four specifications are sufficient. You don't require the material number of the configurable material here, because it is clearly defined in the material master of the material variant.

2.10.3 Routing and Material Variant

Similar to the BOM, you can use the following two options for the routing:

▸ Specific routing

   ▸ Creating a new routing

   ▸ Using a routing as a template

▸ Assignment to the routing of the configurable material

If you want to use a specific routing for the material variant, for example, to map special aspects, you are again provided with two options.

Irrespective of the routing of the configurable material, you can create another routing. For this purpose, the traditional transactions of the routing maintenance and the Engineering Workbench can be used. The second option for creating a specific routing is to use a routing exploded in simulation as a template. This option is provided in all transactions in which routings exploded in simulation are possible. This includes the simulation transaction (Transaction CU50), the configuration simulation in the product modeling environment, or the maintenance transaction of the order BOM (Transaction CU51) and additional transactions, such as the maintenance transactions of the SD documents if routing explosions are possible within the configuration.

To explode routings, navigate from the value assignment interface of one of the transactions mentioned in the previous paragraph to the result screen. You must navigate to the result screen of the routing. For this purpose, you can use the 🗂 icon (Task List) to navigate from the BOM to the routing. The result screen of the routing contains the [⊙ Task list] button, which comprises exactly this function of creating a routing using the explosion result as a template. The [⊙ Task list] button takes you to the neutral transaction for the creation (Transaction CA01) of routings, which provides full flexibility with regard to the key and the routing's reference. What does full flexibility mean?

▶ **Template for the routing of a material variant**
This function is not restricted to the creation of specific routings for material variants. Here, you enter the material number of the material variant as the material master.

▶ **Template for an order routing**
You can also explode routings in simulation with regard to the configuration of a sales order. In this case you can use this function to create an order routing. For this purpose, you must enter the configurable material including the sales order number and item number in the task list header. This option for creating order routings is not tied to any specific transaction. One variant is to implement this directly from the order BOM maintenance and use the transaction provided there. However, you can also implement this from the simulation transactions or the transactions of SD document maintenance.

▶ **Template for a routing of any material**
Because you can specify any material number in the header of the routing that is supposed to be created, you can also use this for routings of any materials — materials that initially have nothing to do with Variant Configuration. You just

need to have a routing that is similar to the routing exploded in simulation (*see* Figure 2.53).

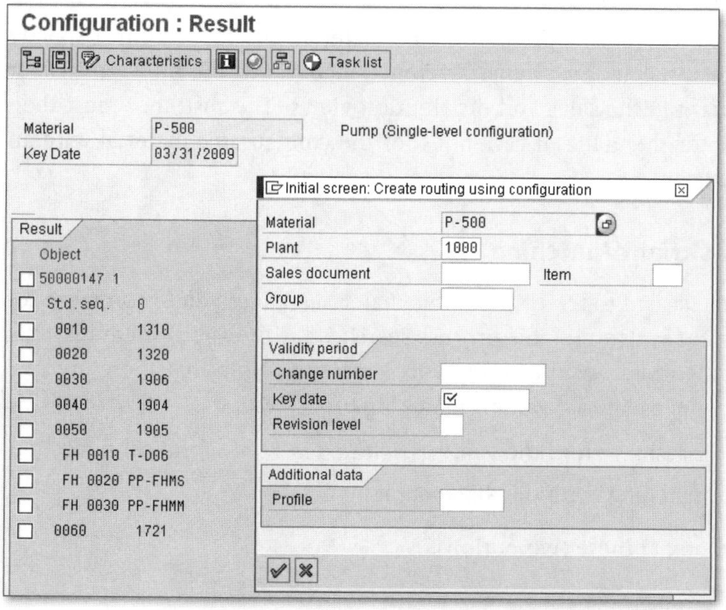

Figure 2.53 Routing Exploded in Simulation as a Template

Similar to the BOM, if you don't have to consider any special aspects for the routing of the material variant compared to the routing of the configurable material, no specific routing is necessary. This is the case if you want to use a routing for the material variant that exactly corresponds to the result of the routing explosion for the configurable material. Again, you can decide for each material variant if you want to use a specific routing or only a link.

In contrast to the BOM, linking a routing to multiple material masters is nothing special here. Therefore, no separate transaction is provided for linking the material variants to the routing of the configurable material. You can create this link in the normal maintenance transactions of the routings via the assignment of materials to the task list.

2.10.4 Pricing and Material Variant

Because make-to-stock production with large lot sizes generally includes cost benefits, you can pass on these benefits to the customers in the form of pricing. This entails maintaining specific pricing for the material variant. You can define separate condition records for the material variant. The material variant can use variant conditions too. In this case, the keys of the variant conditions must be selected in the same way as those of the configurable material. The configuration model of the configurable product also defines the logic for when each condition record is effective.

If you don't want to use specific pricing for the material variant, you must specify the configurable material as the pricing reference material in the material master of the material variant in the sales and distribution views. The customer must then pay the same price for the material variant as for the configurable material with an analogous configuration.

2.10.5 Material Variant Matching

Linking the material variant to the configurable material enables you to configure so-called material variant matching. This process searches for the appropriate material variants in the configuration of the configurable material. Note that there are two completely independent material variant matching processes:

▶ Material variant matching from the value assignment screen of the configuration
▶ Material variant matching from the SD document

Let's have a closer look at these two options.

Material Variant Matching from the Value Assignment Screen

If you navigate from the SD document to the value assignment interface, you make use of the first option: material variant matching from the value assignment screen (see Figure 2.54). The same applies if you call the value assignment interface from other transactions (simulation, purchase order, production order, and so on).

You control this material variant matching process via the settings in the configuration profile. As already mentioned, these settings can be overridden user-specifically and changed dynamically during the configuration. You can choose one of the following options in the Variant Matching tab of the Settings:

▶ Permanent variant matching
▶ Variant matching on request

In the first case, variant matching is started each time data is released ($\hookleftarrow$). In the second option, the value assignment interface additionally displays the 🔄 button. You can start variant matching by clicking on this button. Then you can choose one of the following settings in the Variant Matching tab:

▶ Partial configuration
▶ Complete configuration

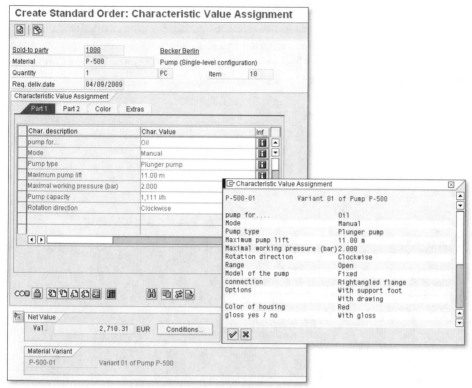

Figure 2.54 Material Variant Matching in the Value Assignment Screen

For the second option, a material variant is only found as the result if the current configuration completely matches the value assignment of the material variant. The first option only compares the characteristics to which values have already been assigned in the current configuration — that is, only this partial configuration. If the system finds several material variants, it displays them all. This material variant matching process is used for information only. As already mentioned, the system displays only the material variants that were found according to the search strategy used (partial or complete configuration). You can also have the system display the complete value assignment of the material variant. Further functions are not available, however.

Material Variant Matching from the SD Document

If you return to the SD document (see Figure 2.55) from the value assignment interface, the material variant matching process just described is no longer relevant.

There is a specific, completely independent material variant matching process for SD documents. This process is not controlled via the settings in the configuration profile but via the Customizing of the item category. The following is only a preview of the corresponding Customizing; Chapter 4, Customizing SAP ERP for Variant Configuration, provides further details.

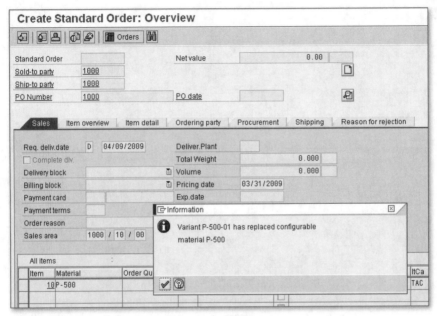

Figure 2.55 Material Variant Matching in the SD Document

In the standard version, you can find the TAC item category for configurable materials in sales orders of the Standard Order order type. As you can see in Figure 2.56, variant matching is selected in Customizing in the Bill of Material/Configuration section. For found material variants, no availability check according to the ATP logic is activated (ATP Material Variant is blank). You can use the Material Variant Action option to make the following settings:

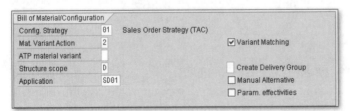

Figure 2.56 Material Variant Matching and Customizing of the Item Categories

► You're only informed about the found material variant via a note

► The material variant automatically replaces the configurable material

You can also decide how often the corresponding action is supposed to be performed:

► Only when an item is newly created

► Also when the item is changed

In Figure 2.56, the automatic replacement option is used. At this point, only the Bill of Material/Configuration section of the Customizing of the item categories is described. For more information, refer to Section 4.3.2, Item Categories and Their Determination, in Chapter 4.

2.10.6 Material Variant Matching at the Header and Assembly Levels

The material variant matching process that this section introduces addresses the header level of a configurable product. The sales order is created with the material number of the configurable product. Material variant matching is implemented at this level, that is, at the main item in the sales order. The process is independent of the configuration scenario that is indicated in the configuration profile.

If it is a complex model with configurable assemblies with the appropriate configuration scenarios, that is, a multilevel, configurable model, material variant matching is theoretically also possible for individual assemblies. Such a material variant matching process is also available at the assembly level. However, for performance reasons, it is not available in low-level configuration.

In a planning run for numerous material masters, for example, integrating such logic would lead to unacceptable performance problems. [Ex]

Therefore, this kind of material variant matching is only available at the assembly level in the high-level configuration, and it is only useful — and thus only implemented — in scenarios in which a result of a BOM explosion can be stored in the high-level configuration. This is only supported for the order BOM scenario. Accordingly, only the order BOM scenario, and only the maintenance of the order BOM in the result screen, provides the option to activate this kind of material variant matching.

► This is also done via the 🔁 button.

► In general, material variant matching is started manually and can therefore not be automated via settings.

▸ Material variant matching usually evaluates the entire BOM. Similar to the material variant matching process described above, the result is generally only displayed in the value assignment screen. The automatic replacement option is not provided.

2.11 How to Create a Product Model for the IPC

When you have carried out the steps described, the configurable product is completely modeled — at least for the traditional configurator of Variant Configuration in SAP ERP. But how does the configurator of the IPC retrieve its model data? As already mentioned, there are two SAP configurators: the configurator of Variant Configuration in SAP ERP and the Internet Pricing and Configurator (IPC).

How do you maintain models for the IPC? One option — in most cases the only manageable option — of setting up a variant model for the IPC is to provide the model that is introduced in this chapter for the IPC. For this purpose, you must basically only collect and format the existing model for the IPC via the knowledge-base object and runtime version. Finally, you copy the model to the database that the IPC uses. In total, there are three options for creating models of configurable products for the IPC:

▸ The modeling in SAP ERP for the traditional configurator

▸ Modeling in SAP CRM

▸ Direct modeling in the IPC with the existing *Product Modeling Environment* (PME)

If you want to use the model for both configurators, you must select the first option. This also applies if Sales and Distribution is generally supposed to work with the IPC, and only the downstream supply chain processes, such as order BOM, order routing, MRP, or production, are supposed to be implemented in SAP ERP. The addressed knowledge base object and the runtime version enable you to convert a traditional model for the IPC in some way; however, this is not possible the other way round. You cannot convert a model that has been created in SAP CRM or in the IPC for the IPC into a model in SAP ERP using a similar knowledge base object and runtime version technology. Accordingly, this section introduces the only manageable option for setting up a model for the IPC.

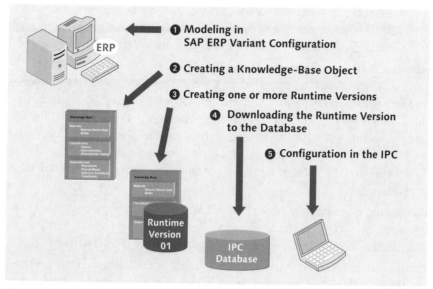

Figure 2.57 Modeling for the IPC

You must perform the following steps in the standard version, at least if the integration with the SAP ERP system is supposed to be fully retained (see Figure 2.57):

1. **Modeling**
   Modeling is done with Variant Configuration in the SAP ERP system, as described in this chapter. You have to consider a few special aspects to use the model for both configurators. These aspects are also discussed in this chapter.

2. **Knowledge base object**
   The knowledge base of the IPC first requires a knowledge base object. This represents the baseline. Here, you define which central object is used to retrieve the necessary data of the variant model. For this purpose, you create at least one (KB) profile associated with the knowledge base object. In the profile, you define the central object, the so-called *OO class*. This is usually the material master of the configurable header material but can also be the variant class. A knowledge base object can contain multiple profiles and thus collect multiple variant models.

3. **Runtime Versions**
   You create runtime versions of the knowledge base object. A runtime version is associated with a specific knowledge base object. A knowledge base object can have several runtime versions. Multiple runtime versions are necessary if different plants are supposed to use different BOM usages or applications or multiple development statuses in parallel. Each runtime version is unique with regard to these aspects. You only need to create several runtime versions for multiple devel-

opment statuses if you want to make them available simultaneously. Apart from that, you can also update existing runtime versions. In this case, the old version is overwritten and the *build* counter (see Figure 2.58, Build# field) is incremented.

4. **Transferring the runtime version to the IPC database**

   The runtime version must be provided on the database used by the IPC. Depending on the system landscape and SAP version, different methods and technologies are available for this transfer. The most elegant method can be used if the IPC is deployed within SAP CRM. In this case, the database of the SAP CRM system is used as the basis, and the runtime version is automatically transferred via the middleware of the SAP CRM system. This requires corresponding configuration of the middleware.

   Another method can be used when SAP CRM middleware is not available. For this purpose, the IPC provides a so-called Data Loader, which can retrieve an existing runtime version in SAP ERP and transfer it to the IPC's database.

   A third method includes a transaction-based export of the runtime version. Transactions that enable you to create the necessary database schema (Transaction CU37) and export the runtime version (Transaction CU36) are available in the SAP ERP system for this purpose.

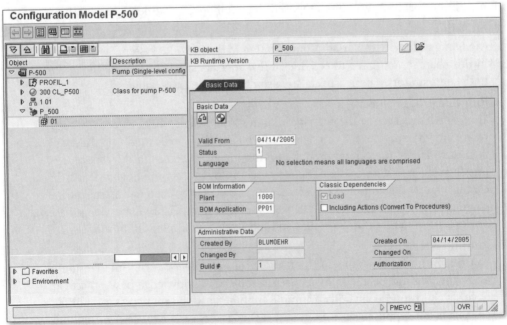

Figure 2.58 Knowledge Base Object and Runtime Version in PMEVC

5. **Configuration using the IPC**

   You can now begin with the configuration via the IPC. You use the configurator of the IPC, that is, the SCE (Sales Configuration Engine, or Application Platform Configuration Engine, as of ECC 6.0), for the configuration in the IPC. The SCE reads the runtime version, loads it, and maps the corresponding configuration in the value assignment interface of the IPC. As a result, a configuration is stored; the runtime version (which is read-only) remains unaffected.

As already indicated, you need to consider some special aspects, known as *deltas,* if you want to use the same model with both configurators. Even though the following list may seem long, most of the restrictions are non-essential "corner cases". Most of the critical aspects have been mentioned earlier.

> You should also refer to a note that describes the deltas and is always up-to-date; it can **[+]**
> be found in the SAP Service Marketplace under *service.sap.com/notes*. See the note on
> page 279 for more information.

▶ **General**

   Please bear in mind the following general notes:

   ▶ The knowledge base for the IPC only knows configurable material masters including the corresponding material BOMs (no networks, general task lists, model service specifications); the IPC only knows master data that is relevant to sales, that is, no routings, for example.

   ▶ An essential attribute of the configurator of the IPC is the increased speed, because the data retention with regard to the knowledge base is leaner.

   ▶ As already mentioned, each runtime version contains exactly one plant, one BOM explosion, and one development status (according to the date, possibly with Engineering Change Management (ECM)). No parameter validities can be considered here.

▶ **Configuration profile**

   With regard to the configuration profile, please note the following:

   ▶ You cannot use configuration scenarios with a single-level BOM explosion specified in the configuration profile, the usual solution is to specify multi-level which subsumes this case.

   ▶ The setting of the availability of the components in the configuration profile doesn't have any effect in the IPC. The IPC doesn't know any availability check associated with ATP logic.

   ▶ You cannot maintain order BOMs in the IPC; that is, you can neither create nor change them. Consequently, no material variant matching is possible at the assembly level, because this can only be implemented within order BOM maintenance.

- In the IPC, you can only start the configuration for the material that has been specified as the OO class ("object-oriented class", an SAP class or material) in the knowledge base. If the configuration model is a multilevel model and if assemblies are supposed to be sold and configured separately, you must create individual profiles for them in the knowledge base object.

- **BOM**
  With regard to the BOM, please note the following:

  - BOM items are considered in the IPC and runtime version context (only) if they are relevant to the configuration or sales process.

  - Runtime versions can include only one BOM for each material.

  - Only material items or class items (no document items, text items, or other types of items) are transferred to the runtime versions for the IPC.

  - The item key in the IPC consists of item number and material and class number. This means that BOM items must not have identical item numbers and object IDs at the same time.

  - For each transferred item, only the key figures and indicators (not the complete item details) that are relevant to sales are transferred.

  - You cannot use the classification of materials as a selection condition (see basic data in the item detail screen).

  - You cannot create and delete items manually.

  - The IPC configurator doesn't support the conversion of units of measurement (this does not apply to the pricing engine). It uses the base unit of measure.

- **Classification system**
  With regard to the classification system, please note the following:

  - For the objects of the classification system, it is critical that class hierarchies don't contain redundancies; that is, a class isn't allowed to inherit via multiple nodes.

  - You cannot use user-specific data types ("UDEFs"). Instead, you should use single characteristics and group them in the user interface design.

  - Value hierarchies are not fully supported; only the leaves are automatically transferred.

  - You cannot use intervals as value assignments. Instead, map this via two characteristics as the upper and lower limits.

  - No class-specific overwriting with regard to the characteristic values and the Not Ready for Input and No Display settings is considered.

  - The IPC doesn't consider application views.

▶ **Reference characteristics**

With regard to reference characteristics, please note the following:

   ▶ Reference characteristics can only be used for field STPO-MENGE, and fields of tables SDCOM, VBAK, and VBAP. Data for fields of VBAK and VBAP, however, is not contained in the runtime version but is implemented via downloads using the sales application.

   ▶ The system only supports a 1:1 (one-to-one) assignment between reference characteristic and table field; that is, you can only use one reference characteristic for each table field and one table field for each reference characteristic.

   ▶ The SDCOM-VKOND reference characteristic is required for both value-dependent pricing methods (object dependencies and direct assignment of the condition record to the characteristic value).

▶ **Class node**

With regard to class nodes, please note the following:

   ▶ Class nodes are only replaced if the IPC finds exactly one allowed material number. If multiple search results are found, the class node is not replaced.

   ▶ Object search regarding class nodes requires that a value assignment has been directly implemented at the class node through a procedure.

   ▶ The configurator user cannot replace class nodes manually.

▶ **Object dependencies and syntax**

With regard to object dependencies and their syntax, please note the following:

   ▶ Some restrictions are also given for object dependencies and their syntax. For example, the IPC can implement a multilevel evaluation for the `part_of` and `subpart_of` syntax elements only within the sales order and planned/production order configuration scenarios.

   ▶ `$sum_parts` and `$count_parts` cannot be used.

   ▶ You have to assign all procedures to the configuration profile or BOM, not to characteristics or characteristic values, which are not loaded to the runtime version. This issue is reported in a warning message when you create the runtime version.

   ▶ You cannot use actions. The download provides the option of converting actions automatically into procedures, but you shouldn't use this option, because an action that has been converted into a procedure doesn't automatically have the same result.

   ▶ All constraint nets of a knowledge base are loaded immediately, rather than on-demand once the material has been selected via the configuration.

▶ Variant functions and user exits must be rewritten in Java. For more information, refer to SAP Note 870201.

▶ The names of variant tables and their columns are only allowed to contain alphanumeric characters and underscores and mustn't begin with "SCE".

▶ Each variant table field must contain exactly one entry.

Finally, note that the SAP ERP system enables you to set up variant models that use features that are only available for the IPC, in which case you use the IPC and not the traditional Variant Configuration tool for the configuration. This is referred to as *advanced mode*, which is activated user-specifically via the SCE value for the CFG user parameter in the user maintenance Transaction SU01. Some settings, fields, and indicators can only be used and are only active in this context, but are also displayed in the standard version. These include:

▶ The SCE Format checkbox in the general data of the object dependencies

▶ The Rule dependency type for object dependencies

▶ Tasks in the knowledge base object

For more information or if you are interested in using the tool, please contact SAP IPC product management.

2.12 Summary

As you have seen, a complete Variant Configuration model is very complex and consists of components from various topics. We hope this chapter introduced all aspects that are essential for your modeling process. The chapter is certainly not suited for the first contact with Variant Configuration. Nevertheless, we've tried to select appropriate topics to provide helpful information for beginners after their first steps and new aspects for "old timers."

While the last chapter focused on the modeling and the master data of a variant model, this chapter concentrates on the business processes in the SAP system.

3   Business Processes in SAP ERP

The topic of Variant Configuration is incorporated into almost all business processes. The last two sections of this chapter discuss some of these topics in more detail. The initial section briefly outlines some points of integration and some terminology.

3.1   Introduction — Variant Configuration in Business Processes

This section serves as an introduction to the chapter. In this context, it is necessary to clarify some terminology. We'll discuss some topics that are relevant for considering Variant Configuration in business processes. This section starts with some definitions and explanations of terminology. The aim is to bring some clarity and systematics into the work of bills of materials (BOMs). From here, we'll move on to a brief introduction to the *Order Engineering Workbench*. In addition, we'll discuss *Advanced Planning and Optimization* (APO) in the context of Variant Configuration before you learn about integrated Product and Process Engineering (iPPE).

3.1.1   BOMs in Variant Configuration

Let's begin with the different types, or categories, of BOMs that were already discussed briefly in Chapter 1, Basic Principles of Variant Configuration, and Chapter 2, Creating a Product Model for SAP Variant Configuration. For example, there are material BOMs, order BOMs, and Work Breakdown Structure (WBS) BOMs (see Figure 3.1). Additionally, there are sales and distribution BOMs and production BOMs, and possibly engineering/design BOMs and other BOMs (see Figure 3.3). And let's not forget about simple and non-simple BOMs (see Figure 3.2). So the following sections discuss the different types of BOMs in more detail:

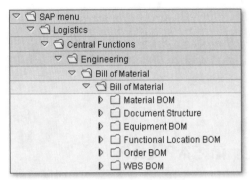

Figure 3.1 Header Object of the BOM

Object Type of the BOM Header

The first difference between BOM types relates to the header object of the BOM, that is, whether it is a material BOM, a document structure, and so on (see Figure 3.1). In most cases that are mentioned in this book, the BOM relates to a configurable material. As already mentioned, not only a material master, but also a document, equipment, or functional location can serve as a header object and provide the name for the BOM. This list should also include the *order BOMs* and the *WBS BOMs*. These two BOMs relate to a material master similar to the material BOM. The specific feature of these two BOMs is that, in contrast to material BOMs, they contain an additional specification, an item of a sales order, or a project.

[+]

Order BOM, Sales Order BOM, and Material BOM

A material BOM is generally valid for a material master. An order BOM is valid for the material master of a specific sales order. Sometimes, there can also be a *sales order BOM*. This term is used as a synonym of the order BOM.

The configuration profile controls to what extent you can create order BOMs for a configurable material. An order BOM is only required if a dynamic explosion of the configurable (super) material BOM is not sufficient, because you need to implement manual, order-specific adaptations. For further details, refer to Section 2.4, Configuration Profile and Configuration Scenarios, in Chapter 2. Next, we'll consider when a material BOM is *simple* and what non-simple BOMs are.

Technical Types of Material BOMs

The *technical types* of material BOMs differentiate simple, multiple, and variant BOMs (see Figure 3.2). Let's discuss these different types in greater detail.

▶ **Simple BOM**

A simple BOM represents exactly one BOM for exactly one material (configurable here). In the standard form, it is used for the configurable super BOM in Variant Configuration.

▶ **Variant BOM**

A variant BOM represents a BOM for similar but different material masters. It is used to summarize BOMs whose components hardly differ.

Variant BOMs in Variant Configuration **[+]**

You can also use variant BOMs for configurable materials as header objects of the BOM. Then, however, all header materials must be configurable (see Section 2.7, Object Dependencies for BOM and Routing, in Chapter 2).

▶ **Multiple BOM**

Multiple BOMs contain several BOMs for the same header material. They are used to map different BOMs for the same material in the context of different production versions, for example.

Figure 3.2 illustrates these three types.

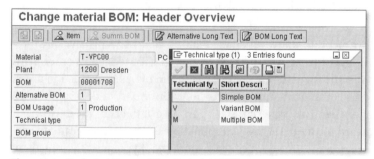

Figure 3.2 Technical Types of the Material BOM

This way, in addition to the simple BOMs, you can also use the non-simple or — better — the *grouped* BOMs of variant BOMs and multiple BOMs.

Usage and Application of BOMs

Let's now discuss sales and distribution BOMs and production BOMs. Here, the *BOM usage*, illustrated in Figure 3.3, is important.

This is another key field; that is, it is used for the identification of the BOM. Consequently, you can create several BOMs for the same material that only differ in this BOM usage. This is an option that you can, but don't have to, use. All enterprise areas that require BOM information can work with the same (universal) BOM.

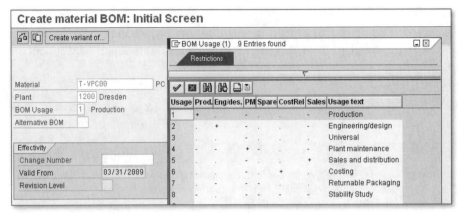

Figure 3.3 BOM Usage

Another concept that is significant in the context of BOM usage is the *BOM application.* It represents the control tool that the enterprise areas or the business processes can use to find the suitable BOM — a universal BOM (if available) or the suitable BOM (from a selection of BOMs that overlap in the BOM usage).

Sales Order BOM and Production Order BOM

Now we know that we have a BOM of material specifically for a sales order: the order BOM. Now we need to know whether or not a BOM specifically for a production is needed or not. The answer is no. Such an explicit production order BOM doesn't exist. If no specific adaptation of a BOM is required for the production order, you use the (unspecific or anonymous) material BOM in the form of a super BOM with object dependencies. This material BOM can be a universal, production, or other BOM with regard to the BOM usage. Production orders in the local environment of Variant Configuration are specific for sales orders. If you need to implement specific adaptations of the BOM for the production order, they are mapped specifically for the sales order, that is, in the form of the previously discussed sales order BOM. This sales order BOM is exploded for the production order as far as this is set plant-specifically in Customizing of the BOM explosion in the material requirements planning (MRP) (see Figure 3.4).

If a sales order BOM exists and if it is considered according to the described Customizing setting, this sales order BOM overrides each anonymous material BOM. The same applies to the BOM explosion for planned orders.

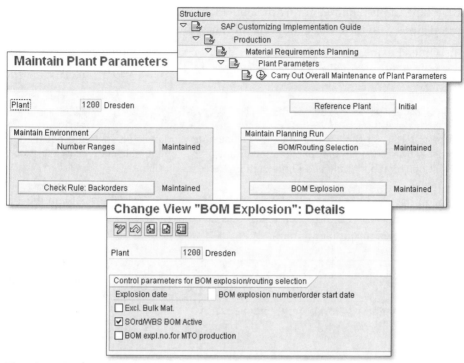

Figure 3.4 Activating (Sales) Order BOMs in Customizing

3.1.2 Order Engineering Workbench

The *Order Engineering Workbench* is a relatively new application that you can use as a supplement to SAP ERP. It is particularly interesting if your enterprise produces specifically according to customer requirements so that doesn't make sense to add all possible characteristics of a product to the product model beforehand. This means the Order Engineering Workbench primarily supports you in engineer-to-order (ETO) scenarios in which you must work, at least partially, with sales-order-specific BOMs and/or routings. You use either Transaction CU51 or Transaction CSKB for the maintenance of order BOMs in Variant Configuration. In this case, the order Engineering Workbench determines which transaction you use.

The Order Engineering Workbench not only supports you in the handling of sales orders, but also in presales activities such as quotations. As early as the quotation phase it is often necessary to go into detail to such an extent that you could also create quotation-specific BOMs and/or routings — at least for some important assemblies. This is important for costing, for example. Because precise information on the costs to be expected is required during the quotation phase, you are better prepared for negotiations with the customer during this phase (for instance, price negotiations).

The Order Engineering Workbench was developed as a supplement for SAP ERP Release 6.0, but it can also be used in combination with earlier releases (up to SAP R/3 4.7).

Order BOM Configuration Scenario

A prerequisite for the use of the Order Engineering Workbench is that you specify the order BOM scenario in the configuration profile of the configurable material (see Section 2.4, Configuration Profile and Configuration Scenarios, in Chapter 2).

In the initial screen of the Order Engineering Workbench (see Figure 3.5), in the left part of the screen, you must enter the Sales Document and the Item (this can be a sales order item or quotation item).

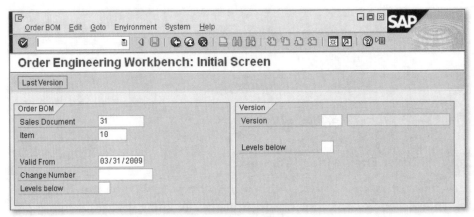

Figure 3.5 Initial Screen of the Order Engineering Workbench

The detail screen of the Order Engineering Workbench (see Figure 3.6) focuses on user-friendly, convenient processing.

This screen provides you with elements such as the tree display of the product structure, a context menu using the right mouse button, and drag-and-drop functionality. You can conveniently create and edit sales order BOMs and sales order routings. You can also implement characteristic value assignments at different stages of the product structure (select the Configure Material option in the context menu). If these characteristic value assignments impact the relevant components of the super BOM (for instance, via object dependencies), the tree display is updated immediately; that is, you directly have an overview of the components that are relevant based on the current characteristic value assignments.

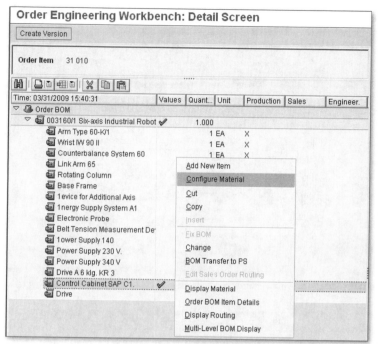

Figure 3.6 Detail Screen of the Order Engineering Workbench

With the Order Engineering Workbench, SAP implemented a concept for *versioning* of order BOMs and characteristic value assignments for the first time (see the Create Version button in Figure 3.7). This function provides several new options that we discuss in more detail in the following section.

Versioning

In the Order Engineering Workbench you have the option to create a version at any time using the Create Version button. This is comparable to a snapshot, and it is saved completely separately. A version is non-MRP-relevant; for example, it is never relevant for subsequent production processes. The version is always displayed as a tree in the right-hand area of the Order Engineering Workbench screen, whereas the left area of the screen always shows the valid BOM structure of the current sales order item. This is depicted in Figure 3.7. Because the use of versions is optional, the tree structure is only displayed in the right-hand area of the screen if you have either just created a version or entered a version in the initial screen.

Using a version, you can save not only the complete relevant BOM structure at this point, but also all characteristic value assignments — also at multiple levels if required. This can be very helpful for documentation purposes, for example.

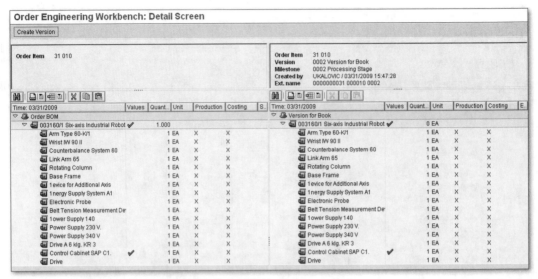

Figure 3.7 Current BOM Structure with Version

Using Versions

For example, let's say you've discussed the configuration in detail with the customer or another department on a specific day and you now want to record this processing stage with a version so that you can refer to it at a later stage.

You can also specify a *milestone* when you create a version. Milestones are freely definable using a Customizing table (for instance, Initial Version, Processing Stage, or Final Version) and can be used for classifying versions to some extent. Although you can create any number of versions for a sales order or quotation item, you should consider the size of the models or BOMs.

It is also possible to define an as-built milestone, for example. Then, the *as-built version* would be created in the Order Engineering Workbench after the completion of a sales order and document the *built condition* of a product.

[Ex] **Using As-Built Versions**

This as-built version can be used, for example, by service technicians or the spare parts sales department to ensure that your customer receives the correct spare parts. This version would still exist if the sales order was archived. This is important because customers are provided with services for delivered products (for instance, machines) over many years, sometimes decades, whereas the corresponding sales orders must be archived after only a few years for performance reasons.

It is also possible to use versioning in the Order Engineering Workbench for documentation purposes only. However, versioning provides many more options than pure documentation:

▶ **Retrieve data from a version**
You have various options to retrieve data from a version to the current BOM structure.

▶ **Import a complete version**
You can import the complete version; then, however, the system overwrites the current BOM structure in the left part of the screen. You should use this option with care.

▶ **Transfer data using drag-and-drop**
Another common option is to drag-and-drop individual parts or entire assemblies from a version to the current BOM structure.

The latter variant enables you to return to the processing stage of last week, for example, for an entire assembly including its characteristic value assignments and multilevel structure because the changes made this week (including the characteristic value assignments and order BOMs) are not supposed to be effective for different reasons. This enables you to respond to change requests and increasing flexibility requirements more quickly and effectively.

Reusability of Similar Sales Orders

The Order Engineering Workbench not only enables the import of data from a version of the current BOM structure, that is, versions that belong to the current sales order item, but you can also use versions of other or similar sales orders or quotations. Of course, this considerably extends the application options of the Order Engineering Workbench.

There are two variants for using such versions of sales orders.

▶ Import the entire version from an old sales order.

▶ Import individual parts from one or more old sales orders.

Prerequisite	[+]
A prerequisite for these two options is that you create at least one version for the old sales order item.	

The next sections discuss these two options in more detail:

For the first variant — the import of the entire version from an old sales order — you start with the status of the old sales order; that is, you must manually rework missing parts. Alternatively, you can import only individual assemblies or other individual parts from the version of the old sales order to the Order Engineering Workbench using drag-and-drop. Consequently, you can use old sales orders or quotations as a source (an assembly from the first sales order, the other assembly from the second sales order, various individual parts from the third sales order, and so on).

Another reusability option is to combine different versions as a kind of *library* using a *dummy customer* and a *dummy quotation*. Via this *version library* you preconfigure assemblies and possibly create order BOMs that occur frequently or can at least form a foundation for future quotations and sales orders. You could use these assemblies in real quotations or sales orders and only adapt a few structures manually.

[+]
Adaptation Using BAdIs

At this point, it's important to mention that you can also adapt the behavior for inserting an assembly (or an individual part) that you've imported from a version individually using business add-ins (BAdIs) in the Order Engineering Workbench.

For example, let's say you want to copy the characteristic value assignments only and re-explode the BOM. Perhaps you only want to copy specific characteristic value assignments, or you don't want to copy the characteristic value assignments at all, but only the BOM of the assembly. You may only want to copy manually changed individual parts from the order BOM or to copy at a single level instead of multiple levels, and so on.

Routings behave like BOMs: For example, you can copy sales order routings of the old sales order if this is relevant. Using this individual adaptation it is also possible, for example, to determine a different behavior for different materials or groups of materials.

Of course, the goal is always to reduce the considerable manual effort, which you have in the make-to-order production during the processing of sales orders and quotations, through optimal reuse of already implemented work from old sales orders and quotations or from the previously described libraries. For this purpose, the Order Engineering Workbench provides the options that were mentioned previously.

Extensions and Adaptations

SAP Custom Development provides the Order Engineering Workbench as a so-called *adaptable custom solution* (ACS), so it is not part of the SAP ERP solution itself, but is a fee-based SAP ERP add-on that is based on the SAP ERP standard.

The concept of an adaptable custom solution contains the term *adaptable*. This means SAP Custom Development is able to provide an additional adaptation or extension of the Order Engineering Workbench according to your specific and individual requirements. SAP also offers maintenance for custom-developed solutions. This can include implementations of BAdIs (as mentioned previously) and additional developments and all other desired adaptations.

The Order Engineering Workbench, an adaptable custom solution developed by SAP Custom Development, is an SAP ERP add-on that particularly supports the engineer-to-order process. In addition to the versioning options of order BOMs and the reuse of existing sales orders or quotations, it supports the remodeling management through a tree structure with an intuitive operation and a drag-and-drop functionality. Here, engineering change management is completely integrated across all levels. The versions that can be created as a kind of nonchangeable snapshot contain the result-oriented storage of components and the characteristic value assignments.

3.1.3 Variant Configuration and SCM-APO

Within the scope of Variant Configuration you can use the planning tools of SAP *Advanced Planning and Optimization* (APO) of *Supply Chain Management* (SCM). For this purpose, you must provide the planning-relevant data of the model to Variant Configuration SCM-APO. Here, you must consider a few specifics for modeling.

Within the scope of the variant class maintenance you activate an organizational area and assign all planning-relevant characteristics to this organizational area. This is necessary because the characteristics are selected during the creation of the integration model for SCM-APO via this organizational area.

You must use the settings in the material master to ensure that the system implements the preferred planning functions in SCM-APO and not in the classic planning of the SAP ERP system. For this purpose, you switch off the MRP in the SAP ERP system in the material master of the MRP views via the MRP type. You also must select the scheduling margin key accordingly.

The MRP type X0 in Figure 3.8 is characterized by its Customizing through the MRP procedure X (Without MRP, with BOM explosion) and the planning method 1 (planned by external system). No further flags are set in Customizing of the MRP type X0.

The scheduling margin key AP1 in Figure 3.9 distinguishes itself by the fact that scheduling margins but no floats are defined. For the planning-relevant components of the super BOM you should set Individual/Collective Requirements Indicator, which you can also find in the MRP views in the material master, to Collective Requirements.

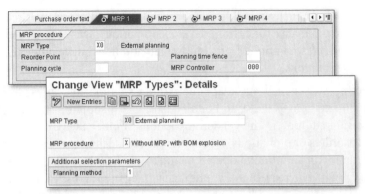

Figure 3.8 MRP Type in the Material Master of the Configurable Material

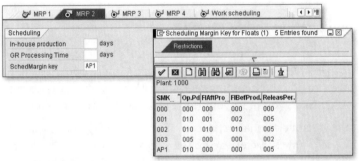

Figure 3.9 Scheduling Margin Key in the Material Master of the Configurable Material

In addition, you require a production version for the configurable material. This production version that you can create from the work scheduling view of the material master contains the key of the super BOM and the super task list of the configurable material. Only after it's been created can the integration model for SCM-APO find the BOM and routing.

You can use Transaction CFM1 to create an integration model that you use to provide all master and transaction data that is required in SCM-APO for the preferred planning activities via the *CIF interface (Core Interface)*.

[+] **Selecting Classes and Characteristics (Organizational Areas)**

Note that you must select the classes and characteristics in addition to the planning-relevant material masters of the model when you create the integration model. You find in the Selection in Figure 3.10, the Material Independent Objects, here in turn via the Classes/Characteristics and here the previously mentioned organizational area and Class Type 300 as the Organizational area indicator.

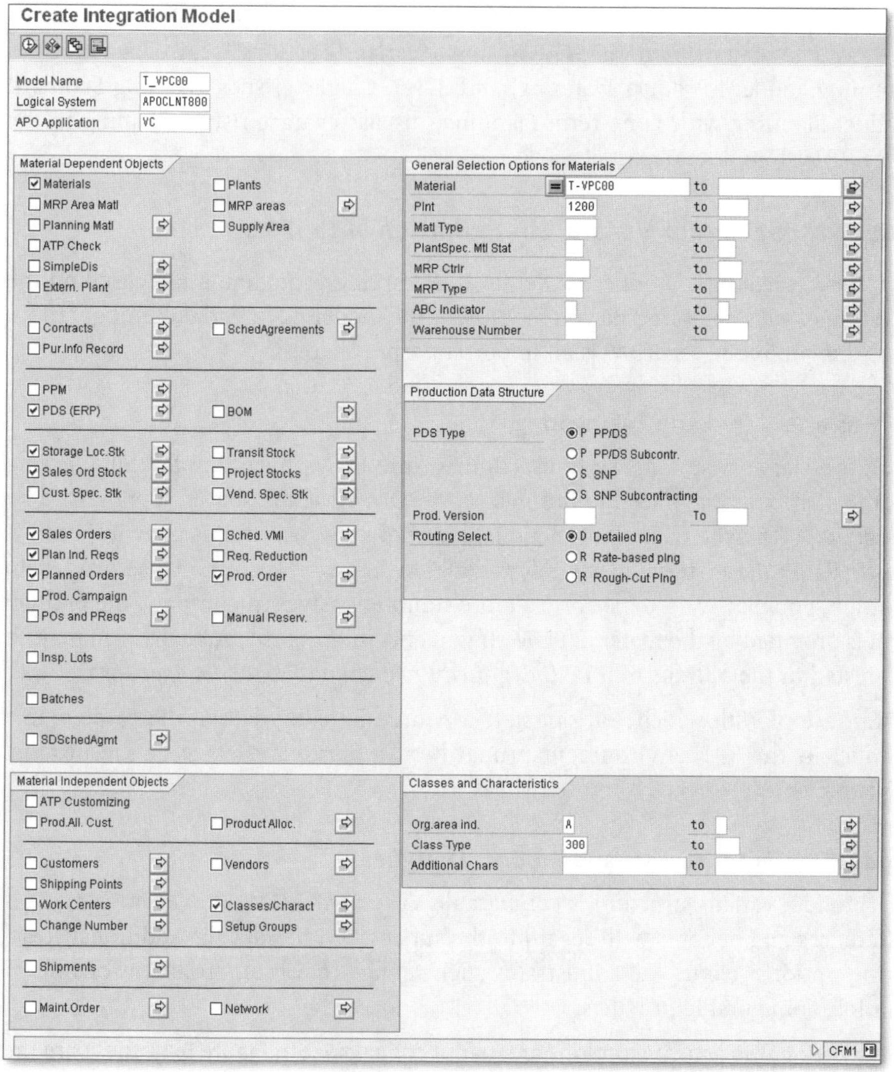

Create Integration Model

Model Name	T_VPC00
Logical System	APOCLNT800
APO Application	VC

Material Dependent Objects

☑ Materials		☐ Plants	
☐ MRP Area Matl		☐ MRP areas	⇨
☐ Planning Matl	⇨	☐ Supply Area	⇨
☐ ATP Check			
☐ SimpleDis	⇨		
☐ Extern. Plant	⇨		
☐ Contracts	⇨	☐ SchedAgreements	⇨
☐ Pur.Info Record	⇨		
☐ PPM	⇨		
☑ PDS (ERP)	⇨	☐ BOM	⇨
☑ Storage Loc.Stk	⇨	☐ Transit Stock	⇨
☑ Sales Ord Stock	⇨	☐ Project Stocks	⇨
☐ Cust. Spec. Stk	⇨	☐ Vend. Spec. Stk	⇨
☑ Sales Orders	⇨	☐ Sched. VMI	⇨
☑ Plan Ind. Reqs	⇨	☐ Req. Reduction	
☑ Planned Orders	⇨	☑ Prod. Order	⇨
☐ Prod. Campaign			
☐ POs and PReqs	⇨	☐ Manual Reserv.	⇨
☐ Insp. Lots			
☐ Batches			
☐ SDSchedAgmt	⇨		

General Selection Options for Materials

Material	▤ T-VPC00	to		⇨
Plnt	1200	to		⇨
Matl Type		to		⇨
PlantSpec. Mtl Stat		to		⇨
MRP Ctrlr		to		⇨
MRP Type		to		⇨
ABC Indicator		to		⇨
Warehouse Number		to		⇨

Production Data Structure

PDS Type	◉ P PP/DS
	○ P PP/DS Subcontr.
	○ S SNP
	○ S SNP Subcontracting
Prod. Version	To ⇨
Routing Select.	◉ D Detailed plng
	○ R Rate-based plng
	○ R Rough-Cut Plng

Material Independent Objects

☐ ATP Customizing			
☐ Prod.All. Cust.		☐ Product Alloc.	⇨
☐ Customers	⇨	☐ Vendors	⇨
☐ Shipping Points	⇨		
☐ Work Centers	⇨	☑ Classes/Charact	⇨
☐ Change Number	⇨	☐ Setup Groups	⇨
☐ Shipments	⇨		
☐ Maint.Order	⇨	☐ Network	⇨

Classes and Characteristics

Org.area ind.	A	to		⇨
Class Type	300	to		⇨
Additional Chars		to		⇨

▷ CFM1 ⊡

Figure 3.10 Selection Criteria of the Integration Model

You then activate the integration model (Transaction CFM2). After the transfer, you may have to supplement the product masters in SCM-APO. Here, you must ensure that no requirements strategy is entered in the product master of the configurable finished product.

You can use the procedure discussed here to include configured sales orders in SCM-APO with planning. Note that characteristics planning is also possible in SCM-APO. Because this topic is rather complex, it is not discussed further at this point (for a more detailed discussion, refer to the book, Dickersbach, J. (2005), *Characteristic Based Planning with mySAP SCM*, Springer-Verlag). Sections 3.3.4, Characteristics Planning/Standard Product Planning, and 3.3.5, Characteristics Planning/Standard Product Planning with Long-Term Planning, discuss characteristics planning in the SAP ERP system in more detail.

3.1.4 Modeling for Variant Configuration with iPPE

Chapter 2, Creating a Product Model for SAP Variant Configuration, focused on the question of *which* master data you require for the modeling in Variant Configuration. The configurable super BOM is an important aspect here.

iPPE as a Tool for Early Engineering

With the classic maintenance transactions of the BOM, including the Engineering Workbench, you may face the problem of a BOM structure that is supposed to be integrated — maybe from the functional point of view — without knowing or having available the material masters for the BOM items or even the BOM header. In addition, you may want to perform a more comprehensive structuring of the product than is provided by the material BOM. If you encounter these problems, you may be interested in the options of iPPE (*integrated Product and Process Engineering*).

iPPE is a tool with which you can start to store modeling data of the configurable product in the SAP environment primarily at a very early stage of the product development.

Product Variant Structure and Product Designer

The product variant structure is comparable to a super BOM. However, it includes differences and benefits with regard to the super BOM besides the additional structuring options. Figure 3.11 illustrates such a *product variant structure* for one or multiple configurable products.

The additional structuring options are mapped via *nodes*. In Figure 3.11 these are the T_CAR car access node and the T_ENG node, for example (see ❶ and ❷).

An *access node* enables you, among other things, to use a product variant structure for different products whose material masters don't have to be known immediately. A *structure node*, like the T_ENG node in this example, enables you to summarize elements, comparable to BOM items, from a functional perspective. In Figure 3.11, this was done for the two possible engines with 150 PS and 200 PS.

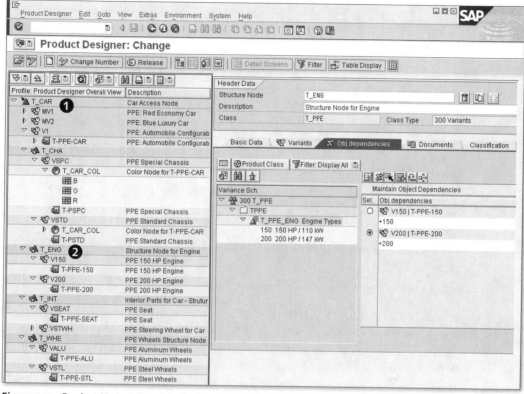

Figure 3.11 Product Variant Structure in the iPPE Product Designer

You can assign these two engines to the product variant structure as *variants* without material masters having to exist for them. Despite missing material masters you can assign object dependencies and test the entire model in this regard.

You can also identify another benefit in Figure 3.11. In the right-hand area of the figure you can work with a considerably shorter syntax that you can define according to a specific enterprise via Customizing. Also, you don't necessarily manually create this shortened syntax that consists of only +150 or +200, but you can create it in the system via a double-click.

iPPE — Modeling from Requirements to Production

iPPE consists of more positive functions than the previously mentioned benefits with regard to structural elements, object dependencies, and nonexistence of material masters. For example, you can store data for the future model at a very early stage of the development process. Figure 3.12 briefly illustrates this.

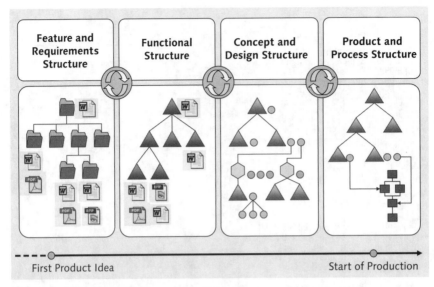

Figure 3.12 iPPE: From Requirements to Production

Within the scope of iPPE you can define your own element types to map structures that the enterprise needs. This can be a feature and requirement structure that structures initial information or contains requirements, customer messages, or specifications about a new product idea. If this information is available as files created in any application, you can assign them as documents to the elements of the feature and requirement structure.

You can derive and map a functional structure from such a feature and requirement structure. It can contain nodes of the previously mentioned product variant structure. Another structure could be the concept and design structure. This largely corresponds to the product variant structure. This function includes the option to continue modeling below the variants, that is, below the elements that correspond to the BOM items. Here you can add and test ideas or different concepts for these elements. You'll decide on one of those concepts and activate it for the element (this element is a variant). The product variant structure, a kind of product structure, is the result of such a development process, so it represents the structure of the configurable product. Similar to this, you can develop and map a process structure to describe the production process.

Applications exist in which you can use this product and process structure directly in production. This particularly applies in the automotive industry and repetitive manufacturing. However, this is not the case in a classic SAP ERP system without industry-specific solutions.

In that situation, you require the classic BOM and the classic routing of Chapter 2, Creating a Product Model for SAP Variant Configuration. For BOMs, you can *convert* a classic BOM from the product structure. For the process structure, such a converter in the direction of the routing doesn't exist. Therefore, the function of maintaining process structures is deactivated in a classic SAP ERP system. In this context the concept of *integrated product engineering* is used instead of iPPE. Also, the workbench for processing in the iPPE in a classic SAP ERP system is not referred to as *iPPE Workbench,* but as Product Designer (see Figure 3.11).

Because this topic goes far beyond the scope of this book, interested readers can refer **[+]** to other literature and sources, such as, Kohlhoff, S., *Product Development with SAP in the Automotive Industry*, SAP PRESS.

3.2   Integration Aspects Along the Supply Chain

This section, on the one hand, focuses on the specifics that you must consider when using configurable products in the individual members of the supply chain. On the other hand, this section shows the high degree of integration and the options available for using it.

Figure 3.13 illustrates a classic process with regard to the integration of Variant Configuration. It also features a more comprehensive scope, which is described in this section based on another example.

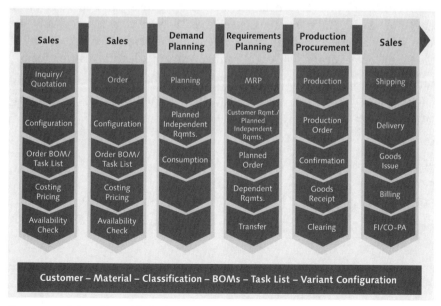

Figure 3.13   Integration with the Logic Business Process

3.2.1 A Classic Process for the Integration of Variant Configuration

Let's begin with a classic — and relatively simple — Variant Configuration integration process, which is shown in Figure 3.13. We'll assume that the required master data (customer, material master, BOM, routing, and master data from the classification system and Variant Configuration) is available.

Sales Activities

The process doesn't start directly with the sales order, but starts with *presales activities*, such as *inquiry, quotation, scheduling agreements*, and *contracts*. Within these activities, you can work with Variant Configuration as accurately as possible for the sales order. More precisely, the variant model works with all object dependencies in all sales and distribution documents, unless explicitly defined otherwise in the syntax of the object dependencies.

[Ex]   **Different Object Dependencies for Different Sales and Distribution Documents**

For example, you may want different requirements on the completeness for the sales order than are valid for the presales activities, or you may want to work with certain default values in quotations.

You implement this by specifying rules within the object dependencies that only become effective in certain categories or types of sales and distribution documents. You can access the sales and distribution document header information via the reference characteristics of the VBAK database table (see Figure 3.14). The VBTYP field enables you to roughly differentiate between the *SD document categories*. In Figure 3.14, this field is accessed using an initial reference characteristic and is therefore available for the object dependencies for configuration in sales and distribution.

In contrast to the display shown of Figure 3.14, you set these reference characteristics as not displayable normally. For instance, the following values apply in the SD document categories:

► A for inquiry
► B for quotation
► C for order
► E for scheduling agreement
► G for contract

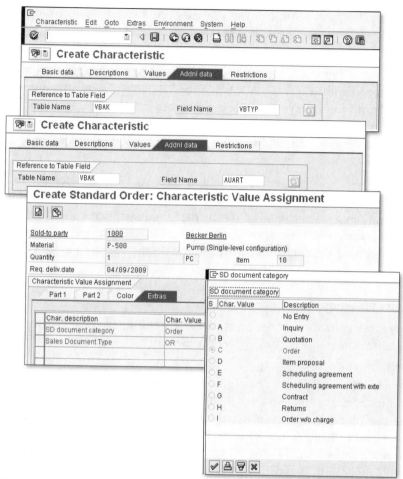

Figure 3.14 Reference Characteristics of the SD Document Category and Sales Document Type

The complete list is displayed if you activate the input help ([F4] help). A more distinguishing option, also used in Figure 3.14, is to query the *order type* via the AUART field in the same table. With a corresponding definition of the reference characteristics, the quotation would look as follows (Listing 3.31).

```
... if VBAK_VBTYP = 'B'
... if VBAK_AUART = 'AG'
```

Listing 3.31 Querying the Sales and Distribution Document Information via the Reference Characteristics

It is also possible to query other fields of the sales document header or the item in the conditions in the object dependencies using corresponding reference characteristics.

In the first column of Figure 3.13, in the inquiry or the quotation you can both specify the configuration and create *order BOMs* or an *order routing*. Normally, however, you don't create this until after you create the sales order and only if required and allowed by the configuration profile. Similar topics include *costing, pricing, availability check*, and *requirements transfer*. The latter is carried out with the sales order in the standard format.

You can create the sales order (second column in Figure 3.13) with reference to a sales and distribution document of such a presales activity or an older sales order. In Customizing you can set in the *copy control* whether and how pricing and costing is supposed to be transferred, for example. This copy control also includes the setting option for transferring the configuration. You also have the option to make a copy or reference.

Requirements Planning

The *demand planning* (third column in Figure 3.13) is to be initially excluded because it is supposed to be detailed separately within the scope of planning and Variant Configuration.

The *requirements planning* (fourth column in Figure 3.13) is based on the planned independent requirements that are created through the sales orders in the standard format. The same would apply to planned independent requirements from planning. The standard is that only one basic date determination is implemented in the requirements planning within the scope of scheduling. For this purpose, you explode the BOM — if applicable, an order BOM — and determine the component list for the planned order. The planned order is used to specify and transfer the dependent requirements. The component list in the planned order is a copy of the exploded BOM and can be changed here. In addition to the basic date determination, you can also implement a lead time scheduling in the requirements planning. The result would be that the system would read and explode both the BOM and the routing.

Controlling the Requirements Transfer

As already mentioned, standard requirements planning is based on the planned independent requirements that are created through sales orders. However, you can also trigger requirements through the sales and distribution documents from other sales activities, for instance, the quotation. Similarly, you can create the *sales order stock segment* with reference to the sales order, but also with reference to the quotation or other sales and distribution documents. Here, the Customizing setting for finding the *requirements classes* and the linked requirements transfer in sales and distribution is decisive. Regarding the Customizing settings, refer to Chapter 4, Customizing SAP ERP for Variant Configuration. Additionally, the next sections briefly discuss the general procedure and specifics from the Variant Configuration perspective (see Figure 3.15).

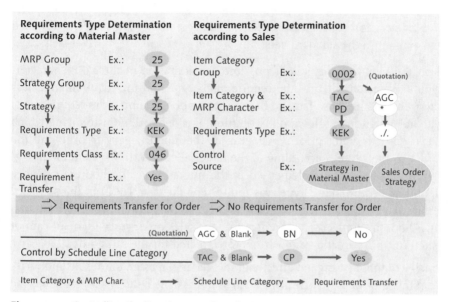

Figure 3.15 Controlling the Requirements Transfer

A general rule is that requirements are only transferred if this is intended according to both the *requirements class* and the *schedule line category*. The system finds these to control factors basically by using the item category or, more precisely, using the following chain:

1. **Item category group**
   The *item category group* is the starting point. It is a control parameter from the material master. In Variant Configuration, the item category group 0002 is used in the standard version. As described in Section 2.4, Configuration Profile and Configuration Scenarios, in Chapter 2, the package or set material in the sales order scenario forms an exception with the item category group 0004.

2. **Item category**
   Based on the item category group, the system finds the *item category*, taking into account the sales and distribution document type. So you obtain the *item category*

AGC for 0002 and the quotation (AG) and, similarly, the *item category TAC* for 0002 and the standard order (TA). (See Figure 3.15, top right, and the descriptions in Chapter 4, Customizing SAP ERP for Variant Configuration.)

3. **Schedule line category**

   Based on the item category under possible consideration of the MRP characteristic, you determine the *schedule line category*. As listed at the bottom of Figure 3.15 and described in the next chapter, this results in a schedule line category without requirements transfer for the quotation and a schedule line category with requirements transfer for the standard order. So, no requirements transfer is provided for quotations in the standard version.

Based on the item category under possible consideration of the MRP characteristic, you also decide whether the requirements type and the requirements class are determined using these factors. In the scope of determining the requirements type via the operation in sales and distribution, it is decided in Customizing which *source* is used with which priority. This can be the item category according to the sales order or the strategy according to the material master. In the example in Figure 3.15, the search was implemented in the quotation using the item category (AGC), and no requirements type was found. For the item (TAC) in the standard order, the strategy is evaluated according to the material master. Because the quotation (AGC) doesn't find a requirements type, the sales order stock segment cannot be created for this quotation and is therefore created for the sales order.

The *strategy* cannot be assigned directly in the material master, but results indirectly from the strategy group. If the *strategy group* is not assigned, the system evaluates the *MRP group*. In the example provided in Figure 3.16, strategy 25, Make-to-order for configurable material, is determined from the MRP group 25 or the strategy group 25 (Strategy field).

A strategy can consist of a requirements type for independent requirements and a requirements type of customer requirement. In the standard version, strategy 25 is not used within the scope of independent requirements, so no entries are made here. Strategy 25 was assigned as the requirements type of customer requirements KEK (Make-to-order configurable material) and therefore indirectly Requirements Class 46. See Figure 3.16. The requirements class plans a requirements transfer.

The last two options in Figure 3.16 indicate that the system generally works with Variant Configuration and no consumption against planning is implemented. The latter is significant in the next section. Chapter 4, Customizing SAP ERP for Variant Configuration, provides further details on Customizing within this environment.

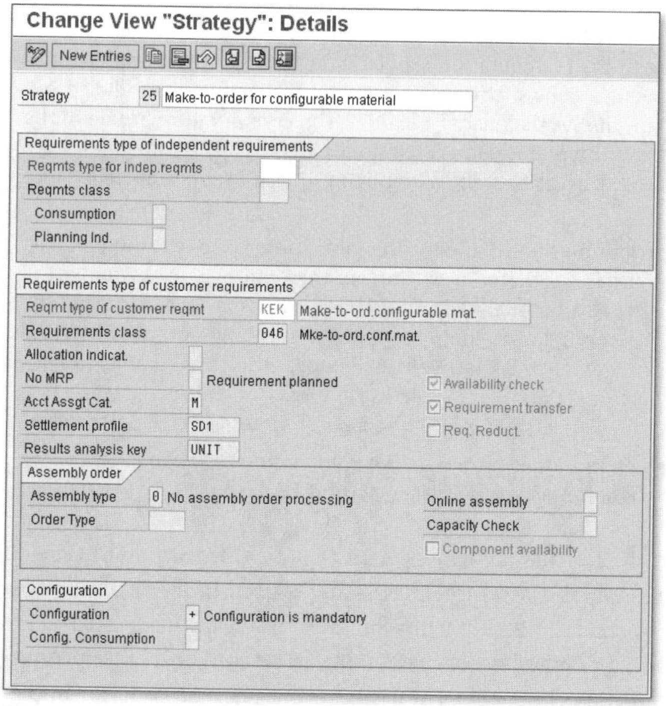

Figure 3.16 Customizing the Strategy

Procurement: In-House Production or External Procurement

We'll only discuss *procurement* (fifth column in Figure 3.13) with regard to *in-house production* as a procurement type. We'll discuss this aspect in detail after an excursus on external procurement.

> **Excursus: External Procurement**
>
> *External procurement* can be used as a procurement type for configurable materials. Both the complete configurable product that represents the sales order item and individual configurable components can be procured externally. These components can be configured in the sales order within the scope of a multilevel configuration.
>
> It is also possible to work without a custom configuration for these configurable components. For example, this is the case if only a single-level configuration is set for the entire configurable product according to its configuration profile. If no custom configuration profile exists for the configurable component to be procured externally, you must query the information necessary for the configurable component from the configuration in the value assignment of the complete, configurable product. The *purchase order* of the configurable component contains a copy of the configuration of the configurable product — the full value assignment provided that no filters regarding purchasing were set via the user interface design.

If the configurable component has its own configuration profile and is configured only at a single level according to the configuration profile with regard to the complete configurable product, you have the same situation as described earlier. In sales and distribution, you can only assign a value to the header material. The configuration profile of the configurable component is ignored; it is considered nonexistent. Only a configuration of the header material is stored. The configurable component doesn't represent an *instance* in this configuration.

Although in sales and distribution the situation is the same, differences exist in purchasing. The difference in purchasing is that you don't copy the entire value assignment of the configuration of the header material to the purchase order, but the unconsidered configuration profile of the configurable component takes effect. In contrast, only the value assignments of the characteristics that were assigned to the configurable component are copied in the purchase order. The value assignments of these characteristics, however, originate from the value assignment of the header material in the sales order.

So you can say that the *configuration profile of the configurable component* acts as a *filter* in the one-level configuration scenario discussed here.

The following sections will consider *in-house production* as a procurement type. If according to the planning strategy, assembly processing is used in the material master, the *production order* is created directly from the sales order. In the standard case with requirements planning, planned orders exist that can be converted into production orders. Here, you can copy the *component list* from the planned order or implement a new explosion of the BOM. The latter generally occurs, for example, if you've activated the header material for *Order Change Management* (OCM) by assigning an overall change profile to the material master in the Work Scheduling view. The routing is exploded for the production order, and a copy in the form of an *operation list* is created like in the BOM.

You can update the component and operation lists retroactively. For this purpose, you are provided with the default function of the production order, Read PP Master Data. This may be necessary in the following cases:

▶ If changes were made to the BOMs and routing retroactively

▶ If an order BOM and order routing were created or changed

▶ If changes were made to the configuration in the sales order that result in changes to the explosion of the BOM and routing

The implementation of the production order, Read PP Master Data, is also possible when the production order has already been released. This release would then be reset, however, but if the production order is created, which is documented via the (partial) confirmations and goods movements, it is no longer possible to read the data. This is where *Order Change Management* (OCM) would be applied. With OCM, even such production orders can be changed with system support. Section 10.2, Change Management, in Chapter 10 contains more detailed information on this.

In the processes of sales that follow procurement (sixth column of Figure 3.13), no specific features need to be considered from the Variant Configuration perspective.

3.2.2 Processes with Extended Integration Aspects

In addition to the considerations in conjunction with Figure 3.13, the following sections address additional aspects of integration. These additional aspects concern:

▶ The integration of the Project System (PS) including the configuration of projects

▶ The integration of configurable model service specifications in the order processes

▶ The integration of Quality Management (QM) including the check against the configuration

▶ The integration of Plant Maintenance (PM) and Customer Services (CS) including configurable general maintenance task lists

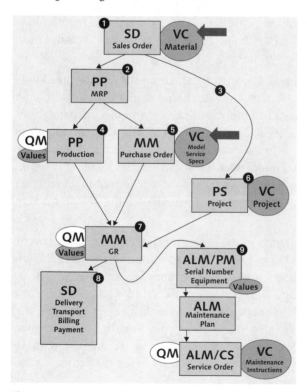

Figure 3.17 Example of an Integration Model with Variant Configuration

So far, you've learned — as shown in the top part of Figure 3.17 — that the process generally begins with a configuration in the sales order in sales and distribution (❶ in Figure 3.17). For this purpose, you create the sales order with a configurable material. The sales order can generate corresponding planned orders via MRP, but also can directly generate procurement elements. For this, an *assembly processing* is triggered via the strategy, requirements type, and requirements class as illustrated in Figure 3.17 with regard to the *project system* ❻.

Project System, Configurable Standard Networks, and Variant Configuration

The sales order directly creates a project for its production using the item with the configurable material. To determine the required elements of the project (work breakdown structure elements, operations, and material components), you can also use Variant Configuration. The *configurable standard networks* are the tool used for this purpose.

The integration can even ensure that the *project* is automatically created from the sales order in accordance with the requirements from the configuration of the configurable material. The configuration in the sales order controls the explosion of the project. For this, you must find a suitable requirements type in the sales order (assembly processing with project system).

For creating the standard work breakdown structure, no special features need to be considered that are relevant for Variant Configuration. Standard work breakdown structures cannot be configurable and cannot be linked with object dependencies. Elements from the standard work breakdown structure are only automatically transferred to the project to be created when operations are assigned to the elements after the value assignment of the object dependencies regarding the *standard networks*. Elements without operations are deleted. The standard work breakdown structure must therefore be created as a superstructure. The same applies to the *standard network*.

Object dependencies can be assigned to the operations of the standard network, however. Both the selection conditions and the procedures can be used. In the procedures, you can implement changes similar to the routings using reference characteristics relating to the PLPO table. The elements of the standard work breakdown structure and the components of the configurable BOM of the material from the sales order can be assigned to the operations. In addition, you can assign the configurable material to the standard network using the corresponding Transaction CN08. A *configuration profile*, which works with the same variant class as the configurable material, is created for the standard network.

Purchasing and Configurable Model Service Specifications

The top area of Figure 3.17 illustrates that an order from MRP ❷ can be transferred to a production order ❹ for in-house production or to a purchase requisition and purchase order ❺ for external procurement. In the purchase order you can generally, that is, independently of configurable materials, work with *services*. You can copy these services from a model service specification. This *model service specification*, in turn, can be configurable. The configuration of the model service specification is generally independent of the configuration of the material in the sales order. *Similar to other configurable object types (material master, network, general task list)*, configurable model service specifications represent super templates from which the elements to be transferred are selected according to the configuration (selection conditions) and possibly changed (procedures). The procedures use reference characteristics relating to the ESLL table for these changes. The model service specification becomes configurable by selecting the appropriate checkbox. Additionally, you require the configuration profile and the variant class.

Quality Management and Variant Configuration

The top part of Figure 3.17 also shows that you can implement *quality management* (QM) measures ❹ during production, for example. The production order can include inspection steps as operations. These inspection steps are then copied from the super task list of the configurable material to the production order. An operation can be indicated as an inspection operation using the control key that is available there. In addition, you can assign *inspection characteristics to such operations.* Also, you activate an in-process inspection in production via the material master in the Quality Management view using the *inspection type.* You can directly link Variant Configuration with the inspection in QM, so it is possible to check against the configuration of the customer request.

To do this, in the material master select the INSPECTION BY CONFIGURATION option in the activation of the individual inspections (see Figure 3.18).

In addition, you can link the Variant Configuration characteristics with the inspection characteristics from QM. This is done by assigning them the corresponding characteristics from the variant class for the configurable material as so-called Class Characteristics in the definition of the inspection characteristics.

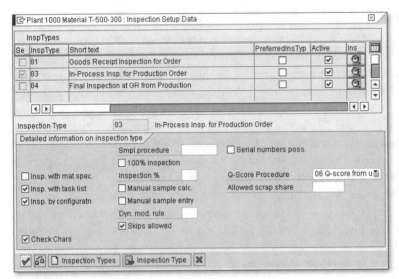

Figure 3.18 Quality Management Settings in the Material Master

[+] In the inspection of the QM, the system displays all inspection characteristics that were assigned to routing operations. This is independent of whether these inspection characteristics were linked with characteristics from Variant Configuration. The system also displays all linked inspection characteristics that were not assigned to the routing. However, this is only done if the corresponding characteristic was assigned a value in the configuration of the sales order; in this case it is displayed as the inspection characteristic for the last operation.

A linked inspection characteristic needs to be assigned a value only if it is selected as a required characteristic and the appropriate characteristic was assigned a value in the configuration. A check against the configuration is only possible for numeric, but not for qualitative, characteristics.

In the standard version of Variant Configuration, you cannot assign an interval to a characteristic so that you can only check against a single value. Multiple-value characteristics that allow for interval values are an exception. However, in the standard version these cannot be evaluated in the object dependencies.

Let's go back to the process illustrated in Figure 3.17. The box MM and GR ❼ is supposed to indicate that the *goods receipt* concludes the procurement process of the material — irrespective of whether the procurement was carried out via in-house production, external procurement, or project processing. Like to the inspection during production, you can perform an inspection of QM within the framework of goods receipt. The settings are also similar to the ones described earlier. One difference is that you work with an explicit *inspection plan* here. In contrast to other task list types, no object dependencies can be assigned to an inspection plan.

The steps following procurement — delivery, transport, invoice, and payment (❽) — don't have any specific features relevant for Variant Configuration, as was already discussed in the context of Figure 3.17.

Customer Service and Configurable General Maintenance Task Lists

Figure 3.17 includes another integration aspect: Plant Maintenance (PM) and Customer Services (CS) ❾. The idea is to provide service for the sold, configured product. In these service quotations and service measures you can directly access the configuration of the product (configuration from the sales order). For this purpose, the product that was delivered to the customer must be maintained in a separate system as a PM-relevant object. This is done using a so-called *customer equipment*. Corresponding settings in the material master enable you to achieve an automatic serialization and hence a linked generation of equipment. If such equipment is created with reference to a configured item of the sales order, it also bears a reference to the corresponding configuration which, in turn, can be used in the service. If you plan service measures for this equipment using a *maintenance plan*, you can work with configurable general maintenance task lists.

Similar to the configurable routings, networks, and model service specifications, such configurable general maintenance task lists are superstructures from which you copy the necessary steps, for example, into a service order (selection conditions) via object dependencies according to a configuration and possibly change them (procedures with reference characteristics for the PLPO table). Because the equipment is connected to the value assignment of the sales order, you can populate the service order with service operations in accordance with this configuration. Here as well, you can work with components from configurable BOMs for plant maintenance. Moreover, you are also provided with QM within the scope of the service order.

Equipment with Custom Configuration	[+]

For example, if during operation the equipment changes owing to service measures in such a way that it no longer corresponds to the original configuration of the sales order, this can also be taken into account. In this case, the configuration of the sales order is copied to a configuration of the equipment and can then be changed, and the reference became a copy. Subsequent service orders would explode configurable general maintenance task lists in accordance with the copied and changed configuration.

These descriptions of the process illustrated in Figure 3.17 show the options with regard to the integration of Variant Configuration with the different object types and logistics processes. They are examples only and don't claim to be exhaustive.

3.3 Planning and Variant Configuration

The term *planning* refers, roughly speaking, to the processes that derive corresponding planned independent requirements from the planning and forecast data. These planned independent requirements, in turn, trigger procurement.

At first glance, planning and Variant Configuration are two contradicting topics. We'll solve this contradiction in Section 3.3.2, Variant Configuration and Planning — An Introduction to the Topic, before presenting the details of this subject. This section focuses on the planning strategies and technologies that were additionally and exclusively developed for planning in Variant Configuration.

3.3.1 Evaluations in the Variant Configuration Environment

Before we present the planning topic in the narrower sense, you should note that in Variant Configuration evaluations are also possible in the *Logistics Information System* (LIS) and in SAP NetWeaver Business Warehouse (BW). For example, a standard analysis relating to Variant Configuration exists in the Logistics Information System under Transaction MCB. You can also use this data source later in planning. This is why the standard analysis is to be presented here.

The standard analyses in Variant Configuration stand out because in contrast to info structures in the Logistics Information System outside Variant Configuration, you must evaluate the configuration in addition to the usual aggregation criteria and key figures. In Variant Configuration, it is not sufficient to know when, where, and how often a product was sold: The preferred properties of the product must also be considered. This calls for custom info structures for Variant Configuration.

Info Structure S138 and Copy Management

For example, the standard analysis (Transaction MCB) uses info structure S138. A special feature of info structure S138 is that it is not synchronously updated automatically. This update is carried out using the *copy management* (Transaction MCSZ). A suitable method within this copy management generates info structure S138 from another info structure, S137. Info structure S137 is not synchronously updated automatically, but is also the result of applying the copy management on info structure S126. For info structure S126, which contains the raw data, you can activate an automatic update.

Within the scope of Variant Configuration, you must meet additional requirements for activating the update. You must carry out the following three steps to activate the update of info structure S126 in the Logistics Information System.

1. **Creating planning profiles and possibly planning tables**

   For the materials that are relevant for evaluation, you must create *planning profiles* and possibly *planning tables*. These two objects are discussed in detail later in this chapter. This first step is specific to Variant Configuration and is only required there.

2. **Setting the update**

   Next, you set the update. For this purpose, you create update groups for customer, material, document type, item type and/or sales area (Transaction OVRP). This is a default step and therefore doesn't consist of any specific aspects of Variant Configuration.

3. **Activating the update**

   The same applies to the last step, the activation of the update (Transaction OMO1).

> At this point, refer to SAP Notes 174758 and 173756 on the topic of info structure **[+]**
> S138.

Info structure S138 can be used as the basis for characteristics planning, one of the planning strategies that is presented in the following sections, where this info structure is discussed again. In the SAP Business Information Warehouse, you can also implement evaluations within the Variant Configuration environment. For this purpose, information from info structure S138 is transferred to InfoCubes using corresponding generic extractors.

3.3.2 Variant Configuration and Planning — An Introduction to the Topic

The basic principle of Variant Configuration is customer-specific production, is which production is triggered by the corresponding sales order. This contradicts to the idea of planning, in which procurement is triggered by planning requirements. Nevertheless, you can still use planning in Variant Configuration (see Figure 3.19).

Of course, you should wait for the sales order with its corresponding configuration for producing the configurable product ❶. No planning is to take place at this level. However, you can use planning strategies for the nonconfigurable components and assemblies (the latter from external procurement or in-house production). This means you generate planning requirements for these nonconfigurable components and assemblies using planning. These planning requirements trigger procurement, and the corresponding materials are placed in storage in the planned quantity. The configurable finished product is not considered until then ❷. If a sales order is created for the configurable finished product, this generates a customer requirement.

Within the scope of MRP, the BOM is exploded for this sales order in accordance with the configuration, and suitable dependent requirements are generated ❸. The dependent requirements must be offset against the requirements from planning with regard to nonconfigurable components and assemblies to reduce the stocks generated there ❹. The planned order that was created in MRP triggers the final assembly of the product.

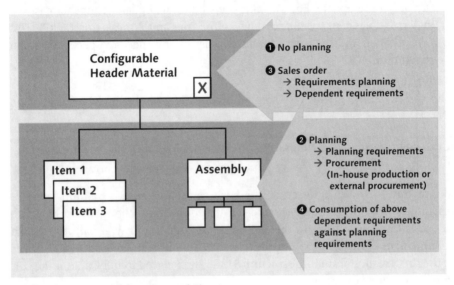

Figure 3.19 Variant Configuration and Planning

3.3.3 Pure Assembly Planning

The pure assembly planning should be known from the planning outside Variant Configuration and can be used in almost the same way here.

For this purpose, a planning strategy, which ignores the planning, is assigned to the configurable product in the material master. Outside Variant Configuration, planning strategy 20 exists for this purpose in the standard version; planning strategy 25 is the analogous strategy available in Variant Configuration. For further information on this planning strategy refer to Chapter 4, Customizing SAP ERP for Variant Configuration.

[+] | **Planning Strategy 70 and the Mixed MRP Indicator**

In the standard version, the nonconfigurable components and assemblies of the super BOM are assigned with planning strategy 70 in their material masters (see Figure 3.20 and 3.22) and require the mixed MRP indicator.

Corresponding planning requirements are created for these components and assemblies within the scope of planning. Here, the consumption indicator in the item screen of the independent requirements must allow for consumption of customer requirement, reservations, and dependent requirements. MRP generates planned orders on the basis of these planning requirements, and procurement can be triggered then. The procurement may include both external procurement and in-house production, like for such nonconfigurable assemblies.

If sales orders are received for the configurable finished product, the BOM is exploded at this level. Planned orders and production orders generate dependent requirements. The customer requirement at the configurable material level is not consumed against planning. Furthermore, the quantities from the assembly planning are not checked in sales and distribution. The dependent requirements that result from the sales order are consumed directly against the requirements from the assembly planning at this level and consequently are consumed exactly. For more information on exact consumption refer to the further descriptions in this chapter.

With this planning strategy you don't require forecasts for the header material. This is both an advantage and a disadvantage: Because you plan at the level of components and nonconfigurable assemblies, you depend on forecasts. These are frequently not available, but exist for the configurable header material at most. Moreover, the recording of planning requirements for large super BOMs is time-consuming.

3.3.4 Characteristics Planning and Standard Product Planning

Using the strategy of standard product planning, you can work without the previously described disadvantages (strategies 25–70; see Figure 3.20):

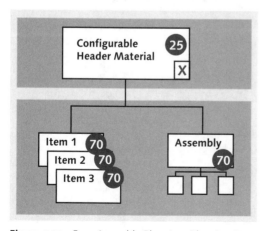

Figure 3.20 Pure Assembly Planning: Planning Strategies 25 and 70

▸ No forecasts at header level usable

▸ High input work at the component level

With standard product planning you create planning requirements at the level of the configurable header material. However, it is not sufficient to only specify the corresponding quantities, but you must determine the properties from the configuration. You must specify how often you must schedule which properties in the specified quantity. The properties to be planned can be mapped through individual characteristic values or value assignment combinations. For this purpose, you require *planning tables* and *planning profiles* as tools.

Before presenting the process for this planning strategy on the basis of Figure 3.22, you assign strategy 56 (standard product planning; see Chapter 4, Customizing SAP ERP for Variant Configuration) via the identical strategy group (see Figure 3.21) to the material master of the configurable header material.

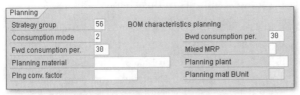

Figure 3.21 Strategy 56 in the Material Master

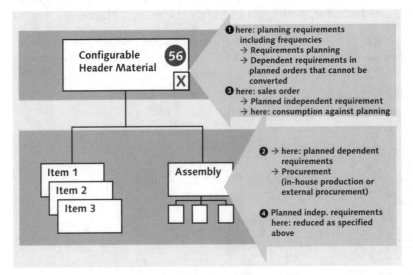

Figure 3.22 Standard Product Planning: Planning Strategy 56

In addition, you must specify the Consumption mode including the time intervals in the material master.

Standard Product Planning and Characteristics Planning   **[+]**

The concept of *standard product planning* is historical and is still in use. However, the strategy contains a characteristics planning for the configurable material. The standard product is the configurable material.

If you want to create planned independent requirements for the configurable header material (see ❶ in Figure 3.21), you must specify the frequency of the properties relevant for planning, as already discussed. To specify these frequencies, you must create a suitable input template using the planning tables and planning profiles (see Figure 3.23). The following sections describe the individual steps for creating such an input template, but let's first take a look at the input template itself.

An example: An input template is to be created for a configurable computer, similar to the example shown in Figure 3.23. This input template has the following planning-relevant characteristics:

▶ Extras with the values mouse, modem, and keyboard

▶ Casing type with the values tower, minitower, and desktop

▶ Processor CPU with the values slow, medium, and fast

Mouse		
Modem		
Keyboard		
Tower	CPU slow	
Tower	CPU medium	
Tower	CPU fast	
Minitower	CPU slow	
Minitower	CPU medium	
Minitower	CPU fast	
Desktop	CPU slow	
Desktop	CPU medium	
Desktop	CPU fast	

Figure 3.23 Example of Frequency Template

The values of casing type and CPU are assigned with values in combination. This way, you can also map dependencies between characteristics. For example, the tower characteristic could be ordered in combination with fast CPUs with a disproportionately high frequency. If you planned the characteristic values individually, this would assume an independency. For example, with 50% tower and 50% fast CPU for the total quantity of planned computers, this would mean that 25% of the computers would have both properties simultaneously.

Via the input template you can map a dependency so that the combination of the two values with 40% is clearly above the calculated value of 25% (50% × 50%). To map such input templates in the system, you must carry out the following steps:

▸ Create the planning table

▸ Create the planning profile

▸ Assign the planning table to the planning profile

Planning Tables

For the rows of the input template that includes the combinations of values and characteristics of casing type and CPU, you require so-called *planning tables*. These are very similar to variant tables and are stored in the same location. They are created from the master data of demand management menu (LOGISTICS • PRODUCTION • PRODUCTION PLANNING) in Transaction MDP1. For the planning table, you specify any name, a description, a status that allows for maintenance, and characteristics as columns (here casing type and CPU). The content of the planning table corresponds to the last nine rows of the input template in Figure 3.23.

Planning Profile

After you've created the planning table, you need to create a *planning profile*. In contrast to the planning table, which is optional, the planning profile is generally required for planning strategy 56.

You create the planning profile with any name, with reference to the configurable material, and the variant class type. All characteristic values of the variant class of the configurable material are then available. Here, you can select the relevant individual characteristic values for the frequency template — in the example of Figure 3.23 these are modem, mouse, and keyboard. In Transaction MDPH of the planning profile you can also assign the planning tables so that the remaining nine rows of the frequency template from Figure 3.23 are activated as well.

You have now implemented all prerequisites for standard product planning, so let's continue with the actual process of standard product planning.

Planning and Consumption for Characteristics Planning/ Standard Product Planning

After you've created the necessary master data, the following sections consider the process of planning with this strategy of standard product planning. This includes:

► Creating planned independent requirements for the configurable material including the configuration supporting point (❶ back in Figure 3.22)

► Planning run and procurement ❷

► Sales orders and their consumption of customer requirement against the planning requirement (❸ and ❹)

Initially, you create the *planned independent requirements* for the configurable material. In addition to the planned quantity, you also require information on the characteristic value assignment. For this purpose, the input template for creating planned independent requirements provides the Configuration supporting point button (see ❶ in Figure 3.24), which you can select to call the frequency template. You can create the frequencies using info structure S138 from the Logistics Information System.

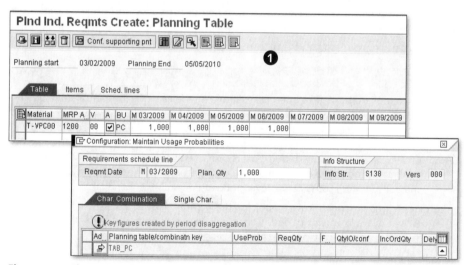

Figure 3.24 Planning Profile and Configuration Supporting Point

---

Configuration Supporting Point **[+]**

You don't need to create a separate configuration supporting point for each planned independent requirement. A configuration supporting point is automatically valid for all planned independent requirements up to the next configuration supporting point. You only need to create a new configuration supporting point if the frequencies change.

Next, you implement a planning run for the configurable material. The result of this planning run is *planned orders* for the configurable material in accordance with the previously specified planned independent requirement. You generate the component list of the planned orders taking into account the frequencies from the configuration supporting points. The planned orders for the configurable material cannot be converted into procurement elements, such as production orders. This ensures that—as requested—no procurement can be triggered at this level, but the system waits for the requirements from the sales orders. For the components and the nonconfigurable assemblies this looks different. The planned orders for the configurable material result in corresponding planned orders for these dependent requirements; they, in turn, can be used for procurement.

So corresponding stock should be available for the desired component and nonconfigurable assemblies so that the final assembly of the configurable product can be started immediately after receipt of the sales orders.

Another topic is the *consumption* of the customer requirement with the planned independent requirement. In planning strategy 56 the customer-independent requirement of the configurable material is consumed with the corresponding planned independent requirement at this level. So the consumption is at the highest level and not at the level of components and nonconfigurable assemblies.

[Ex]

Consumption against Planning

Let's illustrate this based on a numerical example. In the example in Figure 3.24, a quantity of 1,000 pieces is planned for a configurable computer within a planning period. Moreover, it is assumed that a frequency of 30%, that is, a dependent requirement of 300 pieces, results from the configuration supporting point for the tower casing type. Additionally, for this planning period a sales order results in an order of 100 computers that are all requested as towers along with several other properties. As mentioned previously, the consumption starts at the level of the configurable material.

The planned independent requirement for the computer is reduced by 100 pieces to 900 pieces for the planning period. The dependent requirements from planning, for example, the tower casing type, are reduced proportionally.

This means that after a new planning run, the planned dependent requirement for the tower is reduced for the computer in accordance with the requirement quantity of 900 pieces and not according to the requirements quantity from the sales order for the tower.

This means the dependent requirement for the tower is reduced from 300 pieces to 900 × 30% = 270 pieces (not 300 – 100 = 200 pieces).

The consumption is therefore not exact at the component level. The error should be determined using a larger number of sales orders. If this is not exact enough for you, it is advisable to use the next strategy.

Consider SAP Note 68033 on characteristics planning, which also includes some restrictions regarding the use of planning strategy 56. **[+]**

3.3.5 Characteristics Planning and Standard Product Planning with Long-Term Planning

The strategy of standard product planning with long-term planning links the benefits of characteristics planning from the last section with the benefits of pure assembly planning regarding the exact consumption.

Simulative and Operative

The trick of this strategy is to perform the characteristics planning at the level of the configurable material simulatively in the environment of the long-term planning (see Figure 3.25). Let's take a closer look on the basis of this figure:

▶ **Simulative environment**
In the long-term planning ❶ you create the planned independent requirements for the configurable material. Simulative configuration supporting points also exist here. Here, you can enter frequencies, provided that this was prepared using planning tables and a planning profile. These simulative planned independent requirements are subjected to a planning run in long-term planning. The result is planned orders that cannot be converted. They in turn generate the desired dependent requirements ❷ for the component and nonconfigurable assemblies. However, these dependent requirements are initially simulative. A transfer report ❸, which is provided using Transaction MS66 (Copying Simulative Dependent Requirements), enables the transfer to the operative environment of planning.

▶ **Operative environment**
No planning exists at the level of the configurable material in the operative environment. Planned independent requirements, however, exist for the planning-relevant components and nonconfigurable assemblies ❹. Via a corresponding planning run, you can generate the required planned orders, triggering the procurement in the form of in-house production or external procurement. This way, you can build the stock that is requested in planning.

A sales order for a configurable product generates a customer-independent requirement ❺ at this level. This customer-independent requirement cannot be consumed directly with the planned independent requirement because no planning (operative) took place at this level. This looks different at the level of the dependent requirements from the sales order. They can be consumed against the planned independent requirements ❻ — and this time exactly.

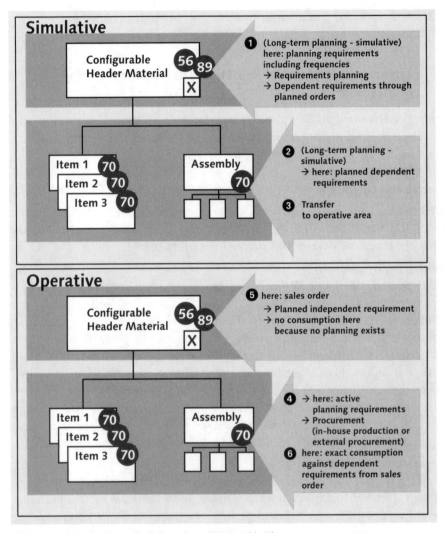

Figure 3.25 Standard Product Planning and Assembly Planning

Planning and Consumption for Characteristics Planning/ Standard Product Planning with Long-Term Planning

How do you proceed in this planning strategy in detail? The following steps are necessary, which are described in detail in the following sections.

- Making settings in the material — assigning the strategy
- Creating planned independent requirements including configuration supporting point
- Planning in long-term planning
- Using the transfer report
- Planning in the operative environment
- Sales order and its consumption

The strategy is assigned to the material master of the configurable material using the identical strategy group. This can be strategy 89, 56, or another strategy, provided that you intend a planning in the standard version using the VSE requirements type and requirements class 103 (also refer to the planning strategies in Chapter 4, Customizing SAP ERP for Variant Configuration). In addition, you must specify the CONSUMPTION MODE including the time intervals in the material master.

Strategy 70 and the consumption mode are assigned to the planning-relevant components and nonconfigurable assemblies. In addition, these components require 2 as the dependent requirements indicator (collective requirement) and 1 as the Mixed MRP Indicator (assembly planning with final assembly).

Create the planned independent requirements for the configurable material. These must not be created in version 00 and must be set to inactive to ensure that the system works simulatively. The configuration supporting points are populated for these planned independent requirements.

You need a planning scenario for the long-term planning (Transaction MS31). You can select any key (greater than 0). You only enter and release the planning period, the planned independent requirements (with the above version), and the corresponding plant. Regarding this planning scenario, you implement a single-item planning for the configurable material (Transaction MS02) in the long-term planning.

At this point, the transfer report comes into play. It is used for the planning-relevant component and the nonconfigurable assemblies. In this process, the data is transferred from the planning scenario of the long-term planning into the operative version 00 of the requirements planning. You can now implement the planning and procurement in the operative area.

Finally, let's have a look at the consumption against requirements from sales orders.

[Ex]

> **Consumption against Planning**
>
> We'll use the numeric example again.
>
> This means that 1,000 pieces were planned for a configurable computer. A frequency of 30%, that is, a dependent requirement of 300 pieces, results from the configuration supporting point for the tower casing type. Additionally, for this planning period a sales order results in an order of 100 computers that are all requested as towers along with several other properties.
>
> At the level of the configurable computer, the requirement of 100 pieces from the sales order cannot be consumed against the planned independent requirement because no such planned independent requirement exists according to the planning strategy. According to planning strategy 70 (assembly planning) at the level of the tower casing, the dependent requirement of 100 pieces from the sales order is (now exactly) consumed against the planned independent requirements.
>
> This means the planned independent requirement for the tower is reduced from 300 pieces by 100 pieces to 200 pieces.

This prevents the inexactness that was described in the last section on the pure standard product planning.

3.3.6 Variant Planning and Planning with Planning Variants

This final section presents a planning strategy that functions on the basis of *planning variants*. This planning strategy can also be motivated based on planning strategy 56, the characteristics planning from Section 3.3.4, Characteristics Planning and Standard Product Planning. Unlike in Section 3.3.5, Characteristics Planning and Standard Product Planning with Long-Term Planning, which focused on a more exact consumption associated with extra work, this section concentrates on the reduction of the planning effort. An inexact consumption with a requirement regarding the sales orders is accepted here.

In the scenario that forms the basis of this strategy and this section it is assumed that the characteristics planning is possibly rejected as too complicated and too complex. The idea is to implement the planning on the basis of a few typical representatives of the configurable material — the planning variants.

Similar to strategy 56, the description *variant planning* of strategy 54 (see Figure 3.26) is historical. It is still based on the terminology used in the SAP R/2 era. With regard to content and in current terminology, it is a planning strategy with planning variants.

Planning Variants

Similar to material variants, planning variants are material masters for configured materials. These material masters are plant-specifically (in the standard version via the MRP view 3) linked with the configurable material master and contain a configuration. The value assignment of the planning variant is carried out like the value assignment of the material variant, but it is a separate entry in the material master.

Planning and Consumption for Planning with Planning Variants/Variant Planning

Proceed in this planning strategy by executing the following steps, which are described in detail in the following sections.

▶ Make settings in the material — assigning the strategy

▶ Create planning variants

▶ Create planned independent requirements

▶ Execute planning run and procurement process

Planning strategy 54 must be assigned to both the planning variants and the configurable material (see ❶ in Figure 3.26). You also must enter the consumption mode and the consumption interval.

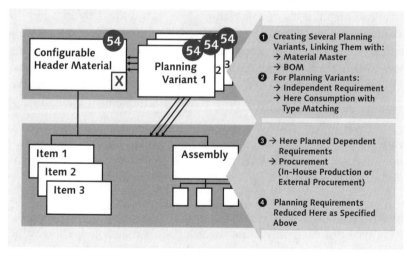

Figure 3.26 Variant Planning — Planning Strategy 54

You require BOMs for the planning variants. However, these BOMs are not explicitly created in the standard version, but are only provided by linking the planning variants with the BOM of the configurable material — similar to the material variants. This is done in Transaction CS40 (Assignment of Configurable Material).

Instead of performing a planning for the configurable material, the planning is carried out for typical representatives — the planning variants ❷. For this purpose, you create a *planned independent requirement* for the planning variants. The planned independent requirement can be created like the "normal," nonconfigurable materials, that is, without variant tables or profiles and configuration supporting points.

A planning run for the planning variants generates planned dependent requirements via the planned orders. These planned orders cannot be implemented the same way as strategy 56. The planned dependent requirements are used to trigger the procurement of the components and nonconfigurable assemblies ❸.

In this strategy, sales orders are received not based on the planning variants, but based on the configurable material. Nevertheless, these customer-independent requirements can be consumed against the planned independent requirements ❷. For this purpose, you set a *type matching*. This function can be found in Transaction MDPV as the last open item in the menu of the master data of the demand or production planning (see Figure 3.27). (The other items of the Master Data menu were discussed in Section 3.3.4 under the topic of planning strategy with characteristics planning.)

In general, you can assume that the configuration of the sales order doesn't completely match the configurations of the planning variants and partly matches all planning variants. This gives rise to the following question: Which planning variant matches the configuration of the sales order best? The settings required for this purpose can be made in the type matching at the level of the Characteristics options. The setting options at the Values level, as shown in Figure 3.27, currently have no functional significance. Under Characteristics, you can implement the following settings:

▶ **Not Relevant**
As the name suggests, the Not Relevant default value involves the value assignment for this characteristic not being considered in the type matching.

▶ **Required**
The Required value has the result that the corresponding planning variant is considered in the type matching only if the characteristic was assigned with a value both in the sales order and in the planning variant.

▶ **Optional**
Optional is those characteristics that are considered in the type matching as far as the corresponding planning variant is still relevant for type matching after the evaluation of the required characteristics.

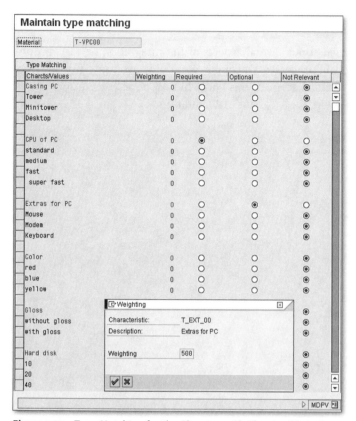

Figure 3.27 Type Matching for the Planning with Planning Variants

For optional characteristics you can also specify the Weighting. The system adds up the weightings of all optional characteristics for which the value assignment of the planning variant matches the value assignment in the sales order. The planning variant with the best match result is the one from the set of planning variants that are still relevant after the evaluation of the required characteristics and for which the total of the weights is the largest. The planned independent requirement from the sales order is consumed against the planned independent requirements of this planning variant. Similar to planning strategy 56, the consumption is at the level of the header material and is not exact at the level of the components and nonconfigurable assemblies. The reduction of the planned independent requirements at the level of the components and nonconfigurable assemblies results from the reduction of the planned independent requirements of the planning variants (see ❹ in Figure 3.26).

3.4 Summary

In this chapter you learned about several aspects of the integration of Variant Configuration with supply chain processes. Of course, we could cover only a small selection of topics to stay within the scope of this book, but we hope you got some ideas and extensions for your work with Variant Configuration.

This chapter provides general instruction for experts and teaches beginners will learn that they don't need comprehensive customizing knowledge in order to use Variant Configuration.

4 Customizing SAP ERP for Variant Configuration

This chapter summarizes the Customizing settings for Variant Configuration. This includes a couple of settings that exclusively concern Variant Configuration, and many settings from other areas, such as the classification system, as well as settings that relate to the processes of the supply chain, including sales and distribution, planning, and production.

The goal of this chapter is to show you that you can use Variant Configuration in the SAP ERP system using the standard Customizing settings. No extensive work is necessary before you can use Variant Configuration.

This chapter details the specific Customizing settings of Variant Configuration, which basically only include questions regarding authorizations, statuses, groups, and default values. In addition, you'll learn about the decisive points for Variant Configuration in the Customizing of the classification system, particularly in the class types settings. The descriptions of Customizing the supply chain processes discussed in this chapter relate to the Customizing of the material master, the item categories, the requirements types and classes, and Order Change Management (OCM).

4.1 Explicit Customizing of Variant Configuration

In Customizing, there is a separate section for the Variant Configuration. Here, you can make minor adaptations, but you can also work with the standard settings. The following sections present the settings that differ from the standard in their details.

We'll discuss these settings in the order in which they're illustrated in Figure 4.1. The description only differs from this sequence when we summarize multiple similar points.

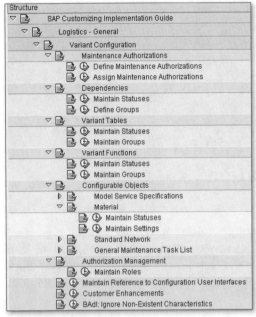

Figure 4.1 Customizing of Variant Configuration

4.1.1   Maintenance Authorizations

The first entry of Variant Configuration Customizing is the Maintenance Authorizations. They fulfill two tasks that — briefly summarized — answer the following two questions:

▶ Who is permitted to maintain?

▶ Where is assignment permitted?

You can use them and the C_LOVC_DEP authorization object to control the maintenance of object dependencies. You can assign this authorization object to authorization profiles and these profiles, in turn, to roles.

Moreover, a maintenance authorization controls to which objects you can assign object dependencies. The maintenance authorization field exists in the basic data of object dependencies. You can use this field to restrict the assignment. You can assign a maintenance authorization that was previously defined with a name and a description in Customizing. Additionally, you must assign the names of tables in Customizing of the maintenance authorization, and you can assign object dependencies to their entries. Among others, the following tables are available:

- Characteristic and characteristic value
- Configuration profile
- Items in bills of materials (BOM)
- Sequence, operation, and production resources and tools assignment in the routing
- Services in the model service specification
- Classes (not available for use in the configuration)
- Events and tasks (only available for use in the *advanced mode* for the IPC)

4.1.2   Statuses

The second entry of Variant Configuration Customizing is Dependencies. Here you can set the status as shown in Figure 4.2. Three statuses exist: In Preparation, Released, and Locked. You only need to make changes if you want to work with *distribution locks*.

View "Dependency Maintenance - Statuses": Overview

New Entries

Stat	Maint.	Release	Locked	Distr.lock	Description
1		X			Released
2	X				In preparation
3			X		Locked

Figure 4.2   Customizing the Status for Dependencies

Distribution Lock   **[+]**

If you assign a distribution lock for an object using the Locked status, the object cannot be distributed, that is, you cannot copy it in a controlled manner to other systems using the *Application Link Enabling* (ALE) tools.

Similarly, you have the option to set the status for variant tables (see Figure 4.3), variant functions (see Figure 4.4), and the configuration profile (see Figure 4.5).

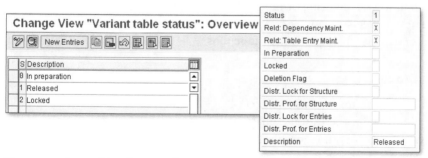

Change View "Variant table status": Overview

New Entries

S	Description
0	In preparation
1	Released
2	Locked

Status	1
Reld: Dependency Maint.	X
Reld: Table Entry Maint.	X
In Preparation	
Locked	
Deletion Flag	
Distr. Lock for Structure	
Distr. Prof. for Structure	
Distr. Lock for Entries	
Distr. Prof. for Entries	
Description	Released

Figure 4.3   Customizing the Status for Variant Tables

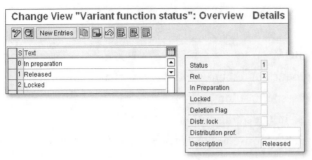

Figure 4.4 Customizing the Status for Variant Functions

Table	Status	In prep.	Rel.	Locked	Chg.no. req.	Distr. lock	Prof. Name
MARA	0	☑	☐	☐	☐	☐	
MARA	1	☐	☑	☐	☐	☐	
MARA	2	☐	☐	☑	☐	☐	
PLKOGMTL	0	☑	☐	☐	☐	☐	
PLKOGMTL	1	☐	☑	☐	☐	☐	
PLKOGMTL	2	☐	☐	☑	☑	☐	
PLKONET	0	☑	☐	☐	☐	☐	
PLKONET	1	☐	☑	☐	☐	☐	
PLKONET	2	☐	☐	☑	☑	☐	
TMPSPECLST	0	☑	☐	☐	☐	☐	
TMPSPECLST	1	☐	☑	☐	☐	☐	
TMPSPECLST	2	☐	☐	☑	☐	☐	

Figure 4.5 Customizing the Status for Configuration Profiles

In contrast to the status for dependencies and the configuration profile, a detail screen exists for the statuses of variant tables and functions. This enables you to make considerably more settings; for instance, for variant tables you can separately control the use in dependencies, the content maintenance, and the distribution of table structure and content.

You can find the Customizing of the configuration profile status in the Statuses for Configurable Objects menu. The status maintenance includes additional control options for variant tables and variant functions.

4.1.3 Groups

You can specify groups for various object types from Variant Configuration and the classification system. These groups enable you to structure the corresponding objects at a single level that ensure another selection and evaluation option. These groups don't have any functionality.

The option to maintain such groups exists for the following object types:

- Dependencies
- Variant tables
- Variant functions
- Characteristics (in Customizing of the classification system)
- Classes (in Customizing of the classification system)

You can define separate independent groups for each object type by specifying only the names and descriptions. No further entries are needed here.

4.1.4 Configurable Objects

Via the menu path CONFIGURABLE OBJECTS • OBJECT TYPE • MAINTAIN SETTINGS, you can preset a status and the indicator for the start logo as a default value for the configuration profile depending on the configurable object type. Section 4.1.2, Statuses, already discussed the status topic.

4.1.5 Configuration User Interface

Via the menu path VARIANT CONFIGURATION • MAINTAIN REFERENCE TO CONFIGURATION USER INTERFACES, you can, as the name of the menu suggests, create a reference to a configuration user interface (configuration UI). This step is necessary if you want to use the Internet Price Configurator (IPC) in the SAP ERP system. As described in Chapter 2, Creating a Product Model for SAP Variant Configuration, you can use the new product modeling environment of Variant Configuration to define for each configurable material whether and when you want to work with the IPC in the SAP ERP system. For this purpose, the detail screen of the material master in Transaction PMEVC provides the option to select from the following three settings:

- Generally classic configurator: Variant Configuration
- Configurator of IPC only in simulation: Variant Configuration for Productive Use—IPC for Testing
- General configurator of the IPC: IPC Only

In this customizing step, only the last step indirectly assigns a configuration UI. Prior to this, you must create this configuration UI within a so-called XCM scenario. This is done using *Extended Configuration Management* (XCM). Here, the configuration UI is defined using a set of parameters. These XCM scenarios are implemented outside the actual SAP ERP system supported by a J2EE server. For this purpose, you must connect the J2EE server to the SAP ERP system. This is done using an RFC destination called IPC_CONFIGURATION_UI.

[+]   For more information, refer to SAP Notes 854170 and 844817.

The configuration UI that you set in this customizing step can be overridden user-specifically by assigning the user a value via a parameter value for the IPC_CONFIGU-RATION_UI parameter ID (user settings).

In addition, you need to run the CFG_ERP_INITIALISE_DB report and assign the COM_SCE_COLLECT_DATA_CHANGE function module (instead of the CRS_SCE_COLLECT_DATA_CHANGE default value) in the TB31 table for the event 00550002 before you can use the IPC in the SAP ERP system. For more comprehensive information refer to the SAP library in the SAP help portal.

The other entries of the Variant Configuration Customizing include standard functions:

- ▶ Role maintenance (Transaction PCFG)
- ▶ Maintain user exits (Transaction CMOD)
- ▶ BAdI Builder for CUDUI_MISSED_CHAR

4.2   Classification System Customizing

The classification system presents important tools for modeling in Variant Configuration (see Figure 4.6), so it's not surprising that some settings in the classification system Customizing are relevant for Variant Configuration.

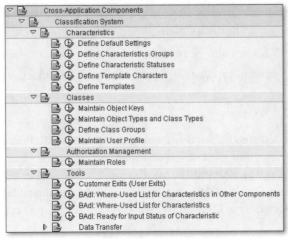

Figure 4.6   Classification System Customizing

> This separate section discusses the classification system as a cross-application component so that the next section can focus on Customizing settings in the environment of the supply chain processes. **[+]**

As shown in Figure 4.6, there are some settings for Characteristics. The first menu entry, Define Default Settings, contains some default values for creating characteristics, specifically, the status and several indicators, such as Multiple Values, Entry Required, and Additional values. You use the second entry, Define Characteristics Groups, to maintain the groups, similar to the groups described in the previous section in the Variant Configuration menu. The third entry, Define Characteristic Statuses, the status maintenance, only contains indicators regarding the release, deletion, and distribution lock including the profile name so that — like the first two entries — the delivery version is also sufficient in Variant Configuration. The same applies to the entries Templates and Template Characters of the Characteristics section. In the standard version, the templates don't play any role in Variant Configuration because they require a free entry with regard to the corresponding characteristics. In Variant Configuration, this can basically only be used in the order BOM scenario because only here can free entry be integrated with modeling in the form of technical postprocessing. You can also work with default template characters in this case. You only need to create templates in Customizing if you want this as a default value in characteristic management.

Customizing of classes starts with an object key. In the menu option Maintain Object key, no changes or supplementations make sense from the Variant Configuration perspective. This table is used not only in the classification system, but also for the object links of documents, for example. It is only extended if additional object types that are classifiable or can be linked with documents are supposed to be defined.

From the Variant Configuration perspective, it is of interest that you specify the key words for the object key in the last column, Table, of Figure 4.7 for the object types that can be accessed in dependencies. So the information provided in Figures 4.7 and 4.8 results in the call of a material master or a document in dependencies (see Listing 4.1).

```
(material)()(nr = 't-vpc')
(document)()(type = 'drm', nr= 't-vpc', version = '00', part = '000')
```

Listing 4.1 Material Masters and Documents in Dependencies

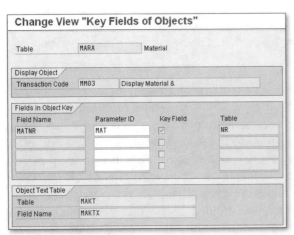

Figure 4.7 Object Key in the Classification System Customizing

The next entry, Class Types, is the most interesting entry of the classification system's Customizing. This step has a hierarchical structure. First, settings on object tables exist at the top-most level that correspond to the object types. You can view details in Figure 4.8. From the Variant Configuration perspective, the Screens section (right) and the Object ID field (bottom right) are of interest. At this general level, you can use the Screens section to restrict the display in the maintenance of the classes for five of the seven possible class maintenance tabs. The remaining two tabs are exceptions. You cannot deselect the Basic data of the class maintenance. The Additional data tab is activated in the detail screen of the class type's Customizing using the Class Node indicator.

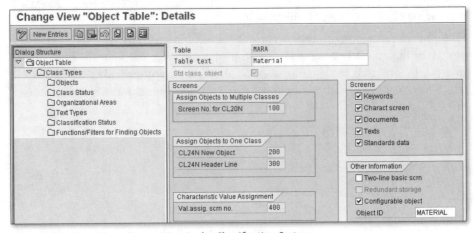

Figure 4.8 Object Type Customizing in the Classification System

As described in the context of Figure 4.7, you use the entry of the Object ID field to access material masters and documents in dependencies.

Below the settings for the object tables you define the class types specifically for the object types in this customizing step. For Variant Configuration, the variant class types 300 (for material master, project system, and plant maintenance) and 301 (for services) and the class node class types 200 (for material masters) and 201 (for documents) are of interest.

Figure 4.9 shows the details of the class types settings. It starts with the permitted screens, similar to the object tables. Here, you can further restrict the permitted screens for each class type. For example, you could hide texts and (DIN) standards data for the variant class type 300. For the classification, you can specify for each class type whether engineering change management or change documents are supposed to be used. This only applies to the classification. For the characteristic and class maintenance both are activated permanently. This doesn't play any role in the actual configuration. It would be possible to use these change documents and engineering change management for the classification in the class nodes or the Variant Class and Configurable Material link. The latter also represents a classification.

Let's take a look at the checkboxes that are displayed on the right-hand side below Functions in Figure 4.9. One class type per object type is the Standard Class Type, a kind of default class type. This is class type 001 for the material masters.

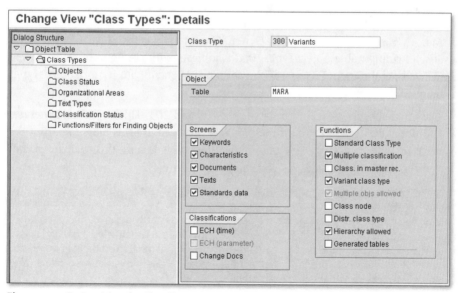

Figure 4.9 Customizing the Class Types

The Multiple classification checkbox enables you to assign multiple classes to a material master regarding this class type. You need this if you want to assign multiple variant classes to the configurable material to compile the characteristics for the value assignment interface. You also use this setting to allow hierarchy networks for the class hierarchy maintenance in this class type. This means that a class can have multiple predecessors.

The Classification in master record checkbox prohibits the central classification functions using Transactions CL20N and CL24N. Consequently, you can classify material masters only from the material master maintenance. This doesn't apply to the variant class type, but to the other class types, for material masters, and to class nodes.

If you select the Variant class type checkbox, a class becomes a variant class. This means that this class is assigned to the configurable object. You can restrict the value set of characteristics as required. Similar to the hierarchy maintenance of classes, dependencies and format specifications, for instance, the single-value attribute, are deactivated for the value restrictions. In addition, the variant class type enables you to select this class type when you create the configuration profiles and consequently generate the value assignment interface. In the standard version, this checkbox is selected for the class types 300 and 301.

The Multiple objects allowed checkbox enables you to assign more object types to this class type. This is done class-type-specifically under Objects (see Figure 4.9, on the lefthand side below Class Types in the Dialog Structure). For class type 300 this is required to be able to use a class type for the three configurable object types, Material Master, Standard network, and General Maintenance Task List. For historical reasons, all configurable object types, that is, objects with configurations, are listed here as well. As shown in Figure 4.9, the checkbox is grayed out if more object types are assigned.

The Class node checkbox is in the class types 200 and 201. This checkbox generates the Additional data tab in the class maintenance. You can use this table to permit the system to include such a class in configurable BOMs. By default, this checkbox is not selected for variant class types, as you can see in Figure 4.9.

[+]   **Combination: Class Node and Variant Class**

This setting to not label variant class types as class nodes at the same time is recommended for performance reasons. However, you must couple variant class type and class node if a class node is supposed to contain configurable materials and be configured at this level. Provided that you don't define a new separate class type, you also select the Class node checkbox in the variant class type.

The Distribution class type is not available by default for class types from Variant Configuration. For material masters, this checkbox is activated for the ALM class type, and it is used for the distribution of material masters across SAP systems using ALE.

The Hierarchy allowed checkbox enables you to set up a class hierarchy within the class type. If you don't select the Multiple classification checkbox, the hierarchy must be strictly hierarchical; that is, a class can only have one superordinate node at the most. With the Multiple classification checkbox, the system also permits hierarchy networks. The restriction of strictly hierarchical doesn't exist then.

> If you don't want to work with class hierarchies in the class type, we recommend that **[+]**
> you don't select the Hierarchy allowed checkbox.

The generated table's checkbox isn't suitable for variant class types and can therefore not be used. To improve performance in very large class systems, you can use this checkbox to permit, in the object search, the system to generate a separate table with the classification data for the search. The information of all classifications (not configurations) of all objects is stored in the AUSP table.

If the number of entries in this table exceeds 100 million to 1 billion, this is referred to as a very large class system. Then, your generated tables improve the performance in the object search. As already mentioned, this isn't suitable for variant class types and is therefore not allowed.

Since SAP R/3 4.5, configurations have been listed not in the AUSP table, but in the tables of the so-called IBase, which are listed in the appendix. From the performance perspective, this doesn't play a significant role in the class nodes. However, there is an interesting use of these generated tables.

If in the dependencies you want to simply evaluate the classified objects that exist in a class node, you can implement this using such generated tables.

Generating Tables for Class Nodes **[Ex]**

Let's illustrate this situation with an example: In the configuration in the sales order, you only want to permit requests that are guaranteed to have a valid assigned value in a class node and can therefore be implemented. Proceed as follows: Create the corresponding characteristics of the configuration as single value and restrictable. After you've fulfilled all prerequisites, generate a DB table that only contains all classified objects of the class node with the value assigned. This is done using Transaction CLGT from the class system. You can take the name of the generated DB table from the tables, GEN_TABLES_MAIN and GEN_TABLES_SUB. Then the system creates a variant table that is linked with this generated DB table. So you can use a table constraint (for instance, created using the corresponding wizard in the PMEVC table) to conveniently restrict the characteristics in such a way that the users can only select the components that are currently available in the class node.

As already mentioned, you can set the entries that are listed in Figure 4.9 in the dialog structure below the class types specifically for each class type. A Class Status and a Classification Status must be created for each class type. Without these two statuses you could not create any classes or classify any object. For the remaining entries, you can specify the use per class type. The Objects item was already discussed in the context of multiple object types in a class type. Organizational Areas can be interesting for the variant class types if you want to use configuration profiles to filter the characteristics that are supposed to be used in the value assignment interface, that is, which characteristics are supposed to be displayed. A C_TCLS_BER authorization object exists for the organizational areas so that you can control who can view and evaluate what in the configuration. The remaining two entries, Text Types and Functions/Filters for Finding Objects, which are specific to the Customizing of the class type are not relevant for Variant Configuration.

4.3 Business Process Customizing Relevant for Variant Configuration

With Customizing of Variant Configuration and the classification system (see Sections 4.1, Explicit Customizing of Variant Configuration, and 4.2, Classification System Customizing), the essential aspects in Customizing for the master data of modeling a configurable product are complete. This section presents some aspects from the Customizing of logistics that are relevant for Variant Configuration.

4.3.1 Configurable Material Master

You've gotten to know the essential aspects in the Customizing for master data of the modeling of a configurable product. The following sections briefly discuss the material types for the configurable material masters.

As shown in Figure 4.11, the Customizing of the material master contains the entry Make Global Settings. Here, you can generally deactivate Variant Configuration, among other things. In this case, the system doesn't display the subscreens of Variant Configuration in the material master.

In addition, you can decide how each material type is used:

▶ **Only for configurable materials**
KMAT is a material type only for configurable materials. In the detail screen of the Customizing of this material type, the Material is configurable checkbox is selected.

▶ **Also for configurable materials**

The FERT material type, which is shown in Figure 4.10, is a material type that can *also* be used for configurable materials.

▶ **Not for configurable materials**

For a material master of such a material type, you can use the field and screen control to prevent it being labeled as configurable.

Figure 4.10 illustrates the FERT material type. If this material type can be used not only for configurable materials, you mustn't select the Material is configurable checkbox as shown here. Additionally, however, you must ensure that a checkbox with the same name exists in the material master and is ready for input. This is implemented via the screen and field control. For this purpose, the Customizing of the material type has a field reference and a screen reference as shown under General data in Figure 4.10.

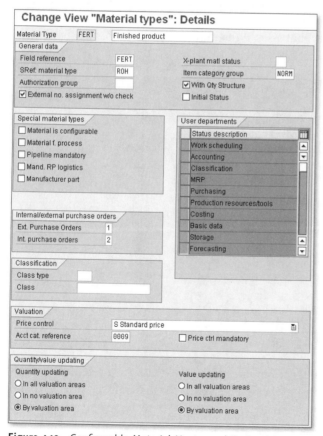

Figure 4.10 Configurable Material Master and Customizing

The screen reference (Screen Reference: material type field) is used in the first menu option of the material master's Customizing, Configuring The Material Master, as shown in Figure 4.11. Here you can control the entire screen layout of the material master using simple customizing steps. For example, this screen reference for finding a screen sequence is supported.

The screen sequence contains the list and the sequence of all material master tables including the additional data. In addition, for each tab it is determined which subscreens are listed in which sequence. Here you must ensure for the material types that permit configurable materials that the subscreen is used with the previously mentioned checkbox, Material is configurable.

In the standard version, this is implemented in the Basic data 2 tab using the second to last subscreen number 2499. You can control this from the material master by selecting SYSTEM • STATUS in the menu from the corresponding tab and the subscreen. The Customizing of the screen control for the material master is also referred to as *customized material master* because the finding of the screen sequence depends not only on the material type, but also on the industry, transaction, and, in particular, the user.

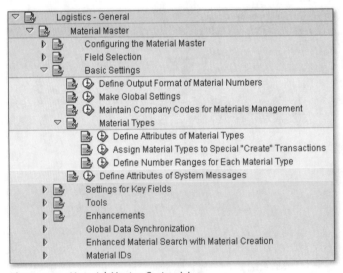

Figure 4.11 Material Master Customizing

In addition to the screen reference, you must also select the Field reference accordingly so that the material masters can be indicated as configurable. For this purpose, in the second menu entry of the material master's Customizing (Figure 4.11), Field Selection, you specify how a field is defined:

▶ As a ready-for-input optional field

▶ As a ready-for-input required field

▶ As a not ready-for-input (gray) field

▶ As a hidden field

In addition to the material master, you can also use other values, such as industry, transaction, procurement type, and plant, to control this.

4.3.2 Item Categories and Their Determination

In Chapter 3, Business Processes in SAP ERP, within the scope of processes along the supply chain, we discussed how some relevant settings exist in Customizing that concern the work with configurable materials in the corresponding processes. Now, let's begin with the processes in sales and distribution. The SD documents have to determine suitable item categories for configurable materials. This applies to the sales order, but also to other sales activities such as request and quotation.

Finding the item category is controlled via the setting in Customizing under Assign Item Categories. For this purpose, the system evaluates the SD document category (first column of the Item Category Assignment in Figure 4.12), the item category group from the material master (second column), and the main item or subitem (fourth column). The third column, Usage, plays only a minor role for material items.

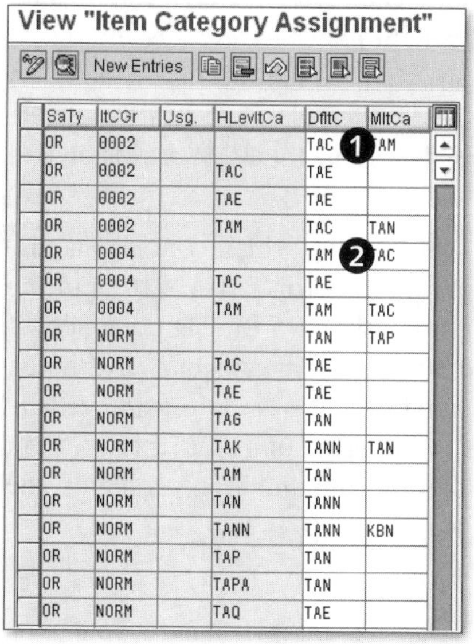

View "Item Category Assignment"

New Entries

SaTy	ItCGr	Usg.	HLevItCa	DfltC	MItCa
OR	0002			TAC	TAM
OR	0002		TAC	TAE	
OR	0002		TAE	TAE	
OR	0002		TAM	TAC	TAN
OR	0004			TAM	TAC
OR	0004		TAC	TAE	
OR	0004		TAM	TAM	TAC
OR	NORM			TAN	TAP
OR	NORM		TAC	TAE	
OR	NORM		TAE	TAE	
OR	NORM		TAG	TAN	
OR	NORM		TAK	TANN	TAN
OR	NORM		TAM	TAN	
OR	NORM		TAN	TANN	
OR	NORM		TANN	TANN	KBN
OR	NORM		TAP	TAN	
OR	NORM		TAPA	TAN	
OR	NORM		TAQ	TAE	

Figure 4.12 Item Category Determination

Figure 4.12 also shows that the system determines the TAC item category ❶ as the main item (fourth column blank) in the item category group 0002 (second column) for standard orders (first column, OR).

In addition, you can see that the system determines the TAM item category ❷ as the main item (fourth column blank) in the item category group 004 (second column) for standard orders (first column, OR). This item category group and consequently this item category is used for the header material in a sales order configuration scenario (SET). For comparison, refer to Section 2.4, Configuration Profile and Configuration Scenarios, in Chapter 2.

Figure 4.12 also shows that the system generally determines the TAE item category, a so-called explanation item, for subitems (fourth column, filled) below a main item with TAC. For subitems below a TAM main item, the same rules apply as for the corresponding main item.

Item categories for other SD documents are determined correspondingly. Similar to TAC in the standard order, the system determines an AGC item category for the quotation, for example.

Section 2.10, Material Variants, in Chapter 2 discussed the settings in the item category as shown in Figure 4.13 under ❶. The Bill of Material/Configuration section is particularly important in addition to the billing, pricing, and schedule line relevance. Here, you must enter the configuration strategy 01 for configurable main items. Additionally, you can use the three fields Variant Matching, ATP material variant, and Material Variant Action to control the material variant matching in sales and distribution. The two fields Structure Scope and Application are significant for the BOM explosion. The entry "D" in the Structure scope field indicates that the settings in the configuration profile, not the settings made here, determine the BOM explosion and the scope in the sales and distribution.

The Application SD01 setting is an exception. This setting, which is not a mandatory field here, overrides the setting in the configuration profile. The setting in the configuration profile for the BOM application is only used in the SD document if no entry exists for the application in the corresponding item category (TAC here). This is not the case in Figure 4.13, however. Therefore, the system would evaluate the application SD01 and first search for sales BOMs in case of a BOM explosion in the sales order, possibly contrary to the setting in the configuration profile on the BOM usage.

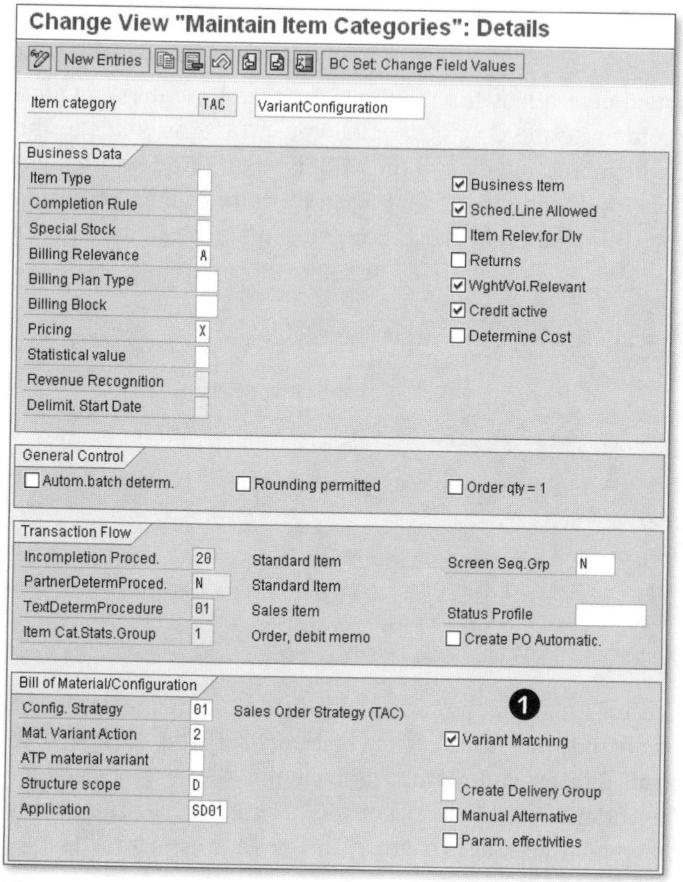

Figure 4.13 Item Categories

4.3.3 Requirements Types and Requirements Classes and Their Determination

Based on item categories, whose determination was described in the previous section, the system determines the appropriate requirements type and thus the requirements class for sales and distribution. To find this requirements type the system uses the settings that are made under the Determination of Requirement Types using the Transaction menu option in the Customizing of sales and distribution (under BASIC FUNCTIONS • TRANSFER OF REQUIREMENTS). The determination of the requirements type is based on the item category (for instance, the three types, TAC, TAM, and AGC, that were described previously) and the MRP type from the material master.

Initially, the fourth column ❶ in Figure 4.14, the source or origin of the requirements type, is decisive.

You can see that the system determines the requirements type from the quotation (AGC item category) according to the Origin Reqmt. Type = 1 (corresponds to the fourth column, Q) using the item category and the MRP type. In this case, the item category and the MRP type, and therefore the entries in the third column ❷, decide on the requirements type. This column is blank, however, so no requirements type is found.

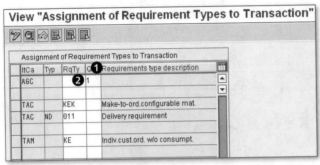

Figure 4.14 Requirements Type Determination via the Process

On the one hand, this ensures that the quotation triggers no requirement. On the other hand, no requirement and stock segment can be created relating to the quotation. Both the requirement transfer and the option including the segment may be required differently. The variant listed here is not the default setting. The setting selected here was used owing to the usage of Order Change Management (OCM).

[+]   Also refer to the descriptions in Section 10.2, Change Management, in Chapter 10.

For OCM it was requested that the system initially create the requirement and the requirement and stock segment using the sales order. Only then does the OCM change process run smoothly, even if the sales order, but not the associated quotation, is changed.

For the two item categories, TAC and TAM, the system initially evaluates the strategy from the material master as the source (blank = 0). In Variant Configuration, the standard version uses strategy 25, that is, if planning or assembly is not used. In Figure 4.16 you can see that this results in the determination of the KEK requirements type (make-to-order configuration) and then in requirements class 046 as shown in Figure 4.17. The requirements type and the requirements class present a 1:1 correlation. There are historical reasons for this separation.

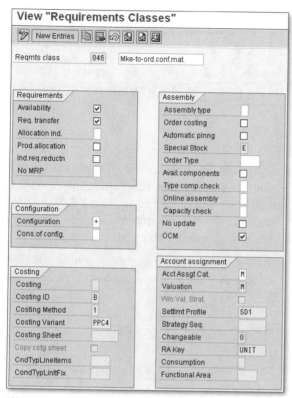

Figure 4.15 Requirements Class 046

Figure 4.15 illustrates the requirements class 046 in detail. The Configuration section in the Requirements Classes view is of particular interest. In general, this requirement class is used only with configuration, as specified by the "+" sign. You could also use requirements classes with "." as the possible configuration. In strategy 25 (requirements class 046) you work without planning and therefore without consumption. Planning strategies with planning and therefore also consumptions against characteristics planning (strategy 56 — requirements class 043) and against planning variants (strategy 54 — requirements class 042) exist.

Additionally, in Figure 4.15, for requirements class 046 you can see that an availability check and the requirements transfer take place according to the setting in the Requirements section. The Assembly section indicates that no assembly is used, which can be used reasonably for configurable materials, however. Also note the OCM checkbox, which you can use to activate the updating of so-called initiating object records for OCM. This checkbox is not set by default.

In addition, the requirements type decides that a TAC item transfers requirement and a TAM item doesn't. In both cases, the system would find the requirements type described previously. That's where the schedule line category comes into play.

In the standard version, a TAC item finds the CP schedule line category (MRP with requirements transfer). In contrast, the system finds a schedule line category CT without requirement transfer for a TAM item.

4.3.4 Planning Strategies

This section briefly presents the planning strategies that are relevant for Variant Configuration including their requirements types and requirements classes based on the corresponding Customizing settings. You can find these settings in the Customizing of production via the menu path PRODUCTION • PRODUCTION PLANNING • DEMAND MANAGEMENT • PLANNED INDEPENDENT REQUIREMENT menu path.

Figure 4.16 maps the Customizing of the planning strategies. A planning strategy can consist of a requirements type for planning and/or a requirements type for sales orders.

Change View "Strategy": Overview

Strategy	Planning strategy description	Reqs-DM	Reqs-Cu.
20	Make-to-order production		KE
21	Make-to-order prod./ project settlement		KP
25	Make-to-order for configurable material		KEK
26	Make-to-order for material variant		KEL
54	Types planning techniques	VSE	KEKT
55	Planning mat. variant w/o final assembly	VSE	KELV
56	BOM characteristics planning	VSE	KEKS
70	Planning at assembly level	VSFB	
80	Project settlement for non-stock items		KPX
81	Assembly processing with repetitive mfg		KMSE
82	Assembly orders with production orders		KMFA
83	Assembly processing netw. w/o project		KMNP
84	Service orders		SERA
85	Assembly processing netw. with project		ZMPN
86	Configuration with process orders		KMPA
87	Assembly orders with production orders B		KMFB
88	Assembly orders with production orders C		KMFC
89	Assembly proc. w. characteristics plng	VSE	KMSE

Figure 4.16 Planning Strategies

Figure 4.17 shows the connection between requirements type and requirements class, which was discussed in the previous section.

Change View "Requirements Types": Overview

New Entries

RqTy	Requirements type	ReqCl	Description
KEK	Make-to-ord.configurable mat.	046	Mke-to-ord.conf.mat.
KEKS	Ord. + cons. of char. planning	043	Mke-ord.cons.StlndRq
KEKT	Order + consumption of variant	042	Mke-ord.consStckType
KEL	Make-to-order, mat. variants	047	MkToOrd.-mat.variant
KELV	Make-to-ord.variant + consumpt	048	Mke->O.mat.var.cons.
KEV	Make-to-order with consumption	045	Stck w.cons.ind.reqs
KEVV	Indiv.cust.with plng mat.cons.	060	Mke->O.Cons. plngMat
KL	Sales ord.manufact.by lot size	041	Order/delivery reqmt
KMFA	Assembly with production order	201	Assembly: prod.order
KMFB	Assembly evaluated w. ProdOrd.	207	Assembly eval. SO
KMFC	Assembly evaluated w. ProdOrd.	ZFA	Assembly: prod.order
KMNP	Assembly with network orders	202	Network
KMPA	AssemOrd. with process orders	206	Process order
KMPN	AssemOrd. with network/project	212	Netplan/proj.settlmt
KMSE	Assembly planned order	200	Assembly: plnd order
VSE	Planning w/o final assembly	103	Plnng w/o assembly
VSFB	Planning for assemblies	105	Assembly planning

Figure 4.17 Requirements Types and Their Requirements Classes

Planning Strategy 25

Strategy 25 is the class planning strategy for Variant Configuration that is basically analogous to strategy 20 outside Variant Configuration. Strategy 25 works without planning and consequently has no corresponding requirements type. The requirements type for sales orders is KEK with requirements class 046, as discussed previously.

Planning Strategies with Assembly

In addition, you can use all planning strategies that work with assembly in sales and distribution. These include the strategies from 81 to 89 (see Figure 4.16).

Characteristics Planning and Planning with Planning Variants

Section 3.3, Planning and Variant Configuration, in Chapter 3 mentioned planning strategies 54, 56, 70, and 89 within the scope of planning. Strategies 54, 56, and 89 use a planning directly at the level of the configurable header material. In all three cases, it is a planning without final assembly; that is, planned orders that were generated owing to planned independent requirements cannot be implemented. The corresponding requirements class 103 contains the following indicators:

- Planning indicator 3: Single-item planning
- Consumption indicator 2: Consumption of planning without final assembly
- Requirements type 1: Planned independent requirement

Strategy 56 can be used for a pure characteristics planning. Strategies 89 and 56 can be used for a characteristics planning in the simulative long-term planning. Here you must transfer the simulative planning-dependent requirements to production requirements for all relevant components, as Section 3.3, Planning and Variant Configuration, described in more detail.

Assembly Planning

These components must be supplied with planning strategy 70. In comparison to requirements class 103, the associated requirements class 105 differs in the following:

- Planning indicator 1: Net requirements planning
- Consumption indicator 1: Consumption of planning with assembly

Strategy 70, the planning at assembly level, can also be used directly. That is, the planned-independent requirements are directly created at this level. For further details refer to Section 3.3, Planning and Variant Configuration, in Chapter 3.

4.3.5 Change Profiles in Order Change Management (OCM)

Section 10.2, Engineering Change Management, presents Order Change Management (OCM). It includes a note that only material masters to which you have assigned an overall change profile participate in OCM. This section also states that this overall change profile is required for checking the determined changes. Here, you decide, depending on the status and the percent of completion of the production order, how the determined changes are to be labeled. You can choose from the following options:

- Uncritical
- Critical
- Disallowed

Uncritical changes correspond to the first two columns shown in Figure 4.18 (None and Info) ❶ and permit an automatic change.

Critical changes correspond to the third column shown in Figure 4.18 (Warning) ❷ and expect a confirmation by the user. You have two options to implement the confirmation:

▶ **Change profile with the "When warning appears, execute and mark changes" indicator**
You can first implement the changes and then confirm the production order.

▶ **Change profile without the "When warning appears, execute and mark changes" indicator**
You can also confirm or reject the changes in the list of determined and checked changes and then implement the changes in the production order.

You can find this indicator in the details of the change profile in the General tab.

In general, disallowed changes cannot be implemented by OCM. For example, already installed components or confirmed operations cannot be deleted. Of course, you can disallow further changes in the change profile.

You distinguish between three change processes (sales order, master data, assembly), for which you can set the checks separately via the corresponding change profiles within the overall change profile.

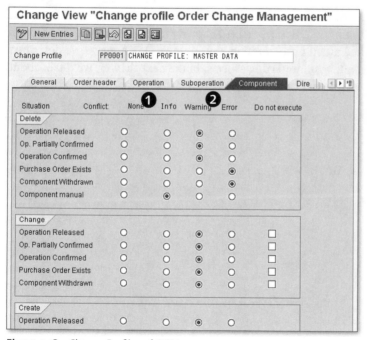

Figure 4.18 Change Profiles of OCM

Figure 4.18 shows the typical appearance of the settings in a change profile. Depending on the objects within the component and operation list of the production order for which the system determined changes, the corresponding tab is used. The General tab includes additional settings. This tab includes the already mentioned indicator that is used for the two critical change procedures. In addition, in assembly you can allow only implementation of individual change steps. You can also decide whether manually added objects may be deleted.

4.4 Summary

In this Customizing chapter you got a complete overview of the Customizing that is explicitly valid for Variant Configuration. Keep in mind that it is not completely correct that the options in Section 4.1, Explicit Customizing of Variant Configuration, are only valid for Variant Configuration. Because, for example, the settings for dependencies, variant tables, and variant functions are also valid if you work with these tools in the classification.

You also learned to what extent the settings in Customizing of the classification system impact Variant Configuration. In the Business Process Customizing Relevant for Variant Configuration section, you learned about some Customizing settings from supply chain processes, which we consider particularly interesting in the Variant Configuration environment. Of course, this chapter mentioned only a few aspects of Customizing in the Variant Configuration environment. We hope we've described the most important and most interesting aspects. In the next chapter, we'll talk about the special features of product configuration in SAP CRM.

This chapter provides insight into Variant Configuration in the context of SAP CRM. You'll learn about the features and differences between SAP CRM and Variant Configuration in SAP ERP, and you'll learn details about the product modeling environment and the use of the IPC.

5   Special Features of Product Configuration in SAP CRM

In this chapter, you'll learn about product configuration in SAP CRM beginning with the different types of product configuration in SAP CRM and the difference in the configuration of products and services. But before we discuss the creation of a product model in SAP CRM, you'll learn about integration with SAP ERP. Finally, we'll introduce you to the special functions and the user interface.

5.1   Product Configuration in Different Channels

One strength of SAP CRM is the multitude of *channels*. You can reach your customers in different ways while always using a uniform infrastructure. The product configuration can be used in all channels offered by SAP CRM. Because the *AP Configuration Engine*, generally better known as *IPC* (see the note on terminology at the end of this section), is always the same, consistency is ensured, and you can develop further channels with minimum investment. Let's briefly discuss the different channels of SAP CRM.

Web Channel (CRM and ERP)

You can use the product configuration in the web channel (also known as e-commerce or internet sales) for SAP CRM and for SAP ERP. It is irrelevant whether you use the B2B or B2C scenario *(business-to-business* or *business-to-customer)*. As described in Chapter 3, Business Processes in SAP ERP, you can adapt the user interface to the appropriate scenario (for instance, in the B2C scenario with screens).

SAP CRM Online and Interaction Center

This includes all office-based scenarios. The product configuration can be used at the early stages of the sales cycle: the opportunities or quotations. Moreover, you can also configure nonphysical products, such as telephone contracts, insurance, maintenance services, and so on.

SAP CRM Mobile Sales

If you want to configure products without a connection to a server (offline), CRM Mobile Sales is the tool of choice. This "miniature CRM" on your laptop provides you with the same product configurator as in SAP CRM (because SAP NetWeaver is not available as the basis, there are adaptations "underneath"). The user interface is identical to the one on the server and is displayed in the web browser. The knowledge bases are replicated to the database of the mobile client. For frequent updates of the models, you must ensure that the clients are synchronized in real time.

SAP ERP

As of SAP ERP 6.0 it is even possible to use the IPC, which was previously only used in SAP CRM, in SD instead of the ERP Variant Configuration (LO-VC). This can be of interest for two reasons:

- You use SAP CRM and want to have a uniform tool to minimize the costs of tests and user trainings.
- You want to benefit from the special functions of the IPC user interface (see Section 5.6) in the SAP ERP system.

[+] **Terminology**

Since SAP CRM 2005, the Java component *Internet Pricing and Configurator* (IPC), which had to be installed separately, has moved to the SAP application platform and is based on the SAP NetWeaver technology (see SAP Note 844816). Since then, the components are formally called AP 7.00 Engines (*AP Configuration Engine, AP Pricing Engine,* and so on). However, in real life — and in this book — the term *IPC* is still used when reference is made to the functions that are programmed in Java, such as product configuration, pricing, tax calculation, and listings. So the original acronym has become an abstract concept.

5.2 Configuration of Products versus Services

Traditionally, the examples in the Variant Configuration area address the production industry. In the SAP CRM environment, further usage areas are available for configured (service) products. Product configuration can be used in the following areas, for example:

- Mail services
- Insurance contracts

- Supply contracts (gas, water, electricity)
- Communication services ("triple play")
- Maintenance contracts
- Tickets

All of these real-life examples have in common that the offered services can be defined according to customer requirements, that these requirements usually influence the price, and that there may be rules that must be taken into account. Using product configuration you can better adapt any kind of product to your customers' requirements.

Therefore, the processes in SAP CRM can be subdivided into two groups:

- **Compatibility with SAP ERP**
  The (physical) product is further processed in the logistics of the SAP ERP system (production, procurement, requirements planning, etc.). SAP CRM is used for the sales process (opportunity, quotation, sales order). From the technical perspective, this essentially means that the compatibility with SAP ERP must be ensured in all process steps.

- **Maintaining master data directly in SAP CRM**
  The SAP ERP logistics is not relevant for the configured product. Either the process is implemented completely in SAP CRM (service contracts), there are industry-specific interfaces (SAP for Utilities), or the further processing is performed in third-party systems. In these cases, a compatibility with ERP Variant Configuration is not relevant, and the master data can be maintained directly in SAP CRM. Sometimes these scenarios are also referred to as *CRM standalone*, which is confusing, however, because an SAP ERP system can still be involved (for instance, for financial accounting).

Distinguishing the Process Groups	[+]
At runtime, that is, in the interactive configuration and in the configuration result, you cannot distinguish whether the master data was created in the SAP ERP or the SAP CRM system. In both cases, the architecture, the user interface, and the result are the same.	

5.3 Procedure for Integrated Production in SAP ERP

Now, let's discuss the first part of our list of processes, the procedure for compatibility with SAP ERP. This description focuses on the most frequently used scenario of the SAP CRM implementations. The famous bicycle is produced, delivered, and sold on the Internet. In particular, this addresses users who use SAP CRM for sales and distribution purposes, but who replicate the quotation or the sales order in SAP ERP, where they trigger the production and further logistic processes using the low-level configuration.

5.3.1 Sales Configuration versus Production Configuration

The SAP CRM system is based on the sale of products and has no information on the production processes. Consequently, the product model should be optimized with regard to the sales cycle. It is decisive which options are presented in which way, which characteristics and components are relevant for the price, and how you design the user guidance (conflict explanations, texts, arrangement). However, you should remove all factors that are only relevant for production, especially a very complex bill of materials (BOM), from the model.

This separation of a sales-oriented and a production-oriented model is not always implemented in real life or is only halfheartedly implemented, although it can considerably contribute to success. From the focus on the sales configuration, some essential restrictions emerge for the configuration in SAP CRM:

▶ *Configure-to-order processes* for which the product model already contains all relevant information are supported.

▶ *Engineer-to-order processes* for which the BOM must be adapted manually in the sales order are not supported.

▶ Routings are not known in the SAP CRM system.

▶ Plant-specific information (for instance, BOMs) cannot be taken into account.

In the SAP CRM system you record all information that is relevant for the sales cycle (*high-level configuration*). After the replication of the sales order into the SAP ERP system, you can access the configuration result in the production-relevant processes and perform a detailed BOM explosion, for example (*low-level configuration*).

Section 1.2.4, Variant Configuration LO-VC, in Chapter 1 discusses in detail this basic separation, which also exists in SAP ERP.

5.3.2 Replication of the Master Data from SAP ERP

To be able to use the master data created in SAP ERP in the configuration using the IPC, you must provide the data to the IPC in a suitable form. In this process, the IPC works with a concept that is completely different from the concept of ERP Variant Configuration (LO-VC).

In contrast to its atomic master data management, all required master data (for instance, variant tables) of the IPC is maintained as *Knowledge Bases* (KB). Different validity statuses are kept in runtime versions, whereas a runtime version is always complete and includes a validity date that indicates the validity start date. If the master data maintenance is carried out in SAP ERP, the system creates a runtime version as a "snapshot" of the atomic master data elements in SAP ERP.

It would go beyond the scope of this chapter to describe this procedure in detail. Therefore, the following only discusses the basic concepts and steps. In principle, the master data replication follows the following steps:

1. **Defining knowledge bases**
   You define a *knowledge-base object* in SAP ERP that lists the products contained in the knowledge base. It can absolutely make sense to bundle multiple similar products (product family) to utilize synergy effects and avoid redundancies. If, for example, a large variant table of 20 products is used, you can create a KB with 20 products instead of 20 KBs with one product each.

2. **Defining runtime versions**
   You then define a *runtime version* (Transaction CU34). Here, you must particularly consider the following fields and specifications:

   ▶ The Valid from field is of central significance: When you create the KB, the system compiles the status of the master data that was valid at the time specified (snapshot), taking into account Engineering Change Management.

   ▶ There is no Valid To field because another change in the future would represent a new version. Because each runtime version on the database is available as a complete object, you can access past, present, and future statuses of your model at any time.

   ▶ Further important fields include Plant and BOM usage because only one BOM is possible for each material.

   ▶ You then generate the knowledge base, which may take some time because the system gathers and compiles all master data.

Simulating the Generation　　　　　　　　　　　　　　　　　　　　　**[+]**

To analyze errors, you can also simulate the generation. For this purpose, you must click on the Check Version button (F6). The generated report is considerably more detailed than the regular generation and indicates possible problems.

Changing Already Used KBs　　　　　　　　　　　　　　　　　　　　　**[!]**

You usually shouldn't make any changes to KBs that were already used in sales orders to avoid inconsistencies. However, if you carry out compatible changes (for instance, adding an additional characteristic value), you can also update the existing runtime versions, that is, overwrite them, in order to save space (delta logic). Of course, the same applies to KBs that are still being developed or KBs that have a validity date in future.

3. **Distributing runtime objects**

The defined runtime objects are distributed using the mechanisms of the CRM middleware (SCE object). Because a knowledge base is complete, you don't need to distribute any further configuration-specific objects, such as BOMs or variant tables. If you update a knowledge base, only the delta information is transferred.

Delta List

Errors can occur when you generate a knowledge base. Besides obvious errors (incomplete data etc.) that would also appear in ERP Variant Configuration (LO-VC), a problem may also occur in the so-called *delta list*.

[+]

Delta List

The delta list contains a list of aspects in which the LO-VC configurator differs from the IPC. You must take this list into account implicitly if you intend to replicate an existing product model from the SAP ERP system to the SAP CRM system.

In principle, the IPC can process the data model of Variant Configuration, but there are some restrictions here. They can roughly be divided into the following three groups:

▶ **Area of use**
As already described, the IPC exclusively performs a sales configuration. Consequently, certain objects or processes are not supported. Moreover, there is no reference to production so that plant-dependent information cannot be used.

▶ **Context reference**
The IPC usually runs in an SAP CRM system, but in principle — a major benefit in comparison to Variant Configuration — it can also run independently because it retains all relevant information in the KB. This is also reflected in the modeling that also requires a certain encapsulation, for instance, for designing variant functions. Like in Variant Configuration, you can transfer context parameters via object characteristics that must be known in the target system, of course.

▶ **Conceptual and functional differences**
Variant Configuration has a high degree of freedom with regard to modeling, and not all options constitute a best practice. Generally speaking, the IPC requires a stricter modeling methodology. A prominent example is procedures on characteristic values that the IPC doesn't support owing to basic considerations (nondefined processing sequence). Another example is unsupported blank cells in variant tables that are created as a transparent table in the IPC.

You can find more information on the differences (the delta list) between IPC/SCE and **[+]**
Variant Configuration in the SAP help portal under *http://help.sap.com/saphelp_crm50/*
helpdata/de/e8/a3fc3739a25941e10000009b38f8cf/frameset.htm and in SAP Note
837111.

The already mentioned function Check Version for the generation of the KB provides
information about possible problems in the generation.

5.4   Creating a Product Model Using the PME

In addition to the SAP ERP modeling of product models for the use of Variant Configuration, SAP CRM also provides a product modeling environment because you
can operate the SAP CRM system independent of the SAP ERP system (see also the
second bullet point in Section 5.2, Configuration of Products versus Services). This
section discusses the specific features of this modeling environment and the modeling objects.

5.4.1   Essential Properties and Differences Compared to Modeling in
SAP ERP

The *Product Modeling Environment* (PME) is a tool in SAP CRM to model master data
for product configuration independently of an SAP ERP system. It is aimed at customers who require no compatibility with ERP Variant Configuration (see Chapter
3, Business Processes in SAP ERP).

Although many concepts of PME in SAP CRM are known from SAP ERP, they differ
greatly in several aspects:

▶ **Local visibility**
All master data for the configuration (for instance, classes, characteristics) is stored
in the *knowledge base* (KB). There is no global master data. On the one hand, this
is a relief because the modeler works in an encapsulated object space, but on the
other hand, it complicates the reuse.

▶ **Own syntax**
Despite the similarities, the product models from the PME and SAP ERP are not
compatible. You can neither display nor even change master data from SAP ERP
in the PME, nor can you load master data from the PME in the SAP CRM system
into SAP ERP.

This doesn't apply to the compatibility of the ERP product model with the implementa- **[+]**
tion in the IPC, which is provided natively taking into account the delta list.

▶ **Radical simplification**
Although the data model contains concepts such as configuration profile, constraints, and so on, they only exist underneath. The user is provided with a highly simplified view to master the common modeling tasks with minimum familiarization effort.

▶ At runtime, the models that form the CRM PME are processed just like models from the SAP ERP system so that you cannot see any difference (see Section 5.2, Configuration of Products versus Services).

5.4.2 Calling the PME

You call the PME from the overview page of the SAP CRM product master. This complies with the natural way of thinking of many users who want to create or change a product model for a product. If the product (material or service) is configurable, the system displays the Product Models assignment block that includes all existing runtime versions. Here, however, the system also displays KBs replicated from SAP ERP that can only be called in the simulation.

From the technical point of view, a KB is not an object that depends on the product master, but it is independent. Because a KB can contain multiple products, the same product model may appear in multiple product masters.

[+]   Up until and including SAP CRM 5.0, the PME used to be an external application with a separate Java Swing UI. As of SAP CRM 2006s the PME has been part of the SAP CRM application and doesn't need to be installed or activated separately. The UI uses the known SAP CRM web client.

5.4.3 Product Models versus Knowledge Bases

Let's discuss the concepts of product model and knowledge base:

▶ **Product model**
Product model is a business term that indicates all objects that are important for the configuration. This includes classes, characteristics, values, object dependencies, BOMs, UI-relevant data (for instance, characteristics groups), variant tables, variant functions, and variant condition keys.

▶ **Knowledge base**
Knowledge base, by contrast, refers to the technical container that contains the product models. A knowledge base (more precisely, a runtime version of a knowledge-base object) contains one or multiple product models.

The reason for adding multiple product models in a KB is the local visibility of the data. The use of classes only makes sense if multiple products can be assigned to a class. This way, you can reference characteristics correspondingly and reuse them. You carry out a change to the characteristics (for instance, adding a value) at one point, and it is then valid for all products in this class.

The other extreme would be to maintain all products in the knowledge base. However, this is problematic because each change results in a new version of the KB. In addition, the maintenance can become very unclear, and only one user can make changes at a time.

It is therefore recommended that you define the KBs along the product families or product groups. Products whose contents overlap (same characteristics, variant tables, and so on) should be in one KB to facilitate reuse, whereas products with minor overlaps should be in another knowledge base.

5.4.4 Version and Status Management

KBs have different statuses (Released/in Process/Locked) and a Valid from date. The meaning and the mechanism to find the valid KB is the same for KBs from SAP CRM and those that were replicated from SAP ERP. In both cases, you use the data set (product ID, product type, logical system, date) for the search. The logical system ensures that for products that were replicated from SAP ERP the system also searches for knowledge bases from SAP ERP. The most interesting criterion is surely the date because it is used to select the different development statuses of a KB. In the default case, this is the date when the sales order was created, whereas the details depend on the sales order's Customizing. Regardless of this, there is a `CRM_CON-FIG_BADI` BAdI that contains the `set_kb_date` and `set_kb` methods, among other things. With these methods you can influence the standard procedure according to your requirements.

The PME also knows the concept of an inactive KB. This means that the changes to an existing runtime version are initially saved as an intermediate state in a new inactive version. This version can then be tested in the simulation, whereas the previous, active status is still used in sales orders (see Figure 5.1).

Only when you accept the changes, does the system activate this inactive version and hence overwrite the previous version. However, if the changes are erroneous or if the runtime version is "defective," the changes can be discarded and you return to the old, active version.

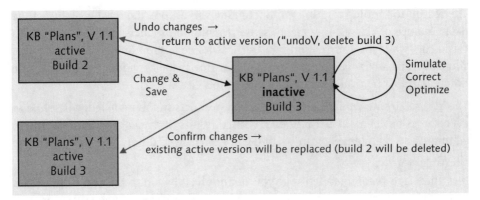

Figure 5.1 Version Management

5.4.5 Classes, Characteristics, and Values

Conceptually, characteristics and values largely correspond to the equivalents in Variant Configuration (see Chapter 2, Creating a Product Model for SAP Variant Configuration). Therefore, the following only briefly lists the specific features of the values and characteristics of the PME:

▶ Types: Characters, numerical, date

▶ Description (132 characters) and optional long text

▶ Cardinality: Single-value or multiple-value (Note: all characteristics can be restricted in the PME by definition. The same applies to the multiple-value characteristics.)

▶ Predefined value list, allowed values (numerical), or free value entry

▶ Attributes (*facets*) that can be defined statically or be changed using object dependencies: Invisible, not ready-for-input, required, not allowed.

▶ Similar to the SAP ERP system, classes are used to summarize characteristics in an object and to then assign products. Moreover, you can build up a hierarchy using automatic inheritance.

Whereas the values and characteristics don't differ greatly, the classification in the PME has some special features:

▶ **An implicit class**
Each product has an implicit class. This way, you can assign characteristics directly to exactly one product, that is, without "detours" via a user-defined (explicit) class. In the simplest case (for instance, only one product in the model), you don't have to deal with classification at all. In other words, a classification makes sense if you have multiple products that share certain properties.

▶ **Assignment to only one explicit class**

In contrast to the modeling in SAP ERP, products can only be assigned to an explicit class; that is, a product is assigned to exactly one implicit and not more than one explicit class.

▶ **Object dependencies in classes**

Classes can contain both characteristics and object dependencies. This enables you to define relationships that are supposed to apply to multiple products in the same way.

▶ **Subclasses**

You can create subclasses that inherit the characteristics and relationships of the superior class. Because you can only assign a product to an explicit class, this results in an inheritance hierarchy for products.

You sell DSL contracts to private and business customers. All contracts have certain **[Ex]** characteristics in common, for instance, the bandwidth for upload or download. You define these characteristics based on the *DSL* superior class, and they are inherited to all subclasses. In the *DSL business* subclass you additionally define the *static IP address* characteristic and assign the corresponding products to this class. Whether you assign the DSL products for private customers only to the superior class or to a separate subclass depends on whether these products are supposed to contain independent characteristics that are not to be inherited to the business products.

Organizing Your Work **[+]**

The use of the class system is very efficient, but changes to the classification in a production model are very tedious. You must therefore come up with a reasonable structuring of your products.

5.4.6 Object Dependencies in the PME

The object dependencies in the PME differ fundamentally from the SAP ERP equivalent. The number of dependency types is considerably reduced, and the display of object dependencies is simplified so that it is relatively clear. In some object dependencies, you can even omit "coding" completely. Internally, the object dependencies are mapped as constraints so that modeling is strictly declarative. Section 2.5.2, The Procedural and Declarative Character of Object Dependencies, in Chapter 2 describes the conceptual difference between the declarative and procedural modeling method. Aside from the academic consideration, at some points this results in perceptible restrictions (see the notes on `specified`), but also simplifications (no processing sequence).

The following dependency types exist, which are discussed in more detail in the following sections:

▶ Formula

▶ Condition

▶ Table formula

▶ Component condition

▶ Component formula

▶ Function formula

All dependencies have some general attributes, that is, an ID or a name, a language-dependent description, a status, and possibly an explanation. The latter can be evaluated by the conflict handling of the configurator at runtime.

[+] **Status**

A new dependency initially has the in Process status. For this dependency to become effective, you must change the status to Released. If a dependency is erroneous (for example, incorrect syntax), it is automatically set to the Locked status.

Dependencies are always created with a reference to a product or a class. This way, you define the context, that is, which characteristics can be addressed. If you define a dependency at the class level, it is valid for all products that are assigned to this class (or one of its subclasses). It is not possible to maintain dependencies directly on characteristics.

Formula

Formulas are used to set values (including variant condition keys) or defaults, to exclude values, and to check the configuration's consistency.

The formula consists of an optional condition part (if) and an execution part (then). The condition part defines when the formula is executed. If this part remains blank or is filled with the true keyword, the formula is always executed. This can be useful, for example, to determine the default value of a characteristic, which is shared by multiple products, for each product individually. The product itself doesn't occur in the condition itself because in the assignment of the dependency to the product, you've already specified to which product the formula refers. The general syntax of the dependency part corresponds to the syntax of the constraints in SAP ERP. You can query values in Boolean logic or use mathematical expressions.

In addition, the PME has two specific features (analogous to constraints in SAP ERP):

► The `specifiable` keyword can be used only in negated form (that is, `not specifiable`). This enables you to check whether a characteristic can no longer be assigned with values. This is the case if it already has a value or if the "not allowed" facet was set.

► The `specified` keyword, by contrast, mustn't be negated owing to the strictly declarative modeling. In its positive form, it can be used to check whether a characteristic already has a value.

Nonexistence of a Value Assignment Cannot be Checked [+]

Unfortunately, in the PME syntax it is not possible to check the nonexistence of a value assignment (analogous to `not specified` in SAP ERP procedures). If this is necessary, you must explicitly check for the nonexistence of all possible values (`not CHARACTERISTIC in ['VALUE A', 'VALUE B', ...]`) or work with help constructions (for instance, by adding the artificial value "no value" to the characteristic to be checked and setting it by default).

The execution part is mandatory, because otherwise this formula wouldn't make any sense. The following activities are possible:

► Assigning a value (`Characteristic = 'Value'`) that also includes the assignment of a variant condition key.

► Excluding a value (`Characteristic <> 'Value'`) or restricting allowed values (`numCharacteristic < number`).

► Assigning a default value (`Characteristic ?= 'Value'`). In contrast to the "fixed" value, the user can change this value.

► Immediately creating a conflict (`false`). This can be particularly useful for complex models in the test phase to immediately identify impermissible conditions that shouldn't occur at all.

The basic rule applies that values that were inferred by the AP Configuration Engine can no longer be changed by the user, and similarly excluded values can no longer be selected. If the user has already set a deviating value before a formula becomes effective, this results in a *conflict*. User entries can never be changed unnoticed, but must be changed actively.

The option Low Latency for Online Games is only available for an upload bandwidth of [Ex] 640 kbit/s. Using a formula, you exclude the Low Latency option if you select another bandwidth. If the user first selects the Low Latency but then selects another bandwidth, this results in a conflict, and the user must decide which input is more important. However, it is also possible to write another formula that automatically sets the bandwidth to 640 kbit/s after the Low Latency option was set.

This example reveals a fundamental question that you must answer: Do you create a "fool-proof" model that excludes conflict situations in principle but is more restrictive, or do you select an "open" model that provides more freedom to the customer and reveals the connections? The latter option is suitable for experienced users.

If a configuration has the inconsistent condition owing to a conflict, the system displays a corresponding message in the UI and — depending on the conflict type — you can perform a controlled solution process including suggestions. If you save such an inconsistent configuration, the status is indicated in the sales order, and a corresponding error message appears there.

Condition

Using a condition, you can dynamically change the attributes (facets) of characteristics. The following attributes are available:

▶ Invisible: The characteristic is not visible, but can have a value.

▶ Not ready-for-input: The characteristic is displayed, but only the IPC can change the value.

▶ Required: The characteristic must be assigned with values so that the configuration is complete.

▶ Not allowed: This facet can only be set using a condition. The characteristic concerned may not have any value; otherwise the configuration is inconsistent.

Similar to the formulas, the conditions have an optional dependency part. The same syntax applies as for the formulas.

In the execution part, you don't need to formulate any syntax. Instead, the system lists all relevant characteristics, and you select the attributes to be set. This way, you can set multiple attributes of multiple characteristics at the same time.

Characteristic attributes can also be set statically. It is confusing and generally not recommended to set static and dynamic attributes in combination for the same characteristic.

Table Formula

Tables are powerful tools to efficiently express a multitude of valid characteristics combinations. In addition, tables have an import-export interface that enables you to decouple the maintenance of the model and its structures from the business content; that is, you can integrate user departments with the modeling and they don't need to deal with the SAP CRM system and the PME.

You can use tables to achieve two essential goals:

▶ Restrict the allowed values and thus ensure consistency

▶ Infer values or default values

The use of tables is carried out in three steps: The structure of the table is defined, and the contents of the table are maintained. Then the table is used.

Step 1 — Definition of the Structure of the Table

In the PME, there is a separate section for tables. In addition to the usual header data, you also specify which characteristics are referenced as columns in the table. Provided that you (also) want to use the tables for inferences, at least one characteristic must be a key field that you can use to find the appropriate rows that are analogous to a relational table.

As soon as a table is filled with content or is used in table formulas, you can no longer change the its structure.

Step 2 — Maintenance of Table Content

You can maintain tables directly in the PME. You can create table rows and enter a value for each field. Blank fields or wildcard characters (*) are not permissible; that is, you must enter an explicit value.

In real life, however, the import-export interface probably has major significance. Here, you can export a variant table as a CSV file (*comma-separated value*) and edit it in any program (for instance, Microsoft Excel or OpenOffice). You can then import this field back to the PME and overwrite the existing content.

You can also export a newly created, still empty table to check which format the CSV file must have. **[+]**

The content of the table is created using the knowledge base in the database. You should therefore store changes to the table's content in a separate version of the knowledge base. **[+]**

Step 3 — Use of Tables

Table formulas enable you to address the tables in the model. The formulas don't have a condition part and have no other coding. You specify the table and select one of the two goals, Restrict or Infer, under Type, that is, the purpose of the table.

The system displays all characteristics of the table. If you want to infer characteristics, the system automatically sets the columns that are defined as key fields to read. For all other columns, you can select whether a default value or a fixed value is set. The effect is the same as for the formulas: A default value can be changed by the engine or the user. A fixed value mustn't contradict another inference or user input because otherwise a conflict is triggered.

[+]   If no characteristics were defined as keys, you cannot use the tables for inferences.

If you want to use the table for restricting allowed values, you can select any number of columns using the Restrict Values option. In the PME you generally restrict in all directions; that is, you can assign values to the characteristics in any sequence and in this way influence the others.

[Ex]   For a car, dependencies exist between equipment lines, types of seat covers, and available interior colors that you define in a table. At runtime, each characteristic can influence the selection of the others. For example, the selected type of seat cover restricts the selection of the equipment line and colors.

For both usages, you have the option to label the characteristics as not relevant (No Value).

[+]   You can create two table formulas with reference to the same table to restrict the allowed values, on the one hand, and to infer specific values, on the other hand.

Component Condition

A *component condition* enables you to define which components are selected from a BOM under which conditions. These components appear as subitems in the sales order. It is also possible to use multilevel BOMs; that is, a component can have a component in turn. The components can be self-configurable or simple products.

[Ex]   Your call rate has the option International Flat. If this option is selected, you create a subitem because you want to manage special cancellation rules for this option.

In contrast to SAP ERP, the viewing direction of the component condition is from top to bottom, that is, it is created based on the superordinate product and not the component to be selected.

▶ The condition part corresponds to the formula or condition.

▶ In the execution part you select those components that are selected if the condition is met. This way, you can select multiple components using the same condition.

If you address a component in multiple component conditions, it is sufficient if one of these component conditions is met.

Component Formula

You can use a *component formula* to express dependencies between components of the BOM; that is, the rules apply beyond the product boundaries.

The International Flat component mentioned previously can have the attributes Standard or Extended. The latter is only available if the customer has selected a particularly high-value call rate.

[Ex]

In the PME, this restricts you to the direct parent-child relations. Dependencies across multiple stages can only be expressed using diversions. Naturally, you can only define component formulas based on products and not on classes. The viewing direction of the component formula is from top to bottom; that is, it is created based on the superordinate product, and you address the component explicitly from there.

The structure of a component formula differs only slightly from a simple formula. You select a component in the header of the formula. In the condition or execution part, the `component` keyword is available to address the selected component in a targeted manner.

Function Formula

In the PME is it possible to address variant functions. Like in the SAP ERP modeling, in the knowledge base you only define the interface and the call of the variant function, whereas their actual coding is programmed in Java and installed in the Virtual Machine Container. In the appendix, SAP Note 870201 describes the corresponding procedure in detail.

To use variant functions in the product model, you must perform two steps:

1. The first step is to define the interface. For this purpose, the PME is a separate area, Functions. Here, you specify which characteristics occur in the interface and which of these characteristics are key fields. They are automatically input parameters, whereas the remaining characteristics are either output parameters to be changed or are not considered at all.

2. The defined and released function can be addressed in classes or products in one or multiple function formulas. A function formula has no condition part, that is, the function is always processed. You only define which of the output parameters are actually supposed to be used by the function.

5.4.7 Transport of Knowledge Bases

Since SAP CRM 2006s it is possible to export or import the knowledge bases that were created using the PME from one SAP CRM system to another. This is done using the XIF interfaces. This way, you can create and test your models in the test or consolidation system and then transfer them to the production systems later on.

[+] You can find further details in the online help for the SAP CRM system (PRODUCT CONFIGURATION • IMPORT OF THE KNOWLEDGE BASE).

5.5 IPC User Interface

It would be too long-winded at this point to go into the details of the User Interface architecture (UI). Therefore, the following sections only discuss some core points: Java Server Pages and the J2EE Engine as well as *Extended Configuration Management (XCM)*.

5.5.1 Java Server Pages and the J2EE Engine

The UI is based on the framework of e-commerce. It uses Java Server Pages (JSP) and Struts. The UI runs on the SAP J2EE Engine (there is a special architecture for Mobile Sales). If you want to use the product configuration in SAP CRM or in SAP ERP with the IPC, you must install the J2EE Engine and the web application on it.

The same UI (uniform code basis) is used in all scenarios, including Mobile Sales and SAP ERP.

[+] You can find details on the general framework and the change concept in the article *SAP E-Commerce – Development and Extension Guide*, which is available in the SAP Service Marketplace (quick link *crm-inst*).

Consultants and developers can find further information on the web sites of the Configuration Workgroup (*www.configuration-workgroup.com*).

5.5.2 XCM (Extended Configuration Management)

A comprehensive Customizing is a special characteristic of the IPC user interface. The idea here is that, depending on the context (user group, channel, and so on) and "taste," different types of display and functions are required without it being necessary to modify the coding.

[Ex]

For example, the UI is to be as compact as possible in the call center, and functions reserved for experienced users or for internal purposes, such as fast entry via the keyboard or XML import-export, are helpful. In the web shop, however, the product is supposed to be presented as descriptively (images etc.) as possible, but the users must not have access to internal information.

The following section discusses only the basic concepts concerning what is maintained in XCM.

▶ Some technical parameters, such as JCo parameters, must be maintained; otherwise the system displays an error message (frequent error source for initial installations).

▶ Numerous settings are necessary regarding the display (traffic light versus text, characteristics groups, images, text types, and so on).

▶ General control parameters, for example, whether a check is carried out by the server after each user input or only upon request. The latter setting enables a quicker input, but is only suitable for experienced users.

▶ Almost all special functions (see next section) can be (de)activated using XCM and require certain parameters as the case may be.

▶ Some very simple enhancements (user-defined buttons) can also be carried out.

There can be several XCM configurations. When you call the UI, a so-called xcm. scenario parameter is transferred in the URL. Then, the UI reads the corresponding settings from XCM. XCM is structured in such a way that the user settings are separate from the SAP standard; that is, the settings don't represent a modification and are not overwritten by upgrades.

SAP provides a different standard configuration in which not all options are activated by default. Therefore, it is worthwhile to have a look at XCM and the documentation to exploit the full potential of the UI. In general, you should note that some special functions (for instance, variant search and large images) may considerably impact the performance.

5.6   Special Functions of the IPC User Interface

Now, let's take a look at the special functions that the IPC user interface provides.

5.6.1   Images and Other Objects

Because the IPC has a modern user interface that is displayed in the web browser, you can easily integrate multimedia objects, such as images, sounds, pdf, and so on.

Common object types are displayed "out of the box"; for more exotic types, you may have to slightly change the coding on the server side to generate the correct HTML output and/or have a corresponding plug-in for the display on the browser page.

[+]   Images can be assigned both to characteristics and to values.

When you create the product models in SAP ERP, you maintain the objects, such as images, via the document management system and assign them with characteristics or values. The files are automatically integrated during the KB generation and are then available to the IPC.

5.6.2   Import-Export of Configuration Results

Normally, a configuration result is stored in the associated document (for instance, quotation or sales order). Templates for frequently used configuration are also based on such documents.

In some cases, it seems to be desirable to be able to maintain the "pure" configuration result. For this purpose, let's take a look at three application examples:

▶ You want to locally store frequently used configurations as personal favorites.

▶ You want to exchange intermediate statuses of a configuration with other processors via email.

▶ You work on a highly complex configuration and want to locally store an intermediate status for good measure because the network connection or the backend system is unstable.

Using the IPC you can locally store the current status of the configuration as an XML file (.cfg file extension). This is possible without any expert knowledge, by clicking a button in the UI. You can import the XML files in the same way and thus start a configuration session with the values stored there. The usual tolerance mechanism takes effect during loading: Characteristics and values that cannot be set (for example, because the KB was changed) are ignored.

[+]   SAP Note 385773 describes the format of the XML file.

[!]   **Externally Managing Configuration Results**

The external management of configuration results principally poses a security risk, because invisible characteristics are visible on the one hand, and you could theoretically manipulate the results on the other hand. For this reason, the import-export is only available if Low was selected as the security level in XCM.

5.6.3 Pricing Overview

The price of a configurable product basically consists of the prices and discounts determined in the pricing procedure and the surcharges and reductions (variant conditions) determined during the configuration. The surcharges, in turn, can be subdivided into two groups:

▶ So-called 1:1 surcharges that are directly linked with a characteristic value and are displayed as a surcharge there

▶ Surcharges determined via relationships

The first group is easier to handle; the second group is considerably more flexible. For example, the surcharge for a leather interior can depend on the selected model type and on the type of seats. This also results in the logic problem of assigning this price, because the price doesn't depend on one factor, but on multiple factors. Only for surcharges that are directly linked with a characteristic value can the system display the price right behind the characteristic value ("leather: 1,000 EUR").

However, the IPC UI provides the option to display for the user an explanation of the currently calculated price during the configuration. The pricing overview displays a subset of the used condition records that was defined in XCM, for example, basic prices and discounts. From the technical perspective, these are rows from the pricing procedure. For variant conditions the system displays the language-dependent text, provided that it is maintained (in SAP ERP using Transaction VK30; in SAP CRM in the CRMC_VARCOND/CRMC_VARCOND_T table).

This text is important to verbally describe the condition's origin, for instance: Premium Edition: Leather for Sport Seats. This way, you can make even complex price inferences transparent to the user (see Figure 5.2).

5.6.4 Better Handling of Restrictable Characteristics

In SAP ERP Variant Configuration, restrictable characteristics may have a serious disadvantage in the handling: As soon as they are assigned with values, other possible values are no longer displayed because the allowed values are restricted to the set value from the engine's point of view (singleton problem). To change the value, you must first delete the set value so that the dynamic allowed value is redetermined; then you can set the preferred value in a second step.

Constraints restrict the list of ten possible colors to three possible colors. If the user [Ex] selects the color red, in Variant Configuration, he would not directly see that green and blue would also be valid. He must first delete red so that the selection red|green|blue is visible.

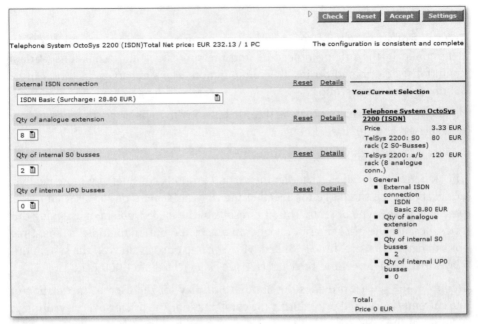

Figure 5.2 Pricing Overview

Since Version 5.0, this problem no longer exists in the IPC. Even if the characteristic is assigned with values, the system calculates the alternatives, and the user can select them as usual by simply clicking on them. This function demands a great deal of computing time (*user domain*) and may reduce the performance; if required, it can be deactivated in the settings of VMC, but this clear improvement of the user-friendliness compensates for the disadvantage.

5.6.5 Search/Set

Experienced users may know the used configuration models, in particular the frequently set characteristics and values, very well and want to be able to navigate efficiently in the model using the keyboard and/or to set values. A prerequisite is that the user knows the language-independent names (IDs) of the characteristics or values. The search/set function provides two subfunctions that can be (de)activated via XCM depending on the scenario — as is usual in the IPC.

▶ **Search**
The user enters the characteristic's name or parts of it. If the system finds a suitable and visible characteristic, the UI navigates to the corresponding characteristic (and across tabs). This is particularly useful for very large configurations.

▶ **Set**

With this powerful function, you can assign characteristic values of the character string type directly and across groups by entering their IDs. The user can even enter multiple values in the corresponding field. If the value is not found or if it occurs multiple times in the model, the system displays a corresponding message. So a prerequisite is that the IDs of the characteristics to be set are unique across the models.

5.6.6 Displaying Long Texts (as of SAP CRM 2006s)

A frequent request is to be able to use characteristic (value) texts with a length greater than 30 characters. Although the data model of the IPC provides for longer texts, this restriction from SAP ERP also applies to SAP CRM if models replicated from SAP ERP are used there, which is frequently the case.

In the IPC it is now possible to bypass this restriction by using a trick. In SAP ERP, you can maintain additional texts with (virtually) any length under Documentation. These texts are then displayed in the IPC if you click on the Details button.

Since SAP CRM 2006s it has been possible to display long texts in the value assignment interface either additionally or even instead of the short texts. In XCM at the characteristics and values level, you can flexibly select whether you want to have the system display the ID, the short text, the long text, or combinations of them. A definable threshold value ensures that very long texts are interrupted after n characters.

5.6.7 Messages Controlled by the Configurator (as of SAP CRM 2006s)

Users frequently request to be provided with additional notes during configuration. These notes can include marketing-driven information, descriptions, warnings, or similar.

Since SAP CRM 2006s, you can use so-called message characteristics in the IPC. The mechanism is rather simple and is therefore downward compatible in terms of modeling. The characteristics are normal characteristics of the character string type, which you can position in the UI as usual. The characteristic (multiple characteristics are also possible) is inferred using the object dependencies. Based on the prefix of the characteristic ID, the system identifies the message characteristic and displays it in a special form (see Figure 5.3). The individual characteristic values contain the message texts. The value itself is not displayed, and a value assignment is not possible, of course. The three message types (info, warning, error) have different colors and the icons and are selected based on the prefix of the characteristic value.

The concrete prefixes are defined in XCM. In the standard SAP configuration, UIMES-SAGE_, is the prefix for the message characteristic and I, W, and E are used for the texts (characteristic values).

[Ex]   The WEXT value is assigned to the UIMESSAGE_EXT characteristic. It is displayed as follows:

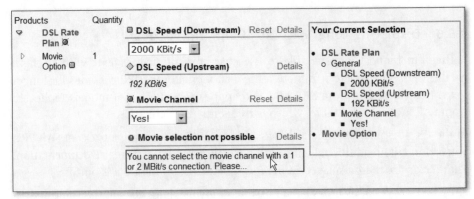

Figure 5.3   Message Characteristics

5.6.8   Configuration Comparison (as of SAP CRM 2006s)

If a configuration has reached a certain degree of complexity, it is not always easy to determine which additional values are inferred through user input. Under unfavorable circumstances, the users find themselves at a dead end at which a consistent state can no longer be attained.

The configuration comparison covers the following functions:

▶ The current state of the configuration can be saved in the memory (snapshot). If required, you can return to this snapshot and undo the entries that have been made since then (see Figure 5.4).

▶ You can compare the state of the configuration using a snapshot. Here, you can view not only the values you set, but also the engine's inferences.

▶ If you change a configuration, you can also compare the current state of the configuration with the state on the database.

Product Configuration (DSL Rate Plan)			
✅ Back			
Return to Configuration			**Configuration Comparison**
			Show All
	Snapshot Switch To 5:33:52 PM, USD 9.90 / 1 PC	Current Configuration 5:34:06 PM, USD 9.90 / 1 PC	Remark
DSL Rate Plan			
General			
DSL Speed (Downstream) △	2000 KBit/s	6000 KBit/s	Value changed by user
DSL Speed (Upstream) △	192 KBit/s	512 KBit/s	Value changed by system
Movie Channel △	Not selected	Yes!	Value set by user
Movie Option △			Product added

Figure 5.4 Comparison of Two Configuration States with the Option to Return to the Snapshot

5.7 UI Designer (as of SAP CRM 7.0)

As described previously, the user interface of the IPC in SAP CRM can be influenced by numerous parameters in XCM regarding functions, display, and behavior. The basic display of the model, that is, the sequence of the characteristics, assignment to groups, images, and so on, corresponds exactly to the master data, however. Figure 5.5 shows the standard IPC UI in the web application.

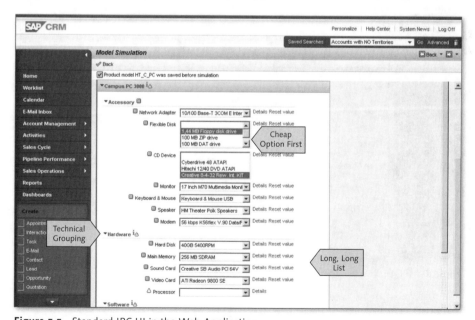

Figure 5.5 Standard IPC UI in the Web Application

As of SAP CRM 7.0, you can use the UI designer to influence the display of the product configuration in SAP CRM scenario-specifically, irrespective of whether it was modeled in SAP ERP or SAP CRM. The following change options exist:

▶ You can change the sequence of BOM components, groups, characteristics, and characteristic values.

▶ You can individually define for each characteristic with a value list whether it is displayed as a compact drop-down list or with selection buttons or checkboxes.

▶ You can decide which additional contents (images, texts, documents) are visible for which characteristic or value. You can even add additional images.

▶ You can arrange the characteristics on pages that are presented to the end user step by step. In contrast to the tabs that can be selected in any sequence, you can use this option to enforce a specific sequence (guided configuration).

▶ Figure 5.6 shows the newly arranged IPC UI in the simulation view.

The display you define is saved in a UI model. Of course, the creation of such a model is optional. If there is no model, the product model is displayed according to the master data.

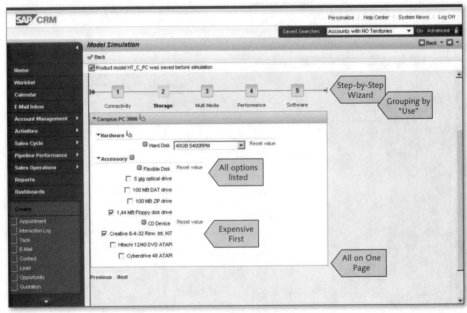

Figure 5.6 IPC UI, Newly Arranged Using the UI Designer

You create the UI model in the CRM PME; this also applies to models that were replicated from SAP ERP (in this case, the actual product model cannot be edited). You assign a *PME scenario* as selection criteria to the model, provided that you also have a user role that serves as a selection criterion at runtime. In doing so, you can ensure that the guided configuration is used in the B2C web shop, for example, whereas you use a compact display for office work.

5.8 Summary

This chapter dealt with the special systematics of product configuration in the SAP CRM system. You learned that in addition to the direct use of the product model from the SAP ERP system (by uploading knowledge bases), the SAP CRM system provides a maintenance environment (PME) to make data available for a configuration. You also learned the differences and specific features, such as Customizing of the UI and the special options in the object dependencies. But because this chapter could not cover all options of the SAP CRM system and because the descriptions of entire processes would go beyond the scope of this book, you should refer to the corresponding literature on SAP CRM.

This chapter provides insight into the world of SAP industry solutions. Of the 28 industries for which SAP develops special software, we chose to cover the Discrete Industries and Mill Products (DIMP), because this area offers some interesting enhancements for variant manufacturers.

6 Enhancements in SAP Industry Solution DIMP

In addition to the pure standard, SAP also provides so-called industry solutions. Most of these solutions are free and provide enhancements in the area of Variant Configuration. This chapter provides an overview of the available solutions. We focus on manufacturing enterprises that are specifically addressed by the industry solution *Discrete Industries and Mill Products* (DIMP). As of SAP ERP 6.0, the industry solutions are delivered together with the standard version and can be switched on using the *switch framework* in Customizing. For further information refer to the corresponding pages at *http://help.sap.com*.

6.1 Overview

Although they've been available for many years — the first industry solution for mill products was delivered in 1999 — there are still customers who don't know about SAP industry solutions at all or only know a little about them. This may be because most solutions are free and are not specifically advertised by SAP. Although the industry solutions (IS) lead a somewhat shadowy existence, it is worthwhile to take a look at this world of industry-specific enhancements, especially because there are many functions in the area of Variant Configuration that customers have developed themselves or have had developed by partners because they simply didn't know about these standard solutions.

> One important reason customers didn't want to use an IS even though they knew about **[+]** the solution was that the industry enhancements were always delivered with a certain delay.

It was possible for six months to pass between the delivery of the current standard release of SAP R/3 and the corresponding industry solution. In addition, not every standard release supported or explicitly developed for the industry solution. So customers were requested to orient on the special release cycle of the industry solution. This deficiency was removed by changing the delivery strategy as of SAP ERP 6.0.

As of SAP ERP 6.0, the industry solutions are delivered in the standard SAP ERP system and no longer need to be requested and installed separately. Nevertheless, interested customers must still pick a specific solution from the various solutions on offer in the switch framework in SAP ERP Customizing and make a commitment because not all industries are compatible with one another. For example, there are major differences between SAP solutions for banks, trade, and the production industry.

SAP provides solutions for 28 industries. Due to this versatility, let's take a closer look at DIMP as an example. This example was selected because experience has shown that it provides the most interesting aspects for you as a prospective customer of Variant Configuration. You're probably in one of the industries assigned to DIMP.

[Ex]

Further Industry Specifications

The Vehicle Management System (IS-A_VMS) for car dealers and Apparel and Footwear Solutions (SAP AFS) for the apparel industry should also be mentioned as examples, but a detailed description of these two and other industry solutions would be far beyond the scope of this book. However, they also address the variant aspect.

6.2 DIMP — Discrete Industries and Mill Products

The industry solution that particularly addresses manufacturing enterprises is called Discrete Industries and Mill Products (DIMP). This solution approaches multiple industries at the same time because it was determined that the basic requirements in these industries are very similar. This includes the following industry branches:

▶ SAP for Industrial Machinery & Components

▶ SAP for High Tech

▶ SAP for Aerospace & Defense

▶ SAP Engineering & Construction

▶ SAP for Automotive

Special functions and enhancements have been developed for the *mill industry*. However, there is not sufficient space here to illustrate this subindustry to its full extent.

> For further information and comprehensive descriptions of the functions, refer to *http://help.sap.com* under SAP FOR INDUSTRIES • SAP FOR MILL PRODUCTS • SAP ECC DIMP 5.0.
>
> **[+]**

In accordance with the title of this book, this chapter only describes those aspects from the mill products solution that are of interest from the Variant Configuration perspective.

6.3 Special Requirements of the Mill Industry

The SAP for Mill Products component as part of the DIMP solution enhances the SAP ERP system with industry-specific functions for the complex requirements of the mill industries that cover areas such as metal, timber, paper, textile, building materials, and cable.

The different branches of the mill industry have in common that they work with a very large number of characteristics and material variants. In addition, the materials used in the previously mentioned industries are primarily oriented toward areas (for instance, paper, cloth, or films) or toward lengths (for instance, cables or pipes). These material properties — described by means of configuration characteristics — must be considered along the entire supply chain.

Owing to this strong affinity to Variant Configuration, this industry solution required enhancements of the SAP ERP standard, and it provides additional functionality that is useful for variant manufacturers. Of course, the original standard functionality is still available and ready for use.

In the mill industry in particular, the variant diversity is practically unlimited because typically the dimensions of coils must be modeled as characteristics, for example. Length, width, and size can have virtually any value within the technical limits. It's almost impossible or not useful to model each of these attributes as a fully configured variant.

In addition to configurable material (KMAT) and "normal" configured variants, you therefore often use changeable material variants.

> **Changeable Material Variants**
>
> **[+]**
>
> The majority of the characteristics are already specified in the changeable variant. After a product has been selected, you then primarily supplement dimensions, additional refinement steps, and certification requests in the sales order; if applicable, you can change already specified characteristics.

The multilevel configuration is rather untypical in the mill industries; the single-level configuration via many low-level codes is used, however. The type of the logistic complexity is different from that of a mechanical engineer.

6.3.1 Sales Order Processing and Production Scenarios

Besides the classic make-to-order production and make-to-stock production, a mixed scenario becomes ever more important in the mill industries: finish-to-order production.

The more you can postpone the decoupling point between the make-to-stock production and the customer-specific production to the end of the production process, the more flexibility the supply chain planer or the planning system has in the assignment of intermediate products for sales orders. This postponement of customer individualization results in considerably shorter lead times.

This mixed scenario is enabled via Variant Configuration in the make-to-stock scenario in the mill enhancement.

6.3.2 Production Discrepancies — Planned Configuration and Actual Configuration

Like in other industries, the customer request is also specified in the configuration. The requirements planning generates purchase orders, planned orders, and production orders that copy the configuration or refer to it.

The mill industries are characterized by production tolerances and deviations that result in the actual characteristics of the finished products deviating (hopefully only slightly) from the customer's specifications. This *deviating actual configuration* is modeled via the batch classification.

[+]
Batch

A batch describes either a single piece of material on stock or multiple similar pieces with a batch number and the associated characteristics: a single roll of paper or an assortment of paper rolls in which the individual rolls are identical with regard to characteristics features, such as length, width, density, and so on.

So the batch is used with the semantics of a configuration to describe the actual value assignment of the configuration characteristics based on the batch classification, rather than in the sense of a production lot like in the chemical industry.

In SAP Advanced Planning and Optimization (APO), which is part of the SAP SCM solution, the characteristics-dependent planning provides comprehensive functions that enable a requirements planning on the basis of characteristics. Here, you have the following options:

▶ **Requirements consumption**
Configured requirements or planned accesses can be compared with and consumed against batch-classified stocks on the basis of characteristics of the same name.

▶ **Block planning**
You can use the block planning to optimize setup times on the basis of characteristics.

In SAP ERP, you can implement such a comparison between the original customer demand and the actual classification only via the batch determination. In the delivery, you can search for stocks with actual characteristics (that is, batches) using the batch selection criteria that are derived from the configuration of the sales order.

Within the scope of a characteristics-dependent planning in SAP APO you can also use configuration and batch selection criteria (as value assignment and requirement) for a differentiated planning. Here, the configuration describes the exact customer demand, whereas the batch selection criteria define an interval or tolerance within which the specific actual characteristic may be.

Configuration and Batch Selection Criteria [Ex]

The customer requests a cable with a length of 3,000 ft. The Individual Length configuration characteristic is assigned with the value 3,000 ft in the sales order.

As agreed and according to the agreement defined in the customer master (lower or upper deviation tolerance of Tl 0.5% or Tu 5%) the customer accepts lengths between 2,985 ft and 3,150 ft. Consequently, the Individual Length characteristic is assigned with allowed values between 2,985 ft and 3,150 ft in the batch selection criteria. Therefore, the system reacts as follows:

Initially, requirements planning is supposed to find stocks that are within the tolerance range. If no stock is found, a production order for the exact lengths of 3,000 ft is to be created. This can be modeled in such a way that the customer request is described in the configuration and the tolerance interval in the batch selection criteria.

6.4   Product Configuration Enhancements in Mill Products

This section describes functional enhancements from a subarea of the DIMP industry solution from which you can benefit in your daily work with Variant Configuration. Particularly remarkable in this context is the fast entry of characteristics, which, as the name suggests, can considerably accelerate the entry of characteristics. The associated inheritance function across order items also enhances this effect. If you then add a customer information record, in which the characteristics are already defined and adapted to the customer requirements, this further reduces the data entry effort.

You can use sales order versions to define and save change requests to the configuration in the source documents without running the risk triggering the logistic and requirements-relevant processes unintentionally and possibly prematurely.

Especially in this industry, Variant Configuration is required in connection with make-to-stock production (see Section 6.4.5, Variant Configuration in Connection with Make-to-Stock Production).

The order combination, however, is of interest to almost all variant manufacturers because characteristics such as color often occur across production orders. Therefore, it would be interesting for the paint shop if, for example, production orders could be resorted and combined beyond the normal sequence of requirements planning.

6.4.1 Fast Entry of Characteristics — Simplified Entry of Configurable Document Items

When you enter a configurable material or a batch evaluation in the SAP ERP standard system, you must always navigate to the configuration mode or the characteristic value assignment screen for each item. This can be very complicated and time-consuming when you enter many sales order items. Therefore, a comparison of the configurations across multiple items is rather difficult.

You can use the fast entry of characteristics to enter the configuration across multiple sales order items in one step without changing the screen.

An inheritance functionality enables the transfer of characteristic values of a super-ordinate item to all assigned items. This way, you can efficiently enter characteristic values across items and materials or change them retroactively. In addition, the system immediately detects entry errors and unwanted deviations in the characteristic values between the sales order items (for instance, a retroactive change of the surface color of a configured kitchen).

The fast entry of characteristics is particularly suited for materials and batches that contain only a few characteristics that still need to be assigned with values. For this reason the fast entry is often used in combination with preconfigured material variants or if further characteristics can be supplemented by entering fewer configuration characteristics using object dependencies. For configuration models with many characteristics to be entered, the additional navigation to the characteristics value assignment screen is possible, but the benefits of the fast entry cannot be utilized.

The Fast Data Entry tab, which becomes available only after activation of the characteristics group in Customizing, is similar to the Configuration tab (see Figure 6.1). However, here you can enter new items and characteristics values in one step, whereas in the Configuration tab you can only have the system display the characteristic values.

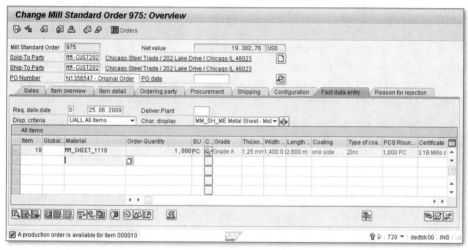

Figure 6.1 Fast Data Entry of Characteristics in the Sales Order

In the Fast Data Entry tab, the system also indicates the status of the configuration. In the Condition of the Configuration column, the system displays icons to indicate whether the configuration is complete or still has inconsistencies. This considerably improves clarity in case of many sales materials for each document because the navigation to check the status is omitted.

The completion of preconfigured materials or material variants in particular is considerably accelerated. Furthermore, you can save the display sequence or the characteristic display user-dependently or even link it with associated material. This way, you achieve a decoupling from the characteristic sequence that was defined in the classification. Direct navigation to the "classic" value assignment interface, that is, Variant Configuration LO-VC, is possible at any time.

The fast entry is available for Variant Configuration in the SD documents (customer request, customer quotation, sales order, customer scheduling agreement, and customer contract), the purchase order, the production order, the corresponding confirmation, and the trading contract. The fast entry for the batch classification is also available in the goods receipt and in the confirmation of the production order.

You can use the Characteristic Display field to select the ten characteristics that are displayed in the overview screen (see Figure 6.1). You define the characteristic display in Customizing.

Fast Data Entry Tab

Note that the Fast Data Entry tab doesn't become visible until you've defined at least one characteristic display in Customizing.

Next, you must assign this characteristic display to an application group. It then provides the connection to the appropriate transaction so that you obtain the correct display in the desired application. The Application group MM consists of the Purchase Order and Goods Receipt transactions, for example.

The characteristic display can be suggested user-dependently or selected manually by the user in the corresponding (sales or purchasing) document depending on the material based on the first document item. You must carry out the corresponding settings in Customizing.

If you want to use the fast entry of characteristics for sales documents, for example, then you must make the following settings in the sales Customizing:

1. **Create a characteristic display that includes the preferred characteristics for the sales order.**
   You define the characteristic display under SALES DOCUMENTS • FAST ENTRY OF CHARACTERISTICS IN SALES DOCUMENTS • DEFINE CHARACTERISTIC DISPLAY FOR OVERVIEW SCREEN (SAP MILL PRODUCTS).

   ▶ **Assign this characteristic display to an application group that consists of the Sales Document transaction.**
   You define the application group under SALES DOCUMENTS • FAST ENTRY OF CHARACTERISTICS IN SALES DOCUMENTS • DEFINE APPLICATION GROUP FOR OVERVIEW SCREEN.

If you don't want to select the preferred characteristic display yourself but want to link it with a material, proceed as follows:

Linking the Material Display with a Material

You want the system to suggest the FLANGE_MM characteristic display in a sales document of the flange material.

▶ For this purpose, you must assign the Default characteristic to this characteristic display in the Set Fast Entry for Characteristic Values IMG activity.

▶ At the same time, you assign this Default characteristic to the 300 class of the flange material and then assign the FLANGE_MM value there.

Of course, you must first create the previously mentioned characteristics and values.

You can find the settings in Customizing under SALES DOCUMENTS • FAST ENTRY OF CHARACTERISTICS IN SALES DOCUMENTS • SET FAST ENTRY FOR CHARACTERISTIC VALUES.

6.4.2 Inheritance in Item Documents — Global and Local Items

Frequently, a document's items differ only in a few characteristics, for instance, the dimensions. Even though most of the other characteristic value assignments of the items are identical, you must separately assign values to all characteristics of different order items in the standard version.

You can predefine these characteristic values that are identical across multiple items in a global item and inherit them to any number of so-called local items that are each assigned to a global item. For this purpose, select the global item and click on the Inherit Characteristics button. The system inherits the changed characteristic value assignments to all items of the selected global item, that is, to those to which you assigned a superordinate item in the Global Item column (see Table 6.1).

Item	Global Item	Material	DIN Certificate	Basic Material
10		Flange 4711	3.1b	St37
20		Flange 4711	3.1c	V2A
30	10	Flange 4712	3.1b	St37

Table 6.1 Global and Local Items

In Table 6.1, item 10 is the global item, and item 30 is the local item. For an inheritance, the local item 30 adopts the characteristic value assignments of the global item 10. The characteristic values of item 20 remain unchanged. Of course, the configuration can be further refined in the local items. A new inheritance from the global item is always possible, but it may not overwrite characteristic values that were changed locally and manually. You can set the exact behavior in the Customizing of sales.

Inheritance   [+]

The inheritance is carried out by means of the identical names of the characteristics in the document items (across materials and classes); that is, you can inherit across different materials as long as you use the same characteristics.

If a specific characteristic of the global item is not assigned to the local item, it is grayed out in the display. Consequently, an incorrect entry is not possible. The use of global and local items is supported both in the sales order and in the purchase order. Of course, you can assign multiple local items to a global item.

6.4.3 Copying Default Values from the Customer Material Information Record

You can further accelerate the fast entry by preassigning certain characteristic values for a combination of material and customer. For this purpose, you classify the material in the customer material information record. The characteristic values from the classification are then copied when you enter new sales order items. The class and characteristics that you want to use for the classification of customer material information must be allowed for the class type 052, Material Customer Information Record.

If you want the system to copy the characteristic value assignments from the customer material information into the configuration of a sales document item, you must permit this in the Customizing of *sales* under SALES DOCUMENTS • FAST ENTRY OF CHARACTERISTICS IN SALES DOCUMENTS • COPY CHARACTERISTIC VALUES FROM CUSTOMER MATERIAL INFORMATION RECORD.

An identical functionality also exists in the purchase order, which is referred to as the purchasing information record. The class and characteristics that you want to use for the classification of the purchasing information record must be allowed for the class type 057, Purchasing Information Record.

6.4.4 Working with Sales Order Versions

You can make retroactive changes to sales orders only to a limited extent, depending on the progress of the requirements planning and production. To document the change request, on the one hand, and to leave the original document unchanged until the feasibility of the change is clarified, on the other hand, the mill industry solution offers sales order versions.

The sales employee can enter the customer's change requests, for instance, changed configurations, directly in the version of the sales order item. After it has been clarified whether the change request is accepted in sales and distribution, MRP, or production, the version can be used as the active sales order item instead of the original item. The main advantage of this functionality is the option to document change requests (multiple versions are possible for each document) and the option to compare individual versions and the original document at the characteristic value level. All of this is carried out without triggering MRP-relevant processes.

The upper part of Figure 6.2 shows an original sales order in which the Coating characteristic was assigned with the One-Sided characteristic value, and the Certificate characteristic with the Certificate 3.1B value for the MM_SHEET_1110 sales material. The lower part of the figure shows the sales order version 1000000003 of this order in which these two characteristics are evaluated with the values Double-Sided and Certificate 3.1A. You can clearly see that the configuration characteristics can only be changed in the version.

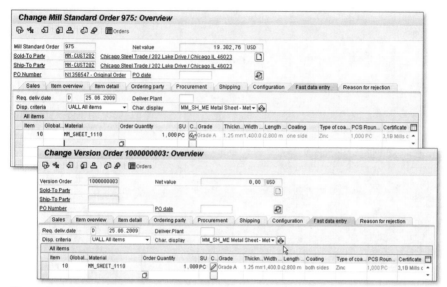

Figure 6.2 Sales Order Version and Original Document

The version comparison in Figure 6.3 uses different colors to indicate the differences from a document item at the characteristic value level. This way, you can easily identify different properties of the products in which characteristics were stored.

Particularly in case of many characteristics, this facilitates the simple search for different value assignments because some value assignments may have been implemented by processing object dependencies, which is not always immediately obvious. For example, characteristics could have been changed that were not visible in the characteristic display of the fast entry.

After you've checked whether the version can be implemented, you have the option to activate the created sales order versions, that is, to overwrite the current value assignment of the material. This automatically triggers the defined logistic processes when you save the SD document. Here, particularly for configurable materials, you should refer to Order Change Management (OCM).

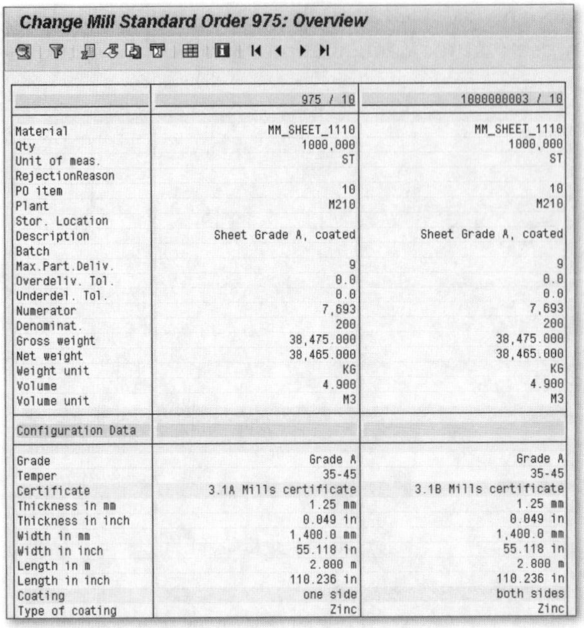

Change Mill Standard Order 975: Overview

	975 / 10	1000000003 / 10
Material	MM_SHEET_1110	MM_SHEET_1110
Qty	1000,000	1000,000
Unit of meas.	ST	ST
RejectionReason		
PO item	10	10
Plant	M210	M210
Stor. Location		
Description	Sheet Grade A, coated	Sheet Grade A, coated
Batch		
Max.Part.Deliv.	9	9
Overdeliv. Tol.	0.0	0.0
Underdel. Tol.	0.0	0.0
Numerator	7,693	7,693
Denominat.	200	200
Gross weight	38,475.000	38,475.000
Net weight	38,465.000	38,465.000
Weight unit	KG	KG
Volume	4.900	4.900
Volume unit	M3	M3
Configuration Data		
Grade	Grade A	Grade A
Temper	35-45	35-45
Certificate	3.1A Mills certificate	3.1B Mills certificate
Thickness in mm	1.25 mm	1.25 mm
Thickness in inch	0.049 in	0.049 in
Width in mm	1,400.0 mm	1,400.0 mm
Width in inch	55.118 in	55.118 in
Length in m	2.800 m	2.800 m
Length in inch	110.236 in	110.236 in
Coating	one side	both sides
Type of coating	Zinc	Zinc

Figure 6.3 Comparing the Item of the Original Document and the Sales Order Version at the Characteristic Value Level

6.4.5 Variant Configuration in Connection with Make-to-Stock Production

A major enhancement of Variant Configuration in the industry solution enables the effective use of Variant Configuration in the make-to-stock scenario. In principle, configuration and make-to-stock production are mutually exclusive because the standard MRP run cannot differentiate the different value assignment of configurable materials in the anonymous stock, leading to the storage in the individual customer segment. However, the mill industries require another procedure for different reasons.

The MTS production scenario is always interesting if the supply chain planning is supposed to be as flexible as possible. For example, you can considerably reduce the lead times if configured intermediate products have already been manufactured without any sales order relation. When you receive a sales order, you should find the correct intermediate product based on the characteristic and produce it according to the customer request in a manufacturing step, for instance, through cutting and refinement processes.

"Reassignment" Process in the Steel Industry

In the steel industry in particular, you can find the process of reassignment. For reassignment, you should have the option to flexibly assign the intermediate material or preliminary material to another sales order using the planning system. In the make-to-order production scenario, however, this is only possible manually through updates between the sales order stocks. Moreover, this procedure leads to some blurring in the material valuation.

So how can you differentiate stocks with deviating configuration in the anonymous stock? In make-to-order production, this is ensured through the sales order stock.

In make-to-stock production you must label the material through batch management to be able to differentiate the stocks with different configurations in the anonymous stock of the same material. If the different configurations lead to different production costs, you should manage the stocks on a value basis. This is possible using the valuation for a single batch. Prior to the goods receipt of the finished product, you determine the value of the batch from the actual costs of the production. In doing so, the costs remain transparent and reproducible across multiple steps in the make-to-stock production scenario.

In the make-to-order production scenario, requirements planning only considers the individual customer segment. This enables you to directly assign the sales order, the production orders, and the purchase orders.

In the make-to-stock scenario, the requirements planning must ensure that the requirement coverage elements, for instance, the production orders, are created with the correct configuration and in the correct quantity. In the SAP ERP system, it is not possible to directly assign the sales order to the production order via the collective segment.

Therefore, SAP APO requires the characteristics-dependent planning. It ensures the correct assignment of requirements and requirement coverage elements based on configuration characteristics.

6.4.6 Order Combination with Configurable Products

In most mill industries, for instance, the steel, paper, and textile industries, and in mechanical engineering, concurrent operations for different products are carried out in a joint production process.

Combination in a Joint Production Process

In the steel industry, this can be the continuous casting and the subsequent hot and cold rolling, for example. In mechanical engineering, this can involve the concurrent painting of multiple components from different sales orders. The combination in painting is driven by high changeover and cleaning costs; in heat treatment this procedure serves for the optimization of energy costs per piece.

Before and/or after these concurrent operations, you implement the different operations in separate production orders for the various products, so the combination is only temporary.

Via the order combination you can now combine such concurrent sequences in a joint production order. You can combine the following order types:

► Sales orders

► Planned orders

► Production orders

You can select the orders to be combined based on joint header material, based on a specific work center, or frequently based on the similar configuration of the original orders. For example, you can combine orders that have the same characteristic value assignment for the Coating characteristic and are produced at the same work center.

The combination can include both individual operations and complete orders. As a basic principle, production orders are always combined, as are sales orders. For this reason, the system creates production orders when combining sales orders or implementing planned orders in production orders. Accordingly, you can only combine orders that are intended for in-house production. You can aggregate production, assembly, and process orders. However, production and process orders cannot be combined with one another. Likewise, not all functions are supported for process orders, for instance, the formulas for quantity calculation are not adapted.

6.5   Summary

As already mentioned, this chapter could only describe a small fraction of the world of additional features and functions of industry solutions. So this chapter focused on the DIMP industry solution and its mill subindustry to highlight facilitations in the area of sales (fast entry of characteristics, sales order versions, customer material information record) and production (order combination, make-to-stock production of configurable products).

Unfortunately, we could not discuss all topics so we refer you to the corresponding help pages of SAP (*http://help.sap.com*; menu path: SAP FOR INDUSTRIES • SAP FOR MILL PRODUCTS • SAP ECC DIMP 5.0). Take a look—it's worth it!

The SAP environment consists of not only the standard SAP system, but also numerous add-ons. Some of these add-ons do not require any modification, while others are programmed. This chapter introduces you to some partner solutions and consulting solutions that add important functions to the basic SAP ERP and SAP CRM functions in Variant Configuration.

7 Enhancements and Add-Ons in the SAP Partner Environment

Within the SAP environment, there are numerous specialists in the area of Variant Configuration. Such partners and freelancers have used their knowledge and expertise to develop their own programs and add-ons for these functions. Because the list of possibilities is extremely long, this chapter provides you with insight into these developments by introducing you to some of the add-ons developed by a few of our SAP partners. Please note that even though we did not choose the following examples at random, we cannot ensure that we have considered every development by every single partner. There is certainly much more available out there.

Our selection is based on solutions that address the shortcomings of the standard SAP system and consists of well-developed add-ons from our various partners.

In addition to their software add-ons, almost all of the partners named in this chapter have a high level of expertise in implementing the basic SAP functions in Variant Configuration, which was yet another reason for selecting these particular partners.

This chapter focuses on the following add-ons:

▶ **Sybit Model Tester**
(company: Sybit GmbH)
This program dispenses with the need for manual testing of product models. You can store test cases and replay them automatically at any time. This not only saves money but also increases model quality.

Since 2001, Sybit GmbH has been involved in Internet Pricing and Configurator (IPC) development and has provided consulting services for SAP Variant Configuration users. Sybit is a Special Expertise Partner for SAP CRM and the manufacturing industry. It is also a member of the SAP Configuration Workgroup (CWG).

▶ **Sybit Configuration Visualizer**
(company: Sybit GmbH)
This software provides a graphical 2D visual image of SAP configuration results.

▶ **VCPowerPack**
(company: AICOMP Consulting)
This is an add-on for the SAP industry solution Discrete Industries and Mill Products (DIMP), that is, for process manufacturers such as the cardboard industry, who need configurations that are oriented toward product creation. AICOMP, which is a Business-All-in-One partner of SAP and a member of the Industry Value Networks Mill, considers itself a services company for customer-focused solutions in the paper and packaging industry. AICOMP is also a member of the CWG.

▶ **it.cadpilot**
(companies: itelligence AG and ACATEC Software GmbH)
This add-on automatically generates engineering and design and production documentation by remotely controlling 3D CAD systems from within SAP Variant Configuration. itelligence AG is one of SAP's largest partners. It provides complete implementations of the entire SAP product range and industry solutions and software add-ons. it.cadpilot has been developed in collaboration with ACATEC Software GmbH, which specializes in CAD Automation and product configuration. Both itelligence and ACATEC are members of the CWG.

▶ **Framework and Variant Engine**
(company: top flow GmbH)
top flow provides options for a newly developed configuration dialog box and process optimizations within the logistical process. In addition to providing highly qualified SAP ERP services, top flow, a software partner of SAP, focuses on developing stand-alone products that enhance the range of functions available in the SAP product range. top flow GmbH is a member of the CWG.

▶ **ConfigScan Validation Suite**
(companies: Fysbee SAS and eSpline LLC)
This add-on addresses common quality and performance testing issues in the context of product models for variant configuration. Fysbee, an SAP small business partner, is the European developer of the ConfigScan Validation Suite.

eSpline LLC is the North American reseller of ConfigScan and drives an initiative called *Managing VC*, which uses ConfigScan plus eSpline products, techniques, and services. Both Fysbee and eSpline are active members of the CWG.

▶ **Managing Variant Configuration**
(company: eSpline LLC)
The applications described in this section support the whole lifecycle of product models, including testing and auditing, from a master data perspective, as well as business analytics and its impact on product model advancement from a business execution point of view.

7.1 Sybit Model Tester (Company: Sybit GmbH)

Now that we've discussed all aspects for creating and maintaining configuration models and their associated business processes, we'll devote this section to testing configuration models.

Configuration models are complex. Multivariant products can easily contain several hundred levels of freedom (in other words, characteristics). For this reason, the associated configuration models are both comprehensive and complex. This is all the more true for mature models, that is, models that have evolved over a period of several years. Consequently, in everyday life, it is often impossible to correctly anticipate all of the effects associated with changes to a model.

The complexity of the models goes hand-in-hand with the importance attached to having correct and accurate models. If some of the possible effects of a changed model are overlooked, this can easily lead to a situation in which it is no longer possible to configure frequently ordered product versions (that is, best-sellers), resulting in a considerable loss in revenue. On the other hand, the configuration model could permit configurations that are not possible from a technical perspective or are only feasible at a higher price. If a customer were to purchase such a legally binding configuration, a financial loss or, at the very least, customer dissatisfaction would be inevitable.

For this reason, correct configuration models are critical to an enterprise's success. What makes things even more difficult is the fact that the standard SAP system does not provide any support for configuration model testing, nor are there plans to implement such a function. Therefore, when performing quality assurance testing on models, most enterprises resort to the only instrument available to them by default: manual testing.

7.1.1 Manual Testing: CU50

It is almost inevitable that the traditional approach to testing configuration models will lead you to Transaction CU50 (Material Configuration Simulation/Modeling). As mentioned in earlier chapters, Transaction CU50 provides a user interface for evaluating the individual characteristics of a configuration model. Here, you can test the object dependencies between characteristic values, on the one hand, and the BOM explosion, on the other hand. There is no option to run tests that have already been entered, so if you want to repeat a test, you have to manually reenter the configuration and then manually check the result. The following disadvantages are associated with manual testing in Transaction CU50:

▸ **Cost**
In an ideal scenario, you would authorize each model change in an extensive test series before activating it in the production system. You would also ask experts from the user department of the relevant product to perform the tests, thus increasing costs.

▸ **Quality**
Manual testing is a tedious process that demands a high level of concentration. It is impossible to rule out human error when entering a huge number of configurations and visually checking the results.

▸ **Time to Market**
You have to perform extensive tests when creating and enhancing new configuration models. Such tests can considerably delay the time to market for a product or new feature.

Therefore, Transaction CU50 is an inefficient quality assurance tool for comprehensive configuration models.

7.1.2   Sybit Model Tester

The *Sybit Model Tester* minimizes the manual interaction required during testing. This not only reduces costs and the time to market, but it also improves the quality of the tests performed. The core concept behind the model tester is to make it possible to run tests repeatedly and to automate the results check.

When compared with manual testing via Transaction CU50, we can assume that the time and effort required for testing is reduced by 60% to 90% when you use the model tester. The more complex the test, the greater your savings potential is. The greatest savings potential is achievable in scenarios where extensive testing takes place on a regular basis (for example, after every model change).

To enable you to effectively reduce costs and the time-to-market as a result of using the model tester, particular emphasis is placed on user support when creating and maintaining tests. The model tester makes it possible to perform continuous *quality checks* throughout the whole model lifecycle because all of the systems in a multisystem landscape can be incorporated into the tests performed. It also supports the use of *SAP Engineering Change Management*. You can reproduce the tests performed in the model tester at any time because both the relevant test version and the test result are stored in a database.

The *test case* and *test run* are the core elements of the model tester.

Test Case and Test Run

The test case is the basic test element in the model tester. You can use it to accurately test a configuration, that is, exactly one value assignment for the characteristics of a configurable material. In the following sections, the terms *test case* and *test* are synonymous. As shown in Figure 7.1, a test case consists of an input and an expected result. Similar to Transaction CU50, the input consists of a configuration, that is, a series of characteristic value assignments. The expected result consists of the following areas:

▶ **Error Area**
In principle, you can test both valid and invalid configurations in the model tester. If you expect an input configuration to be invalid, you can enter this information in the error area or you can even specify the expected error itself. Because you cannot expect a result for an invalid configuration, the other areas are not filled in this case. However, if you expect an input configuration to be valid, the error margin remains empty, and all other areas can be filled.

▶ **Configuration Area**
You can specify the expected configuration in the configuration area. This area is used to test the object dependencies used by the system to fill the characteristics. If, for example, if you want the characteristic value assignment Large and the value XXL to trigger the characteristic value assignment Length with 35.5 within the system, Large = XXL would be part of the input configuration, and Length = 35.5 would be part of the configuration for the expected result.

▶ **Areas for Bill of Materials (BOM) and Price**
In the BOM area, you can specify the expected BOM items resulting from a BOM explosion and the expected price (price testing will not be supported until a later version).

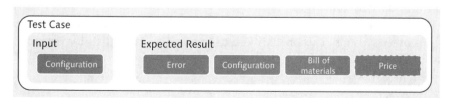

Figure 7.1 Test Case Consists of an Input and Expected Result

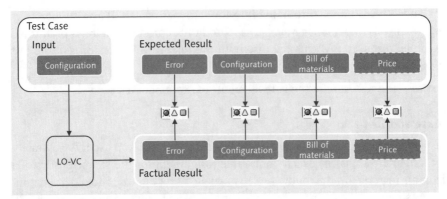

Figure 7.2 Test Run Consists of a Test Case and an Actual Test Result

Figure 7.2 shows the relationship between the test case and the test run. The upper area shows the test case from Figure 7.1. Similar to Transaction CU50, the input configuration is transferred to Variant Configuration during the test run. The result created, that is, the *actual* result, is logged here. You can then compare the areas for the expected result with the corresponding area for the factual result. Here, the Model Tester distinguishes between error messages (red traffic light), warning messages (yellow traffic light), and success messages (green traffic light).

Creating, Maintaining, and Organizing Test Cases

The Model Tester has numerous tools for creating and maintaining test cases in the most simple and efficient manner.

▶ **Import Tools**
The fully automated import is the fastest and easiest way to create test cases. The Model Tester has two import tools:

▶ *Import from sales documents*
With live configuration models, you can extract available data from sales documents and store this data as a test case in the Model Tester. This makes it easy to ensure that configurations that have been sold to customers are still possible after model changes.

▶ *Import from external sources*
The Model Tester has an XML interface that can be used to transfer test cases from other systems. This enables you to import test cases that, for example, were automatically generated from a formal definition of the relevant configuration model.

▶ **Test Case Tool**
You can use the test case tool to edit individual test cases. This includes the input

configuration and all areas for the expected result. To simplify the input in the individual areas, you make the system generate default values with the click of a button, which you can then accept or correct.

▶ **Test Group Tool**
The test group tool is used to organize test cases. To provide a clearer overview, you can combine any number of tests into one test group.

▶ **Test Run Tool**
The test run tool is used to run tests or execute test groups. During the test run, all of the tests within a test group run in sequential order. The actual result is then saved and compared against the expected result.

▶ **Evaluation Tool**
You can use the evaluation tool to analyze test runs. The correlation between the expected result and the actual result for each area is displayed separately as a traffic light signal. At a glance, you can identify problematic test areas and analyze the problems further with the click of a button.

▶ **Mass Maintenance Tool**
The mass maintenance tool supports you when maintaining tests while enhancing or changing the configuration model. If, for example, you add a new mandatory characteristic to a model, you can open any number of test cases in the mass maintenance tool and, at the same time, enhance the value assignment for this characteristic.

Processes and System Landscape

Many enterprises implement a multisystem landscape in the SAP ERP system whereby you distinguish between the development system, the test system, and the production system, known as the *three-step model*. Frequently, only some data is kept up-to-date in all systems. For example, BOM data may be missing from the development system.

To get the best support in this scenario, it is possible to specify the relevant destination system in each test run and perform tests in every system in this manner. If the object dependencies for a new configuration model are created in the development system, which is the case in our example, but the BOM is not maintained until the test system, you can, from the outset, create test cases whose expected result contains a configuration. After the object dependencies have been transported to the test system, you simply have to add the BOM area to the expected result for the test cases.

The system architecture for this scenario is shown in Figure 7.3. Here, the Model Tester runs in the test system, for example, and connects itself via a remote function call (RFC) to the development and production system. The Model Tester runs as an application in the ABAP stack of SAP NetWeaver. The web-based user interface is implemented in accordance with ABAP Web Dynpros and is therefore fully web-, portal-, and SAP GUI-compatible.

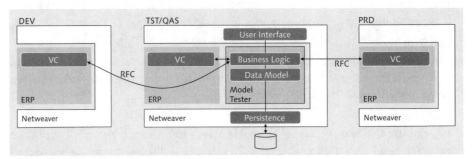

Figure 7.3 Architecture in a Multisystem Landscape

7.1.3 Summary

The Sybit Model Tester checks the quality of configuration models. It contains a range of useful coordinated tools for creating, maintaining, and organizing test cases, thus minimizing the manual interaction required during testing. Compared with manual testing, the Model Tester reduces the costs associated with the testing process and increases test quality. At the same time, it reduces the time required for testing, which has a positive effect on the time to market. For a comprehensive quality check, you can perform both positive and negative tests in the Model Tester. Here, you can incorporate all of the systems within a multisystem landscape into the tests. The model tester is an important and helpful tool when creating and maintaining configuration models.

7.2 Sybit Configuration Visualizer (Company: Sybit GmbH)

To configure a product, you must fill a number of characteristics with values. Frequently, this solely text-supported user interaction with the relevant configuration model is not very intuitive. For inexperienced users, in particular, it is often difficult to anticipate the effects of their characteristic value assignments.

This section describes a lightweight solution that visually supports you during the configuration process. It can be seamlessly integrated into the SAP component *Internet Pricing and Configuration* (IPC), which is used for configurations in all SAP CRM scenarios.

7.2.1   Problem

In practice, the purely text-based interaction by the user with Variant Configuration has proven to be difficult and the user is frequently faced with major challenges. Of course, these challenges increase not only with the inexperience of the user, but equally with the complexity of the model. In particular, if object dependencies are used to derive many characteristic values from other characteristic values, untrained users cannot assess the effects of their characteristic value assignments, let alone anticipate or control these effects. Instead of a goal-oriented configuration process, this scenario has a trial-and-error process whereby the user must, step by step, test the behavior of the configuration model. This empirical testing process is not only tedious, but also slow and expensive, and inevitably leads to user uncertainty, especially with configuration models that are subject to frequent changes. At the same time, practice has shown that the likelihood of incorrect configurations increases with the complexity of a model.

Incorrect Configurations	[+]
Please note that incorrect configurations are not impossible configurations, because impossible configurations are prevented by the configuration model. Rather, they are possible configurations that do not reflect the true wishes of the customer. The resulting financial implications of such incorrect configurations can be huge.	

Essentially, the problems discussed here can be traced back to the fact that you cannot obtain an overview of your current configuration during the configuration process. Besides a consistency and completeness check based on traffic light colors, the standard SAP system does not have a mechanism for supporting the user in this regard. Now, the Configuration Visualizer closes these graphical gaps.

7.2.2   Sybit Configuration Visualizer

The Sybit Configuration Visualizer (SCV) supports you during the configuration process. It provides an up-to-date image of the current product configuration after each characteristic value assignment. This enables you to visually check the effects of each configuration step at any time. The advantages associated with visualizing the configuration process are obvious:

▶ **A considerably faster configuration process**
The ability to visually check a configuration affirms your actions and accelerates your work.

▶ **Higher quality as a result of visualizing the configuration process**
Inexperienced users, in particular, are effectively prevented from creating incorrect configurations.

The Configuration Visualizer has two parts: On the one hand, the actual visual image is shown on an additional tab in the IPC. On the other hand, the SCV and its Visualization Modeling Environment (VME) contain a web-based tool for creating and displaying the display logic.

7.2.3   User View

The SCV adds a Visualization tab to SAP IPC. Figure 7.4 uses the example of a mobile crane to show this tab in the enhanced UI for SAP IPC.

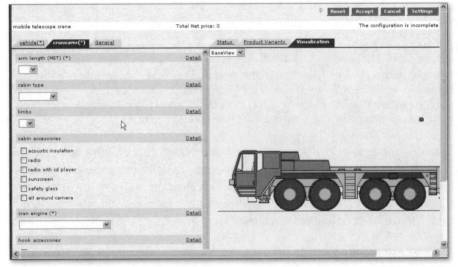

Figure 7.4 Adding the Visualization Tab to SAP IPC — Mobile Crane Displayed Without a Crane Arm

You can perform the value assignment for characteristics ❶ on the left-hand side of the screen, which is usually the case in SAP IPC. After each assignment, the image on the right-hand side of the screen is automatically adjusted to reflect the product instance currently configured ❷. If parts of the product are not specified, they will be missing from the image. Figure 7.4 shows the configuration for a mobile crane in the user view of the Configuration Visualizer. The crane arm is missing because it has not been specified.

Immediately after making the relevant value assignment for the characteristics associated with the crane arm, SAP IPC displays the image shown in Figure 7.5.

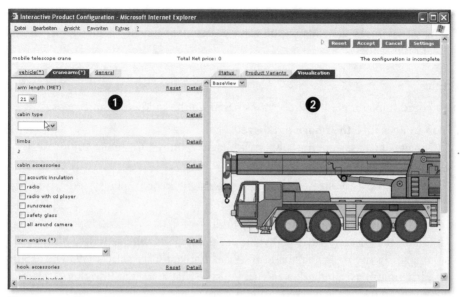

Figure 7.5 Adding the Visualization Tab to SAP IPC — Mobile Crane Displayed With a Crane Arm

In practice, the IPC user interface delivered by SAP is rarely used unmodified. When you implement SAP IPC, you'll almost always make enterprise-specific adjustments to the layout and sometimes even to the operating logic. To make it as easy as possible to integrate the additional tab into the various attributes of the user interface, the visualization component is located on a Customer Tab.

Customer Tabs   **[+]**

SAP's *Customer Tabs* technology enables third-party vendors to add their own components to the IPC user interface.

7.2.4 Modeler View — The Visualization Modeling Environment

Some preparatory steps are necessary to provide the above graphical support to an SAP IPC user. To make work easier for the operator of the configuration model, the Configuration Visualizer contains a *Visualization Modeling Environment* (VME).

This VME is primarily used to manage the images displayed and their links with the characteristics and values in the configuration model. However, to ensure that you do not have to create and maintain separate images for each possible configuration, the VME can layer partial images, thus generating the images displayed in IPC.

Therefore, if components are added to a partially specified material (as shown in Figure 7.4 and Figure 7.5), the VME simply incorporates an additional partial image into the visual image for the existing configuration. The image incorporated into the existing visual image depends on the value assignment for the characteristics of the configuration model.

Method

The method implemented in the VME is based on the following three steps:

1. **Load images into the image database**
   The image database contains all partial images required for the layering effect.

2. **Scale and align images**
   To produce correctly layered images, a preassigned coordinate system is used to align the partial images for a product.

3. **Create rules for modeling the display logic**
   Rules link the characteristics and values of the IPC Knowledge Base with the partial images in the image database. If, for example, a user selects yellow as color and driving cab as standard, a rule can ensure that a standard driving cab is displayed in yellow in the layered visual image.

User Interface

In the VME, all three steps are grouped together in a user-friendly interface. Figure 7.6 shows the view in which you can scale and position partial images. Both the size and position of the driving cab in the mobile crane are determined by clicking on the chassis. To create correctly layered images, you have to assign a layer number to the driving cab in the table in the editing area to ensure that the driving cab is not displayed *behind* the chassis.

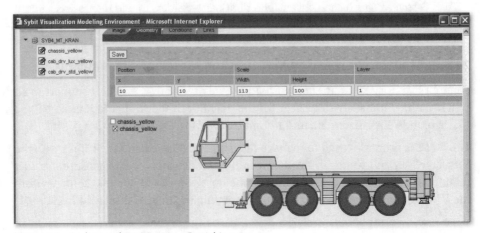

Figure 7.6 Scaling and Positioning a Partial Image

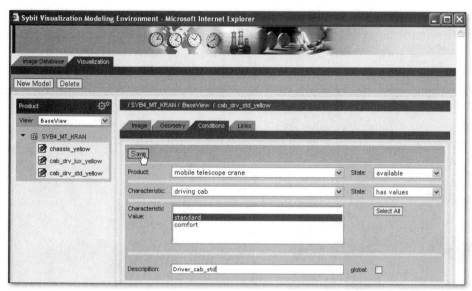

Figure 7.7 Creating a Rule Condition

Figure 7.7 shows you how to create a rule condition. Rule conditions are always linked to possible characteristic values. The figure shows a condition that is fulfilled as soon as the value standard is assigned to the driving cab characteristic. In the next step, the partial image of the (standard) driving cab, shown in Figure 7.6, is linked to the condition created here. Therefore, as soon as you assign the value standard to the driving cab characteristic, the correct image is incorporated into the overall image. Furthermore, this image is sized correctly and occupies the correct position.

7.2.5 System View

Figure 7.8 provides an overview of the system landscape. Section A shows the standard IPC scenario. The IPC user ❶ communicates with the IPC Web UI via a web browser ❷. The IPC Web UI runs in SAP NetWeaver Application Server Java (SAP NetWeaver AS Java) and communicates with the IPC Server ❸. Section B shows the enhanced scenario, which contains the Configuration Visualizer. For the IPC user ❹, the only visual change is the additional tab in the Web UI ❺, which displays the current configuration at all times.

As already mentioned, the visual image is displayed on the Customer Tab. The visualization modeler creates and manages partial images, the display logic, and therefore the layered visual images. The visualization modeler ❻ interacts with the VME ❼ via the web browser ❽. Similar to the IPC Web UI, the VME runs in SAP NetWeaver AS Java ❾. The model information (that is, the images and logic) are stored in the standard Web AS database.

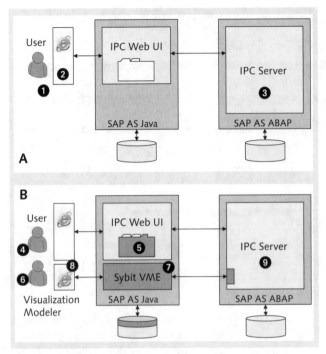

Figure 7.8 Standard IPC Architecture (A) and SAP IPC with an Integrated Configuration Visualizer (B), Additional Tab, and VME

7.2.6 Summary

The Sybit Configuration Visualizer visualizes the product currently configured by the user. The visual image is displayed on an additional tab in the IPC, and it is updated after each characteristic value assignment. To ensure that you do not have to retain images for each possible configuration, the overall image displayed consists of layers of partial images. The Configuration Visualizer has a Visualization Modeling Environment (VME) in which you can maintain partial images. Here, you can manage partial images and reconcile them with each other with the click of a mouse. You can then use a simple logic rule to link the partial images with one or more characteristic values in the configuration model. For a relatively low maintenance effort, the Configuration Visualizer generates helpful overview images and provides valuable graphical feedback during the configuration process.

[+]   Employees from Sybit GmbH *(www.sybit.com)* kindly contributed the above information to this book.

7.3 VCPowerPack (Company: AICOMP Consulting)

In this section, we'll explain the advantages of VCPowerPack — an application developed specifically for the SAP industry solution Discrete Industries and Mill Products (DIMP). This native SAP software add-on has built rapid momentum in industries in which the product configuration is the cornerstone of design, costing, pricing, and shop floor control and the usability of the configurator, internally and externally, is of prime importance. It promises to vastly improve the response time and accuracy of customer communication, along with improved planning, scheduling, and forecasting. New industry or branch-specific solutions within SAP that were once considered too risky, too small, too irrelevant for revenue, or too costly are now within reach by taking advantage of this innovative software solution. This is of particular interest for small and large businesses wanting to leverage the power of SAP, but with restrictive budgets.

Until recently, configuration for products of any kind has been done using two different methods:

▶ Using LO-VC/IPC for product description, costing, and pricing by working around or living within the current boundaries of LO-VC/IPC

▶ Using third-party ISP for configuration outside SAP and then interfacing relevant data back to SAP

Owing to the complexity of configuration and its specific requirements, many manufacturers of products rich in variation choose the second method, using external configuration software. However, this process is very time and cost prohibitive. Large amounts of master data must be gathered and structured outside the system, calculations must be processed, and the resulting data has to be funneled into LO-VC and processed through a custom SAP-internal interface.

VCPowerPack is radically changing the current concepts and bringing the configuration back home where it belongs: SAP ERP. The system permits the construction of a complex configurator in SAP ERP, with many advanced functions: complex calculation possibilities, custom screen layouts, faster system performance, and more detailed error analysis with instructions for troubleshooting. This advanced functionality therefore represents a useful alternative to the standard solution provided by LO-VC/IPC. VCPowerPack allows for the smooth implementation of custom configuration solutions and out-of-the-box-configurations for an increasing number of industries.

7.3.1 How VCPowerPack Works

VCPowerPack consists of four distinct modules: CoreVC, SmartVC, SmartMD, and SmartPR. These modules are designed from the ground up to work seamlessly together or as individual components based on the complexity of customer requirements. The following sections explain how to use modules.

7.3.2 VCPowerPack — CoreVC

Common architecture built using one central core leads to standard concepts used across the whole enterprise. Modularized components within that architecture provide:

▶ Seamless integration of each requirement into the SAP flow of information to meet specific demands.

▶ Data integrity within one model for different industries, distinguished by customized master data — not by additional programming. This ensures a solid design base without legacy data structures.

▶ Adaptable system setup of the CoreVC allows the configuration of complex industry requirements through master data — not custom programming.

▶ CoreVC allows the quick and structured development of standard interfaces with SAP and non-SAP systems.

7.3.3 VCPowerPack — SmartVC

Using the Core VC module, VCPowerPack is designed for user-friendly configuration based on the customers' business processes – see Figure 7.9. The tree structure replaces the characteristic name–characteristic value restriction of LO-VC/IPC. ABAP–class-oriented screen and menu controls provide a new way to configure products; from a tab-oriented navigation of screen controls to a production- and assembly-oriented tree configuration, the interface is very intuitive. In addition, real-time consistency checks and validations of all entries keep the information flow high and the frustration level low. Consistency validation can be compartmentalized, whereas specific error messages and troubleshooting tips provide a clear user interface that improves the overall configuration experience.

Smart VC provides a full set of configuration objects to take advantage of the native tree structure, beginning with the free definition of levels and manufacturing details such as machines and work centers, consumption materials, and design structure. The use of an object-oriented, modular configuration allows for rapid response to changes in business requirements.

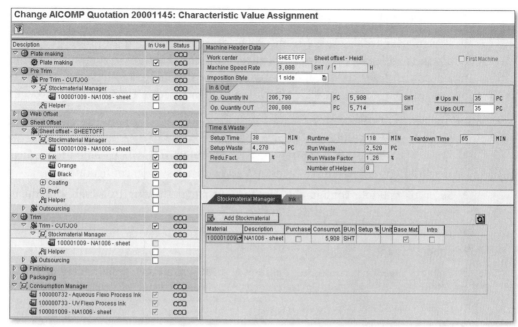

Figure 7.9 Smart VC – Configuration Screen

7.3.4 VCPowerPack — Industry Solutions

Although AICOMP was originally spawned by the needs of the paper and packaging industry, VCPowerPack was designed and built from the beginning to be used in multiple industries and market segments. With CoreVC and SmartVC, AICOMP offers two extensive configuration enhancements, providing SAP ERP with a more powerful product configuration solution. VCPowerPack's CoreVC and SmartVC support a wide range of configuration requirements, whether design, material, or routing driven.

From a technical perspective, VCPowerPack shares a common architecture of classes, user interfaces, and structures across different modules for different industries. This provides for minimal custom development, encapsulated in ABAP classes, for screen processing, and for business-specific calculations. By encapsulating these customer-specific functionalities in their own classes, the core functionality of the system is untouched. This allows for much simpler unit testing and debugging along with increased customer confidence.

Additional industry-specific solutions are being developed for VCPowerPack on an ongoing basis to further improve the speed with which it can be implemented within these industries.

7.3.5   Project Acceleration

VCPowerPack enables project teams to perform multiple complex product configuration tasks with ease. VCPowerPack enables the configuration of common business functions that leads to logical documentation, less training, and less custom development. Using VCPowerPack reduces the implementation time for the following reasons:

▶ Less custom development owing to highly flexible system configuration.

▶ Automatic creation of technical SAP objects and master data based on functional specifications entered directly in the VCPowerPack system configuration.

▶ Setup of multiple configurable products in parallel owing to a highly modular system with little or no programming required.

▶ Increased response to business process changes owing to flexible system configuration.

▶ The VCPowerPack configuration and the ERP system are directly integrated out of the box.

7.3.6   Summary

VCPowerPack enables the SAP ERP system to meet your industry-specific needs. Moreover, it provides a cost-efficient implementation without the burden of extensive custom development. VCPowerPack enables your company to meet your individual configuration needs by controlling the processes from acquisition to quotation to production, processing all relevant calculations and automating the flow of master data. If you have a restrictive project budget or timetable, VCPowerPack is the solution to your complex configuration requirements.

[+] This section was edited by Jens Hennecke, Sascha Rauhe, and Rainer Förster of AICOMP Consulting (*www.aicomp.com*).

7.4   it.cadpilot (Company: itelligence AG and ACATEC Software GmbH)

Products are often essentially described using *Computer Aided Design* (CAD) models and drawings. This is also the case for configured variants within a product range. Enterprises require CAD models for these variants to respond to inquiries from potential customers. Later, they require CAD-based production drawings and assembly drawings for their own order processing or the service itself.

Variant Configuration can be used to determine and derive the following product data in a rule-based and automated manner: prices, BOMs, routings, inspection plans, and therefore also the quantity structure for the cost estimate.

For a long time now, users have been able to manage CAD documentation such as 3D models and drawings in SAP Product Lifecycle Management (SAP PLM). These documents have been fully integrated here, enabling you to obtain a professional check-in and check-out process, releasing, and version creation with integration into SAP Engineering Change Management. However, product configuration is requiring the most automated, rule-based creation of a new product variant, in other words, the CAD content. Despite using a configuration system, many enterprises still adopt the tedious approach of order creation when manually creating CAD documentation for configured variants.

In the following sections we'll show you how a partner solution makes it possible to fully integrate CAD Automation into SAP Variant Configuration. This solution is a product called *it.cadpilot*, which was developed by itelligence AG and ACATEC Software GmbH.

7.4.1   CAD and SAP — Two Configuration Worlds?

Because SAP does not provide a stand-alone CAD or graphics application, there has been a long-standing myth that SAP and CAD must occupy two different worlds. However, if we simply consider the numerous successful customer projects that exist, it's easy to refute this claim. Even 10 years ago, it was possible to make graphical configurations and to integrate them into SAP software (consider the kitchen and furniture industry). However, this was generally an individual customer project, which is not comparable with implementing a standard solution. Thanks to SAP's shift toward an open service-based platform, it is now possible to integrate standard products that provide such options.

7.4.2   Structure of Modern 3D CAD Systems

A modern, parametric CAD system has a complex and comprehensive data structure that describes the structure of a model, starting with basic sketches for which parameters have been set and on which CAD modeling steps are built, as well as mating conditions and CAD auxiliary geometries such as planes, axes, and frames.

A geometry-based product structure is created by grouping CAD parts into CAD assemblies. Within the CAD system, CAD drawings are derived associatively from the models. The drawings represent scaled views and sections of models and are therefore valid product documentation from a mechanical engineering perspective.

For the following reasons, however, the CAD product structure frequently differs from the BOM structure:

▶ Depending on the configuration, some components, including their mounting parts and, for example, drill holes may be positioned differently in the CAD model only. Of course, this has a significant effect on functionality and the whole assembly, but it does not affect the BOM.

▶ The BOM structure may contain electric or electronic parts in addition to the mechanical parts. This may also be the case for any associated software controls (for machine tools, for example).

▶ Non-CAD-modeled parts are required in the BOM structure (for example, adhesives, mounting parts, or complete variable-size item BOMs).

▶ On the other hand, CAD auxiliary objects (for example, skeletons and planes) should not have a process-relevant impact on the BOM.

For these reasons, it is not always possible to have a 1:1 assignment between BOM components and CAD structures, nor is this necessary, because you can derive many dependencies from characteristic value combinations.

7.4.3   Controlling CAD Systems

The user interface of a CAD system allows the CAD designer to create and change his models. He can use CAD Automation tools to control and therefore automate the CAD operations required for each program (see Table 7.1).

Control Type	CAD Operation	Example
Component	Add, replace, suppress, retrieve, delete, reroute, rename	Configuration-controlled installation of alternative parts, even for configuration-dependent CAD references
Dimensioning	Set or change value	Controlling dimensions, tolerances, distances, and so on
Parameter	Set or change value	Basic material, color, and so on
Form element	Add, replace, suppress, retrieve, delete, reroute, rename	Controlling CAD features such as drill holes, beading, undercutting, and so on

Table 7.1 Controllable CAD Operations from SAP Variant Configuration

A *master model* is the starting point for the CAD configuration. The master model is a minimum product version, from which all predefined product variants can be automatically derived through the use of CAD operations. The standard functions in the CAD system are used to model master models, which are suitable for CAD automation, but this is requiring that users always adhere to an appropriate modeling methodology.

7.4.4 Super BOM in Variant Configuration

In SAP ERP, the solution space in the product structure in the standard system must be predefined in general, even if a dynamic component could be specified via class items and by controlling the number of parts and adding new items. Therefore, the basic approach in SAP ERP comes from a maximum approach whereby such a CAD model immediately becomes a corresponding maximum model that frequently can no longer be loaded and becomes extremely complicated.

From a conceptual perspective, it is therefore necessary to reconcile the maximum approach and master approach with each other.

7.4.5 Automatically Creating CAD Data from Variant Configuration

The graphical derivation out of SAP Variant Configuration is often wanted and, in many cases, implemented within a project. In addition to the aforementioned 2D presentation, however, a real CAD drawing is often required for engineering and design, sales, or services. it-cadpilot fulfills this wish.

7.4.6 Architecture

To automatically generate CAD data from SAP Variant Configuration, the master data must fulfill the following three prerequisites:

▶ In the CAD system, you must create a CAD master model for a product variant area.

▶ In the CAD-related part of the add-on solution, you must determine which CAD operations are permitted for the master model.

▶ Finally, in an SAP control table within the add-on solution (see Figure 7.10), you must define which CAD operations are to be performed, depending on the SAP configuration result.

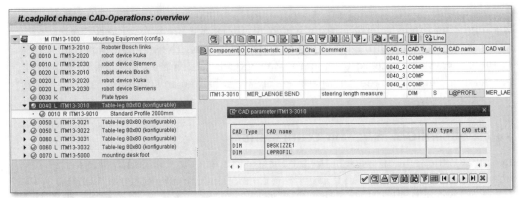

Figure 7.10 Example of a Control Table

Figure 7.10 shows an example of this. Here, you define how the characteristic value from a configuration result is used to determine dimensions in CAD control. When defining the content of the control table, you have direct access to both the Variant Configuration master data and the permitted CAD operations.

Master models, including the exchangeable parts, are also consolidated, with their control information, in SAP Documentation Management System (DMS). This makes it possible to use SAP Engineering Change Management controlling release versions and the validity of dates. If you're using SAP-supported CAD model and drawing management in addition to DMS, the master CAD originals are integrated into this master data management. Figure 7.11 provides an overview of the entire additional solution architecture: master data in the upper section and transaction data in the lower section.

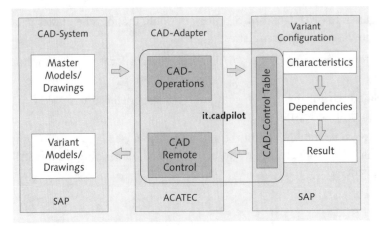

Figure 7.11 Architecture of CAD Control in Variant Configuration

7.4.7 CAD Configuration

If the master data has been defined, the process for a variant is as follows:

1. **Configuration in SAP ERP**

   As usual, a variant is configured in SAP ERP, for example, in Transaction VA02, and CAD creation can be called via an additional button (see Figure 7.12).

2. **Load the CAD master and create the CAD variant**

   Load the CAD master. You apply the CAD operations determined from the configuration result and control table to the master model by remotely controlling the CAD system in the add-on solution. Consequently, the CAD model is created for the variant. All of the required derivations for the variant are also automatically created: drawings, conversions into neutral formats such as JPG or TIFF, conversions into interchange formats such as STEP, IGES, or DXF (see Figure 7.13). Derived product images and drawings can be automatically checked-in.

Figure 7.12 Starting CAD Creation from Sales Order Processing

The CAD descriptions for the variant product are simultaneously created and added to the BOM and, if necessary, the routing. Of course, with such an approach, the BOM structures can also be more comprehensive than the pure CAD-modeled structures (e. g., variable-size items and bulk material).

If you are using SAP-supported CAD model and drawing management, the CAD results can be checked-in and are therefore available for further engineering and design steps, depending on the scenario (also in conjunction with the order BOM that is created at the same time).

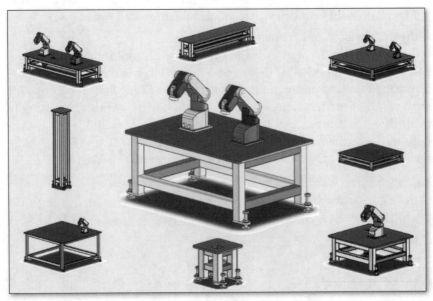

Figure 7.13 Range of CAD Model Variants

7.4.8 Advantages of a CAD Configuration Integrated into SAP ERP

There are numerous advantages by consolidating the three main types of product configuration (sales, order processing, and CAD) into one SAP application platform:

▶ The resulting BOM and the associated CAD documentation are closely related to each other.

▶ One rule set can be used for all configuration types, thus avoiding time-consuming and risky redundancies.

▶ Object dependencies, super BOMs, and the related CAD master models are managed on one platform.

▶ Engineering Change Management provides end-to-end functions for release management.

▶ Everyone involved in configuring a product in an enterprise can work together more easily, without integration gaps.

▶ The creation of CAD data is considerably faster.

7.4.9 Application Scenarios

The SAP-based CAD configuration has many use cases. Essentially, the scenarios can be divided into order acquisition and order processing. In addition, depending on the order processing type, you can use the CAD-based documentation for the configured variant in different ways.

Automatic Creation of Drawings in the ATO Scenario

The *assemble-to-order* (ATO) process is a closed configuration: The characteristic value assignment fully describes the product. Furthermore, the production and assembly process can start immediately without the need for order engineering processing. In addition to the SAP production papers, it is now possible to immediately and automatically create the associated assembly and production drawings (see Figure 7.14).

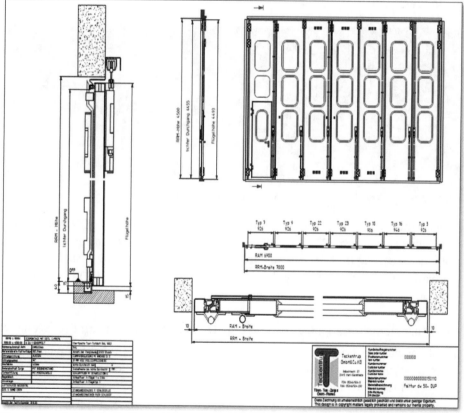

Figure 7.14 Automatically Created Variant Drawing (Source: Teckentrup)

Teckentrup, a German manufacturer of doors and gates, has automated their procedure for sales order processing in this way. The drawings are created at the same time as, and in conjunction with, the manufacturing papers for the configured materials. There is no need to create any other drawings, nor is there a need for engineering postprocessing. The time and effort associated with creating the CAD configuration pays off within a short space of time, and enterprises immediately benefit from order acceleration and better-quality documentation.

Order-Related CAD Design in an MTO Scenario

Sometimes the *make-to-order process* continues to require order-related manual engineering. In commercial terms, it is not possible to predefine a rule set that allows for every conceivable customer preference. The benefits are considerable if you apply the *80:20 rule*: On average, above 80% of customer requirements can be rule-based pre-configurated, whereas the remaining special features require manual postprocessing.

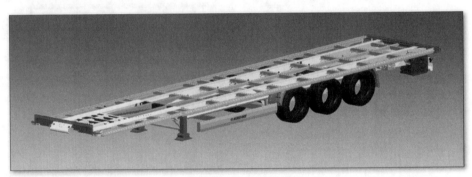

Figure 7.15 Configuration of Semi Truck Trailer (Source: Krone)

Krone, a manufacturer of semi truck trailers, uses the integrated CAD configuration to automatically control Pro/ENGINEER models, thus achieving considerable savings potentials – see Figure 7.15. The major advantage here is that CAD designers, which are hard to find, can completely devote themselves to their actual task, namely the development and maintenance of a new product range.

CAD Documents for Engineering and Construction Assembly in an ETO Scenario

The automatic creation of CAD models and drawings is a key factor in an engineer-to-order scenario represented by engineering and construction. Here, control can also be achieved by adding engineering and construction components to network activities. Once a characteristic value assignment is made there, project BOMs and the associated CAD documentation are created automatically, checked into SAP-supported CAD management, and automatically linked with plant-related work breakdown structure (WBS) elements.

CAD Documentation for Retailers and Architects in a B2B Scenario

Enterprise customers like partners and architects require high-quality CAD product variant documentation with the request-for-quote (RFQ) process and the quotation process. This includes the following:

▸ Native CAD models, if necessary with details that are automatically removed in advance (to protect intellectual property)

▸ Fully scaled, standard-compliant drawings as binding sales documentation

▸ CAD interchange formats such as STEP for 3D or DXF for 2D

All of this documentation can be generated fully automated (embedded into the SAP E-Commerce web shop for ERP or SAP CRM Internet Sales).

7.4.10 A Look Ahead

The methodology can also be applied to other documentation, for example, Microsoft Word and Excel formats and other viewing formats associated with a configurable product structure.

Like CAD, a type of master is also defined for Microsoft Word, for example, for automated quotation processing, including allowed changes and additions to the Word master (this is done at the placeholder or text module level).

Values from the SAP database can be applied dynamically, for example, the customer number, customer, document number, product for which the potential customer wants to receive a quotation, characteristic value assignment, prices, and true-to-scale 3D images of the product configured from CAD Automation. Therefore, the creation of quotations and product specifications can be automatically generated from Variant Configuration.

Another area is the control of viewing formats, which can be derived from the CAD documentation. However, they are more compact and faster to load and therefore are often better suited to illustrating the variant product on the Internet in real time during configuration. Once again, a similar methodology consisting of a master and permitted operations can be used here.

As part of SAP's SOA initiative, there are more and more possibilities of integrating third-party products, especially in terms of delivering Variant Configuration services. This presents opportunities to incorporate related areas into SAP integration, for example:

▶ **Guided selling**
Solution-based support during the product determination process. Graphical representations of products or product areas are often essential here. SAP Variant Configuration can be controlled in external, graphical, freely configurable user interfaces.

▶ **Solution configuration**
A space requirement is geometrically preassigned by customers or interested parties (for example, the layout of a plant). Both the overall plant engineering assembly and the detailed configuration are graphically possible under these conditions.

[+]   Dr. Hans Joachim Langen from itelligence AG *(www.itelligence.de)* and Dr. Reiner Kader from ACATEC Software GmbH *(www.acatec.de)* kindly contributed the information about it.cadpilot for this book.

7.5   top flow-Framework and top flow-Variant Engine (company: top flow GmbH)

The ability of the configuration dialog box, enhancements in the object dependency logic, and the options for make-to-stock production despite the use of SAP ERP Variant Configuration are the highlights of the optimization package available from the company top flow for the configuration application.

Enhancing ERP-based Variant Configuration (LO-VC) gives rise to the following three optimization potentials:

▶ Optimization of the configuration dialog box beyond the standard system

▶ Enhanced options for creating and using a dependency logic

▶ Process optimization with the top flow Variant Engine: Option for make-to-stock production despite the use of SAP ERP Variant Configuration

The top flow Framework does not represent a stand-alone system with interface issues. Rather, it is fully integrated into the SAP ERP environment as a result of its implementation in ABAP. These new functions enable users to make changes and adjustments more efficiently, without having to limit the standard SAP functions.

7.5.1   Optimizing the Configuration Dialog Box

The first potential for optimization is aimed at the usability of the configuration interface and the option to integrate additional functions. This point is noteworthy in view of the fact that this aspect of the SAP ERP configuration application receives the most criticism. To achieve improvements here, top flow has standardized the communication between the standard SAP Variant Configuration and custom-developed, customer-specific interfaces; that is, the interfaces automatically respond to the entries made in the standard configuration logic.

If, for example, object dependencies are used to control the display of the FINISHING characteristic, the custom-developed interface responds in exactly the same way as the standard interface for the characteristic value assignment (see Figure 7.16).

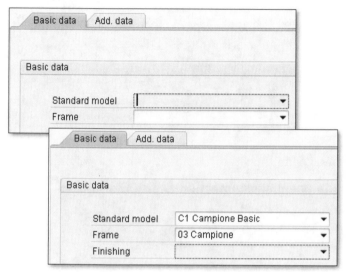

Figure 7.16 Enhanced Dialog Box that Displays a Characteristic Subject to Another Characteristic Value

This figure shows the enhanced dialog box described above. The configuration editor contained in the standard SAP system is not used here. Nevertheless, the response to the standard object dependency is identical: A characteristic is displayed subject to another characteristic value.

Thanks to this standardization, it is possible to use all SAP NetWeaver development tools to build configuration interfaces.

▶ Screen-based (Transaction SE51)

▶ Business Server Pages (Transaction SE80)

▶ Web Dynpro (Transaction SE80)

Sample Dialog Box with an External Interface (Screen-Based)

The following example shows a screen-based interface that is automatically displayed instead of the characteristic value assignment defined in the standard system (see Figure 7.17).

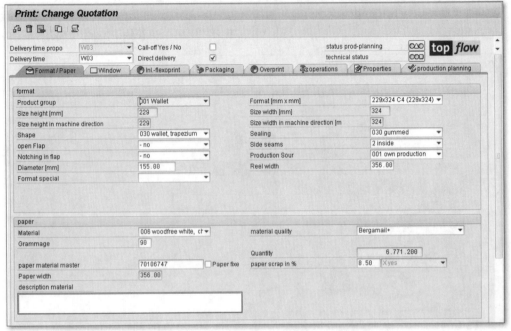

Figure 7.17 Newly Designed, Screen-Based Interface

The difference is immediately apparent. Users are no longer bound to the rigid sequence of characteristic values. In addition to improved usability, for example, as a result of using list boxes to select characteristic values, the open interface design makes it possible to integrate any additional functions into the configuration dialog box.

Sample Dialog Box with Web Dynpro or BSP

Let's consider additional examples for a screen-based interface (see Figure 7.18).

The following are two examples of SAP web applications that use Web Dynpro or Business Server Pages (BSP).

Figure 7.19 shows the example a standard list of characteristic values, similar to what you see in Transaction CU50. However, a static product image is displayed during configuration.

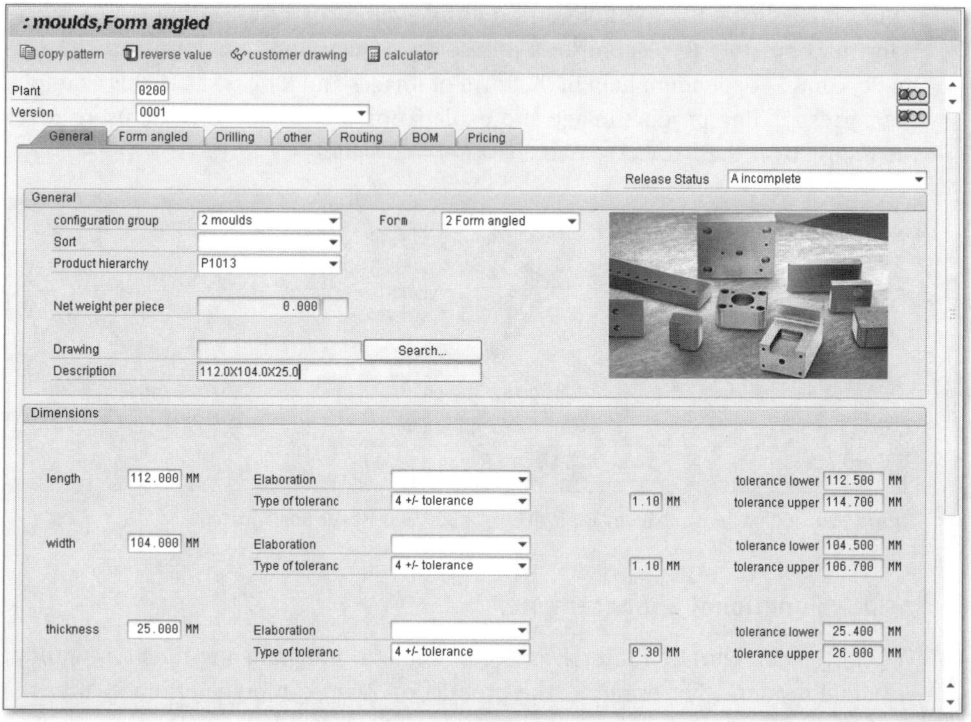

Figure 7.18 Additional Example of a Screen-Based Interface

Figure 7.19 Characteristics Value Display, Including the Static Product Image

Figure 7.20 shows the example of a multiple-value characteristic. In web applications, it is possible to respond to a mouse-over event and therefore map an automatic, context-dependent help in the form of images and long texts. In this example, the corresponding product image and explanation are displayed if the user moves the mouse over the PROCRAFT ALU characteristic value.

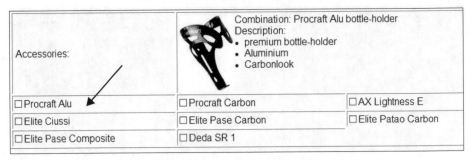

Figure 7.20 Characteristic Value Display Using a BSP and Hover Box Function

7.5.2 Functional Enhancements

Within the configuration interfaces, it's possible to integrate any function without any modifications. For example, the production plant can be determined after the configuration has been made. In the standard system, the plant is determined subject to the sales material. As a result of a functional enhancement, the production plant can now be determined and changed subject to a successful availability check. When you use the top flow solution, this and other information can be displayed and automatically or manually overwritten by the user.

Sometimes, owing to the complexity of the configuration, you cannot or may not want to define the complete configuration logic in the system. However, you nevertheless want to use the configuration as an initial aid or shortcut for dynamically correcting the routing determination result, for example. You may want to add or interactively change additional processes or components. All downstream processes such as costing or creating planned orders and production orders then automatically inherit this additional information.

Other examples of functional enhancements include integrated pricing within the configuration logic, displaying images as an auxiliary function for the user, workflow integration, or simply displaying additional information.

7.5.3 New Object-Dependency Logic Options

The enhancement of the object-dependency logic is based on the idea of storing all object dependencies directly, without any diversions, in transparent tables.

All of these functions can be accessed via an initial screen (see Figure 7.21).

Table Structure

To simplify table maintenance, a transformation provides users with a tool for creating, changing, and deleting tables via drag and drop. In addition to creating the database tables, the maintenance dialog box is structured in such a way that the tables can be maintained directly (Transaction SM30) or can be uploaded with a tool from Microsoft Excel.

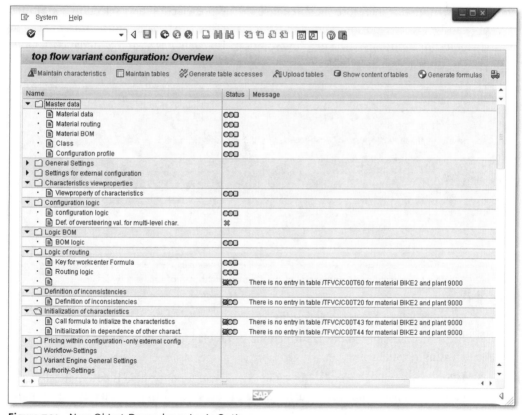

Figure 7.21 New Object-Dependency Logic Options

Configuration Logic

The configuration logic describes sequential processing of user-defined tables and formulas. This logic is used not only for the configuration dialog box itself, but for determining the BOM, its components, and the routing. The search for production resources and tools is also provided.

To interpret the tables, you have to define how to access the tables and which table content to transfer. It is also possible to define multiple accesses for a table and therefore define a type of access sequence.

Maintaining the Display Attribute for Characteristics

In extensive configuration projects, the definition of display attributes can be rather complex, that is, when to display or hide a particular characteristic subject to another characteristic. Therefore, maintenance of the preselection and selection conditions can be a very laborious and confusing task. This is particularly true for changes to object dependencies. The top flow solution also facilitates clear maintenance in tabular form (see Figure 7.22).

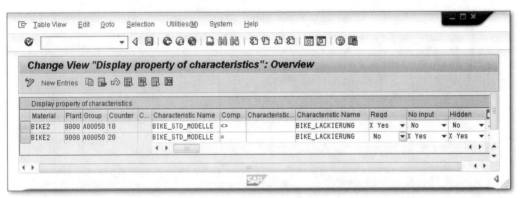

Figure 7.22 Simplified Maintenance of the Display Attributes of Characteristics

Support When Creating Formulas

If it is not possible to display functions in tabular form, you can also use a tool for creating formulas.

When you create a formula, the standardized call for a formula is automatically generated within an editor. To actually map or process the formula, you can use all of the logical configuration data at your disposal. When you save the formula, it is created as an ABAP routine that can be entered in the configuration logic. Known standard functions such as the syntax check and debugging support you in your work.

7.5.4 Process Optimization with the top flow Variant Engine

Frequently, you want to benefit from the advantages associated with Variant Configuration, but you nevertheless want to use make-to-stock production for certain orders or, at the very least, you want to automatically generate material variants for certain orders. This is often necessary for complex Warehouse Management settings or specific electronic data interchange (EDI) processes involving customers and suppliers.

The top flow Variant Engine makes it possible to easily create material variants. On the basis of a configured sales document item, the top flow Variant Engine can start automatic creation of all master data in the process with the click of a button (see Figure 7.23).

The top flow Variant Engine can automatically execute the following steps:

▸ Creation of the material master with field control either via formulas or subject to the configuration

▸ Copying of the characteristic value assignment from the sales document item into the new material master (thus creating a material variant)

▸ Assignment of the material master to the relevant super BOM

▸ Assignment of the material master to the relevant super routing

▸ Costing of the material, including the standard price update

▸ Creation of a customer material info record

▸ Creation of SD condition records

▸ Material classification

▸ Creation of master data and the documents (RFQs and purchase orders) for commercial parts

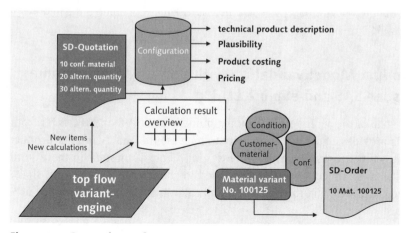

Figure 7.23 Process for top flow Variant Engine

Therefore, all of the master data required for a subsequent process is available in a single step. Consequently, it is possible to automatically create an order with reference to a quotation.

7.5.5 Solution Map for the top flow Variant Engine

The solution map shown in Figure 7.24 provides a clear overview of the possible functions.

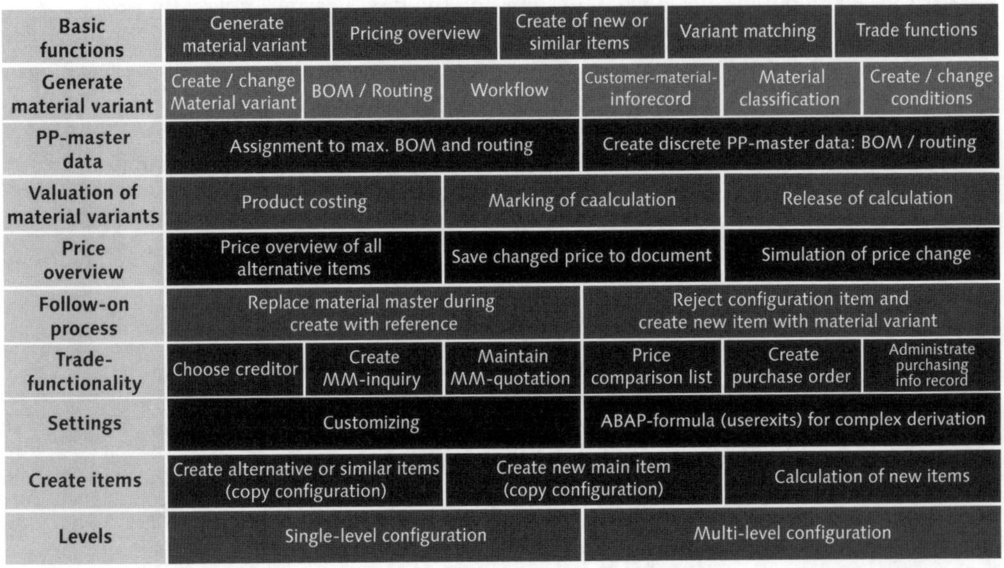

Basic functions	Generate material variant	Pricing overview	Create of new or similar items	Variant matching	Trade functions
Generate material variant	Create / change Material variant	BOM / Routing	Workflow	Customer-material-inforecord	Material classification	Create / change conditions
PP-master data	Assignment to max. BOM and routing			Create discrete PP-master data: BOM / routing		
Valuation of material variants	Product costing		Marking of caalculation		Release of calculation	
Price overview	Price overview of all alternative items		Save changed price to document		Simulation of price change	
Follow-on process	Replace material master during create with reference		Reject configuration item and create new item with material variant			
Trade-functionality	Choose creditor	Create MM-inquiry	Maintain MM-quotation	Price comparison list	Create purchase order	Administrate purchasing info record
Settings	Customizing			ABAP-formula (userexits) for complex derivation		
Create items	Create alternative or similar items (copy configuration)		Create new main item (copy configuration)		Calculation of new items	
Levels	Single-level configuration			Multi-level configuration		

Figure 7.24 Solution Map for top flow Variant Engine

[+] Employees from top flow GmbH *(www.top-flow.com)* kindly contributed the information in this section.

7.6 Product Model Validation with ConfigScan (companies: Fysbee SAS and eSpline LLC)

The current state of the art in testing product models for LO-VC and IPC is far from optimal. This provides a golden opportunity to increase the quality of your product configurator, optimize your internal processes, and help your staff do more with less. To capitalize on this opportunity, you need to check your assumptions, shake up established practices, find the right tools and techniques, and put optimized processes in place.

If your company uses LO-VC and/or IPC, chances are high that you are using more than one *anti-pattern*: processes that contain significant business risk, inefficient use of human resources, "keeping your head in the sand," that is not measuring performance and not using metrics to drive continuous improvement.

Fortunately, by borrowing proven ideas, best practices, tools, and techniques from the software engineering world, you can make significant improvements to your current practices.

7.6.1 Business Scenarios that Motivate the Need for Change

Most companies with configurable products seem to be in denial about their testing problem — and they're right; there is no *testing* problem. Instead, there are real, serious, costly business issues and broken processes like the following situations, frequently encountered by eSpline and Fysbee on customer projects over the past decade:

▶ **Inconsistent orders**
You implemented Variant Configuration to detect incorrect orders as early as possible, but you still get incorrect orders. Fixing these inconsistencies requires expensive and time-consuming intervention by staff, who should be more productively employed. The later in the process a defect is caught, the more costly is its repair. Example worst-case: A production order is sent to the plant, inconsistencies are detected, internal communication is started to negotiate possible manual solutions, interaction starts with the customer to get an acceptable alternative, a discount or sweetener is added, and the shipment is delayed.

▶ **Incomplete sales offering**
Your product configurator users often can't find, or are prevented from ordering, some of your less-common options; that is, you can't present the entire offering in the configurator. This makes your company hard to do business with, and puts you at a competitive disadvantage, but very likely your competitors have the same problem, in which case fixing the problem will give you a competitive advantage.

▶ **Discrepancies among multiple configurators**
If you're required to create and maintain your product models not only in SAP ERP, but also in a parallel or legacy system or two, or if you're migrating a third-party product configurator into or out of SAP, you need a way to ensure that LO-VC, or its cousin, IPC, behaves the same way as the third-party configurator. You may even have this problem within your SAP landscape, for example, because you have two parallel chains of "development, quality assurance, and productive" systems that may not be perfectly synchronized.

351

▶ **Inefficient software factory, trial-and-error in production system**
A similar scenario occurs if you make changes, and manually test the changes, directly in production — a practice more common than it should be — and if you rely on production refreshes (client copies) to keep your development and test environments up-to-date. Sometimes the refreshes don't occur for several months or more. It is time-consuming to manually ensure that product models in your development or test system are in sync with production and to make sure they behave the same way. Your product modelers will have the luxury of up-to-date data and models, but they are hampered by the restrictions of your production environment: the business imperative to not break your quote-to-cash process, change controls that slow experimentation that could otherwise rapidly converge on a solution, no place to try realistic performance tests, business disruption whenever your staff turns over. Worst-case: frequent disruptions in your order management processes, significant delays in releasing product models, and requiring staff to be on high alert to resolve emergencies.

▶ **Missing metrics to drive continuous improvement**
You're responsible for quality and speed of development of configurable product models, but you have no way to measure the quality of, and progress toward, a new product introduction or update release, or upgrade to a new SAP software version. You need quality and progress metrics, at a manageable cost.

▶ **Manual testing is slow, costly, or incomplete**
Owing to combinatorial explosion of the number of possible configurations, both valid and inconsistent, manual testing can only reach a tiny fraction of most product models, it introduces serial delays in time to market, and it requires the same investment each time you test. If you try to reduce the cost, or increase the tiny fraction of coverage, by employing offshore resources, there is no guarantee of diligence — that all tests will be run, run correctly, and reported correctly — except by measuring failures that should have been caught by your tests.

7.6.2 Anti-Patterns in Common Use

Do the following practices seem risky? What are the negative consequences to your business? Are you making best use of your human resources? Are you realizing close to the full potential of the SAP solution that your executive team and sponsors expected when they decided to implement Variant Configuration? Are there better ways to run your back office configure-to-order (CTO) processes?

Although the following practices are questionable, they are quite common with many implementations.

▶ You create, maintain, and test configurable product models directly in your production environment.

▶ You have no written specifications for your product models, or the ones you have are obsolete.

▶ Test plans are informal and undocumented, are focused on releases, and rely on manual testing of a small fraction of allowed combinations.

▶ You have no way to verify that your entire product offering can actually be ordered using LO-VC or IPC.

▶ Governance and change control are great for ABAP code — user exits, Business Add-Ins (BAdI), RICEF (Reports, Interfaces, Conversions, Enhancements, Forms), and transactions — but not so advanced when it comes to controlling master data changes, for example, to configurable product models. Standard tools provided by SAP such as Engineering Change Management, and change documents on single objects of your product model provide an audit trail, but don't directly address your need to manage the product data required by your CTO business.

▶ It is difficult to collaborate on changes to your product models, to apply consistent naming conventions, to stick to coding guidelines for configurator rules, and to reuse successful design patterns.

▶ You have no or few metrics in place to monitor the quality and performance of your configurator applications.

▶ When it is time to release a new or updated configurable product, you have no way to justify even a limited budget for manual testing. That budget and the associated delay in the release of the product are seen as necessary evils.

▶ Some of your testing is done offshore, manually, but it is difficult to verify that a successful test report means the test was actually run correctly — let alone run at all.

▶ Some percentage of the defects in your configurable product models is caught by your internal salespeople, some by partners or agents, some on the production floor, and some by your customers — sometimes only after they receive your product. Your manufacturing processes are set-up according to the Six Sigma business management strategy — but not your CTO processes.

7.6.3   How ConfigScan Addresses these Issues

There are better ways, including techniques, processes, practices, and software tools that support a more rigorous, higher-quality result. The *ConfigScan Validation Suite* addresses many of the quality and performance testing issues listed in the previous subsection. Figure 7.25 illustrates the modeling process, typical test steps, and involved roles.

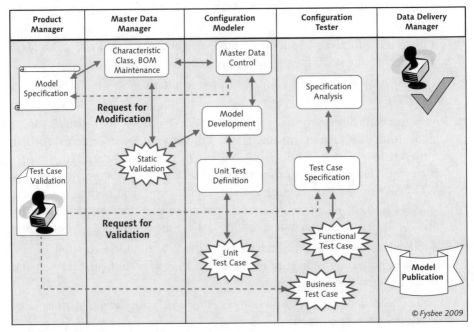

Figure 7.25 Modeling, Testing, Release Process, and Roles

ConfigScan Validation Suite provides effective ways to create and maintain functional and performance test cases in SAP ERP that can be centrally stored as a library or catalog of test cases and executed automatically in LO-VC and IPC (as part of SAP ERP and SAP CRM) product configurators.

Benefits include:

▶ Using best practices borrowed from software engineering including Test-Driven Development, which makes product modelers more productive.

▶ Testing runs automatically, which reduces the cost and time to release new or changed configurable product models.

▶ Automatic testing enables more frequent testing so defects are caught as early as possible when the cost of repair is lower.

▶ Test cases can be created in a model-aware development environment (Test Case Editor) and created automatically by importing sales orders with configurable products on line items.

▶ A Test Engine enables test cases to be specified declaratively, rather than by writing code, which significantly lowers the cost of creating and maintaining test cases.

▶ Ongoing investment in test cases can be applied toward creating more tests instead of rerunning them manually, thereby increasing quality and shortening time to market.

▶ Troubleshooting, including communicating a problem and collaborating on its resolution, is accelerated because all parties have access to a test case that explicitly and precisely describes how to reproduce a problem and can easily execute the test case to parachute into the exact trouble zone in the product model and configuration.

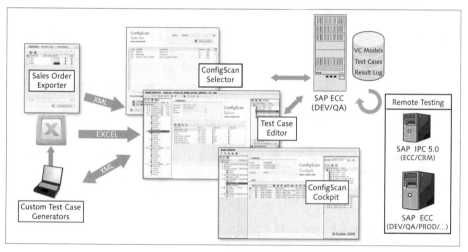

Figure 7.26 ConfigScan Validation Suite — Landscape

ConfigScan Validation Suite supports and guides you in taking a big step forward to address any of the above or similar scenarios that may apply to you and your company. Figure 7.26 illustrates the integration of ConfigScan in the system landscape.

7.6.4 ConfigScan Validation Suite — The Basics

ConfigScan applies time-tested techniques from the wider software engineering world to product configuration.

It is compatible with *Test-Driven Modeling* (TDM), in which tests are created in parallel with model development. As the modelers develop models or implement specification changes, tests that initially fail start succeeding, giving quick, vital feedback to both the modeler and the project manager. Typical questions that can be answered include:

▶ *Do the tests for which I just implemented product model changes succeed now, or do I need to keep at it? Is the performance adequate?*

▶ *Are we on target to complete these new product models or changes by the date required by the business?*

You can apply Test-Driven Modeling gradually by starting with new models and by adding a test case each time a defect is discovered in existing models to "lock in success" on all future releases. TDM is helpful not only post-go-live, but also during the implementation phase. You'll measure the progress of your Variant Configuration team, which for at least the first half of a typical implementation project tends to be on the critical path.

ConfigScan runs directly on your SAP instances and requires no extra or dedicated hardware. It can be configured to fit your existing system landscape. Typically, you install the Test Case Editor on the same system and client where you master product models, for example, a GOLD VC client of your development system. You can install Test Case Execution and Sales Order Export on other clients and instances as well. With the IPC add-on, you can also remotely run tests against the IPC on SAP ECC or SAP CRM. A typical installation takes less than a day; your biggest challenges will be explaining the benefits to persuade your staff to change their existing processes, and deciding how to adapt ConfigScan's flexibility to meet your requirements and maximize your return on investment, quality, and productivity improvements.

7.6.5 Working with the Test Editor

You use the Test Case Editor as your development environment to create, run, analyze, and troubleshoot test cases and their associated models (see Figure 7.27). All of the model object names are at your fingertips if you want to manually enter test cases, or copy and modify them, according to your test plans. Other sources of test data can also be valuable: You can import your sales order history and turn valid order line items — and rejected items — into test cases automatically. You can import test cases that are generated from a custom program into ConfigScan using an intermediate XML, or Excel, test case format. The resulting library of test cases can be stored in the SAP Document Management System (DMS) or externally if you have a reason to master the test cases outside SAP.

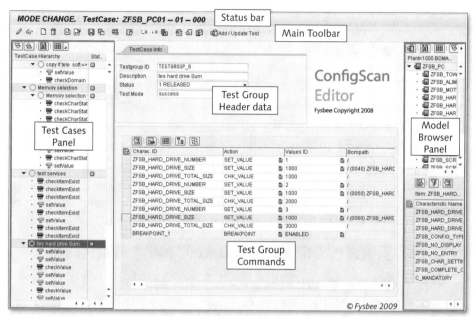

Figure 7.27 ConfigScan Test Case Editor

Analysis of a problem depends on stopping the configurator at important inspection points in the test case script, which ConfigScan supports by pausing at the first error, or pausing at a breakpoint, allowing the modeler or tester to inspect the current configuration to determine the cause of a problem, whether it's an undesired change to a model or a test case that needs to be updated.

A test case is like a script or sequence of "moves" in the game of configuration. Set a few characteristics to certain values, and test other characteristics to ensure the configurator inferred their values. The straightforward test case language covers all of the important moves you and the configurator can make manually, including setting, unsetting, and testing of characteristic values, restrictions by rules to a smaller set of possible values, and testing characteristic attributes such as visibility, required, defaulted, and inherited — without programming.

The test case language fully supports multilevel configuration with bills of materials (BOMs). Questions such as the following are covered: Does an item exist in the configuration? Does it have at least or at most a given quantity? Is the status of the instance at an item consistent and complete?

You can test the entire configuration, including variant conditions and pricing factors, to see if it's consistent and complete.

The ConfigScan Cockpit enables scheduling of recurring test executions, for example, at a time when system load is expected to be low, to run as background jobs. It uses the SAP Catalog to flexibly organize test plans by, for example, product type, product line, model release, or completion milestone dates.

Test case results are stored in standard SAP structured logs, which provide comprehensive messages about successful and failed test steps (see Figure 7.28).

When you examine and monitor the test results, you'll see red and green lights to quickly find failed tests. Regression (loss of functional or performance correctness) is detected and avoided by comparing test case logs over time.

The administration of test cases allows you to copy or modify test cases, create a variation, reuse test cases across multiple materials, import and export test cases or test logs, and find the materials tested by a test case ("where-used").

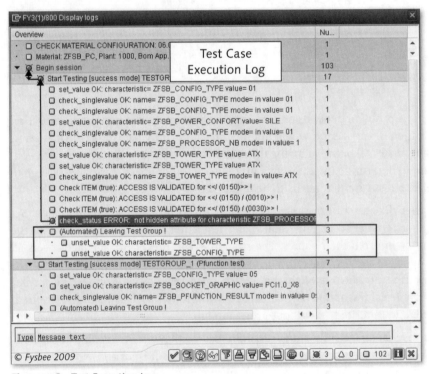

Figure 7.28 Test Execution Log

7.6.6 Summary

You have a golden opportunity to optimize and automate your company's configurable product processes, improve the quality of your product configurators, and make your staff more productive. To achieve this, you need to change today's inefficient, entrenched practices for testing, developing, and releasing your configurable product models. This requires making a business case: You'll need to detect and quantify the problem, compare it against tomorrow's best practices, become a champion for change, get commitment from management, and find the right tools, techniques, and partners to make it happen. ConfigScan Validation Suite, which has been used in production since 2007, can help you in this quest by providing you with automated, integrated, and efficient testing of LO-VC models.

> The section on the ConfigScan Validation Suite was provided by David Silverman from **[+]** eSpline LLC and Christophe Faure from Fysbee SAS (*www.espline.com*, *www.fysbee. com*).

7.7   Managing Variant Configuration (company: eSpline LLC)

As a product modeling expert or the manager of a Variant Configuration project, you're probably aware that there are only a few software tools outside of the SAP modeling environment that can help you manage the lifecycle of your Variant Configuration models. How do you control and approve changes and easily move them between development and production? What if a customer escalation leads to an audit of your LO-VC model? How do you know your LO-VC team is operating at peak efficiency, and what can you do if it is not?

Many LO-VC users have developed their own tools to help support translation of requirements to model specifications; most are making at least some use of two functions provided by SAP: Engineering Change Management (ECM) to control model changes and Application Link Enabling (ALE) to transport them between SAP instances. Virtually all LO-VC users have had to bridge many gaps to fit the LO-VC model into their configure-to-order process.

Project experience shows that managing the LO-VC model lifecycle is a time-consuming, hit-or-miss process. For product models that are changed frequently, or for large, new product models, it is a challenge to get a complete and understandable overview of the model, or to just find all recent changes.

eSpline LLC addresses this aspect of the LO-VC lifecycle management problem by combining several existing and some new software application tools and providing a methodology that injects best practice patterns into both LO-VC model lifecycle and LO-VC transactional order fulfillment processes.

At the heart of the solution is a process and data integration hub called *Avenue Orchestrator,* which is capable of pulling LO-VC master and transactional data from multiple SAP instances, performing detailed comparisons, and synchronizing approved changes across system boundaries (see Figure 7.29).

Avenue Orchestrator is a server application accessible over the simple web user interface depicted in Figure 7.30.

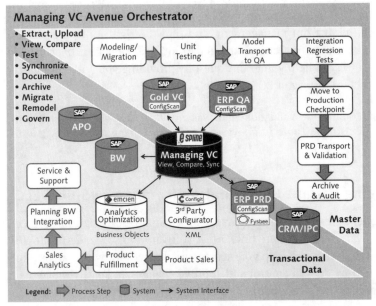

Figure 7.29 Avenue Orchestrator — Processes, Related Systems, and Operations

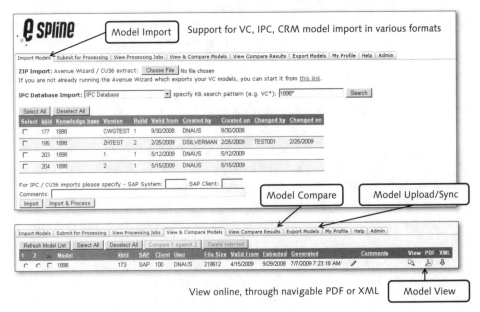

Figure 7.30 Avenue Orchestrator User Interface

Avenue Orchestrator can be deployed on your premises within your SAP infrastructure, but it is also offered as a hosted Software as a Service (SaaS) environment.

7.7.1 Managing the LO-VC Model Lifecycle

The LO-VC model lifecycle process is depicted in the top-right corner of Figure 7.29. It starts with business requirements leading to creation of LO-VC models on the Gold VC Client SAP instance. These models can be created by hand or generated by a one-time or ongoing migration from other sources, for example, legacy configurators.

Next, all LO-VC models should go through unit testing on the Gold VC Client, either manual or tool-assisted with the ConfigScan Validation Suite from Fysbee SAS as described in section 7.6, Product Model Validation with ConfigScan. At the successful completion of this step, the model and test results are imported into Avenue Orchestrator and approved for transport to the Quality Assurance (QA) SAP instance.

As an incremental, low-risk way of introducing automated unit and regression testing, Avenue Orchestrator includes tighter integration to and from the variant configuration testing solution. The compare process recognizes model changes and passes that information to the testing solution to build test cases.

Upon approval, Avenue Orchestrator automatically performs the model transport to the QA system by using ALE and Uniform Packaging Service (UPS). QA is an ideal environment for integration and regression testing, because it allows massive creation of simulated sales documents. This testing step can be manual or tool-assisted with ConfigScan. Once the integration test is complete, results are again imported into Avenue Orchestrator.

The next step is a formal move-to-production checkpoint. In addition to LO-VC team resources, it often includes representatives of infrastructure, IT management, and business. Avenue Orchestrator facilitates an efficient checkpoint by first extracting the existing production model and then running a series of detailed reports highlighting all of the changes that are subject to the checkpoint approval. No changes are missed; even minor description changes are highlighted.

The first of these reports is the Model View, which provides a single, complete, and easily navigable document describing an entire LO-VC model in either web or PDF format. Figure 7.31 shows an example.

The second important report is the Model Compare, which shows all of the changes performed on a model's objects and their fields between two points in time, including timestamps and users responsible for the highlighted changes. This feature works well in conjunction with ECM, but the process does not force you to use ECM; all changes are highlighted nevertheless.

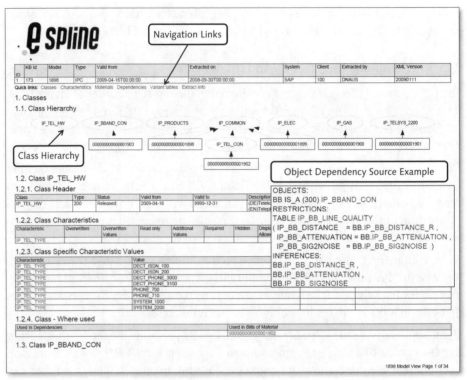

Figure 7.31 Navigable, Printable Model View Report

[+] **Example Model Compare Report (Figure 7.32)**

The two revisions of the model are identified as #1 and #2, and changes are noted both by this ID and in color: information that was deleted is shown in red. Added information is noted in green. Information that is in both models but is different is displayed in blue. In this way, changes are easy to identify and monitor.

Additional reports can be generated as required, such as the Model Compliance report (indicating any violations of defined naming conventions or other agreed-upon conventions, guidelines, or standards), and the Model Health Check report (including recommendations, suggestions, or warnings if the model contains modeling techniques that are not recommended).

Changes can be individually approved or rejected, because business conditions often change between the original model specification and the move to production, and some model features might not yet be ready for release to the supply chain. After approval, Avenue Orchestrator moves only approved changes to the SAP ERP productive instance (see system "ERP PRD" in Figure 7.29).

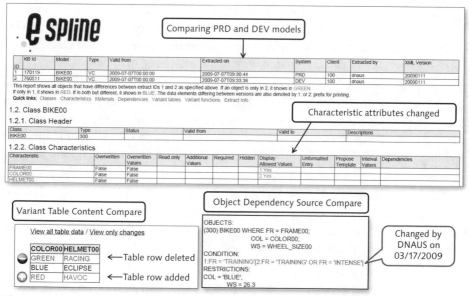

Figure 7.32 Navigable, Printable Model Compare Report

Additionally, Avenue Orchestrator provides the capability to have multiple iterations of testing and changes that feed ECM by generating engineering change masters for the approved part of the product that will be moved to production. This feature alone can accelerate model changes and possibly eliminate the reason to hold up other engineering change number–related transactions.

The last production verification step is executed to confirm that all of the changes were implemented successfully in all production environments (including SAP CRM/IPC), and if wanted, a detailed record of the documents and decisions made in the move-to-production checkpoint is archived to provide a complete audit trail. This record is saved in Avenue Orchestrator by default, but can instead be stored inside the Document Management System from SAP.

This LO-VC model lifecycle process is estimated to save about 40% of the LO-VC model release effort by focusing on automating burdensome and error-prone "busy-work" steps of model change control and transport.

Automated LO-VC Testing Solution

Experience with many LO-VC and IPC projects over the past 10 years has allowed eSpline to examine how a variety of SAP customers are managing their product models. Informal surveys over the past couple of years support the same observations: Testing of LO-VC models is most often performed manually, perhaps supported by spreadsheets, often run directly in productive systems, and focused only on recent changes. Meanwhile, the software engineering world has progressed to use proven techniques, such as *test-driven development* of automated, repeatable test cases, which have been shown to increase productivity while increasing quality.

Using the theory that a product model represents a product definition, a test plan can be created in advance of a proposed change and created in the automated testing software prior to any change. This process also incorporates and reuses previously created tests. Hence, the quality increases substantially, and the turnaround time for a change to an LO-VC product model is a fraction of the time required for a manual process.

SAP doesn't currently offer comprehensive variant configuration testing solutions that support regression testing and the ability to build test plans, so eSpline LLC partnered with Fysbee SAS to offer a comprehensive testing solution to complement the Variant Configuration solution from SAP.

The benefits of using an automated solution are numerous. Using test cases and test plans, performing regression testing, and using the ability to test the entire model every time a change is made improve the productivity of the LO-VC modeler and the testing process and at the same time improve the reliability of the final product models.

LO-VC and IPC Knowledge Base Extraction and Upload to and from XML File

A *knowledge base* (KB) is a container for all of the objects in a product model that are required for running product configuration outside LO-VC, specifically in the IPC. An *extractor* collects the required objects, given a top-level material or class in SAP ERP; this helps avoid the need for dual maintenance of configurable product master data. The same concept can be used to supply knowledge bases to third-party configurators.

However, knowledge bases are also useful as master data documents that can be input to, or output from, LO-VC management processes and tools running outside of SAP, such as those integrated into Avenue Orchestrator. These tools use an application-independent interchange format called Avenue XML interchange format, which is similar to the Knowledge Base Interchange Format (KBIF) published by the CWG, and compatible with the full LO-VC data model and the IPC.

The module used by the Avenue Orchestrator to extract a KB from either LO-VC or IPC is called *Avenue from VC*. It is in productive use, providing knowledge bases to cross-industry configurators such as Configit (see next section) and industry-focused configurators that require tight knowledge-base integration with LO-VC or IPC.

The reverse, "upload," direction is accomplished by another module, called Avenue to VC.

Product Model Migrations, Conversions, and Remodeling

New SAP customers, or LO-VC customers who acquire other companies, often have proprietary or third-party configurators that require safe conversion into LO-VC and IPC. A module called *Avenue Migration* includes a custom, pattern-based conversion of configurable product model structure and rules into a format compatible with the Avenue Orchestrator LO-VC data model. Many conversions can be accomplished with a mapping effort using the Structured Query Language (SQL), rather than the more-intensive programming language conversion. The approach is pragmatic, meaning that conversion is automated, pattern by pattern, if data volumes justify the incremental effort. Then the models are iteratively uploaded into LO-VC using Avenue to VC, iterating until the converted data loads without errors, meets guidelines and standards such as naming conventions or best practices, performs well, and passes all ConfigScan test cases. This technique of repeatedly generating the LO-VC models has been used on multiple projects. It minimizes any time windows during which development of product data must be frozen, allows experimentation to optimize and refine the models in an environment where mass changes and restructuring of the models is well supported.

Movement of product models can be accomplished in several ways, depending on the data source (SAP ERP, IPC from files or database, SAP CRM) and data model (LO-VC, IPC, SAP CRM): via BAPI, Application Link Enabling (ALE), or Product Data Replication (PDR).

Owing to scoping or risk-mitigation considerations, project management may choose to limit conversions during a migration to the minimum required to reproduce identical functionality. In this case, optimizations may be postponed to a future step or project. Optimizations can have the goals of easier maintenance, including unification of design technique across models, as well as changing to best modeling practices, and performance optimization. In this case, a conversion from LO-VC into new LO-VC models may be required. It may also be required when there is a business need to convert stable product models, which is called *remodeling* and is supported by the module Avenue Remodel.

Examples of conversions that may be part of either migrations or remodeling projects include:

▶ Simple renaming, for example, to add or change a prefix on all characteristics or classes, including automatic conversion of rules to match

365

▶ Change to best practice use of constraints, variant tables, and restrictable characteristics from preconditions on characteristics and values

▶ Unify models created by different modelers using different modeling techniques

▶ Replace many slightly different configurable or non configurable materials by a smaller set of configurable materials

7.7.2 Managing the LO-VC Transactional Processes

Once an LO-VC model is released for production use, Avenue Orchestrator provides a bridge to the order-to-cash and planning and fulfillment processes. This is depicted in the lower-left corner of the overview diagram (Figure 7.29), and the steps are described in more detail as follows.

Configurable Quotes and Orders Integration

Configure-to-order products and processes are an outstanding asset and often give your business a significant competitive edge, if you succeed in making them available throughout your company's value chain. In particular, Internet sales and mobile sales scenarios have a critical need for up-to-date and pervasive configuration capability. You want to enable your sales force and external partners to generate accurate quotes and accept orders — even for your most complex configurable products.

However, integrating external applications with the Sales and Distribution module of SAP ERP for configurable products is notoriously difficult. You need to master often poorly documented intricacies of Business Application Programming Interfaces (BAPIs), Web services, and recently also enterprise services. Heterogeneous and distributed environments (Java, Microsoft .NET inside or outside of the company's intranet) need to be integrated.

Avenue Orchestrator offers a Quotes & Orders module that supports multiple integration techniques with SAP ERP, and abstracts the cumbersome lower-level interfaces with an easy-to-use XML-based interface. This module has the capability to transparently synchronize simultaneous independent changes made to the same document in both SAP ERP and Mobile Sales.

Mobile Sales and Offline Product Configuration

For those LO-VC users who do not have or do not want to continue using existing custom sales applications for configurable products, eSpline partnered with Configit A/S, a Copenhagen, Denmark, configurator application company that offers a very close emulation of the LO-VC behavior with the ability to do the following without requiring dual maintenance of product models outside SAP ERP:

- Customize the user interface to make it more user friendly
- Operate outside SAP ERP on laptops and on the Internet
- Integrate with .NET applications

Avenue from VC works with the quote tool and product configurator from Configit to supply knowledge bases from LO-VC.

Characteristic and Characteristic Value Analytics Reporting

Comprehensive analytics for complex configured products have long been sought after by many SAP configurator customers. There have been some initiatives to provide this capability, but none became a commercial product.

Emcien, Inc. an Atlanta, Georgia–based software development organization is focused on providing EmcienMix™ and EmcienMatch™, a unique SaaS analytics reporting application that imports SAP sales orders and quotes and provides comprehensive reports, dashboards, and forecasting of characteristics based on history.

Why add this type of solution to managing Variant Configuration? Today most product managers are in the dark about what combination of options or characteristics are sold. They understand their ever-expanding product offering, but there is little feedback available that can help them forecast by option or combination of options, or report what was sold or not sold, and where these products and product options are sold. Usually, this is a manual, extensive, time-consuming effort.

Once information is known about what options or characteristics are sold or not sold, a company's product management is very likely to fine-tune the product offering and product mix to match sales patterns and trends. Without that information, the product offering usually continues to grow, along with parts inventories and delivery lead times. To improve product offerings by geographic region, and to minimize the number of combinations of options in the sales definition, it is important to streamline configurable product offerings.

Companies with complex products and services historically have found that although product management teams attempt to identify the best-selling products and eliminate the nonperformers, multiple feature and option choices make this extremely challenging and virtually impossible when a configurable product has thousands, millions, or even billions of combinations and beyond.

It is possible at a feature and characteristic level to:

- Easily make adjustments to options of a product offering as the market changes
- Identify and help standardize the most profitable feature bundles

- Report in easy-to-read customizable dashboards
- Provide feature-level snapshots and supply chain planning and forecasting

Quote and order data including product features and options are collected, analyzed, and presented as a series of dashboard displays illustrating combinations of product features. Users have the opportunity to view product options selected and sold for the top five combinations of product options, for example. The output is controlled by the user and product manager. This is done daily or weekly or when needed, so the analysis is current.

The Emcien analytics application examines, on an as-needed basis, what product options are sold, what option combinations are best-sellers, and what options and combinations of options are the poorest sellers. It identifies optimal product combinations and includes what was sold by region. Thus it is possible to determine what combination of features to offer, in which regions to offer them, and at an improved return on investment, because cost and price are part of the equation. Buying habits vary by region; so should product offerings. If product option combinations are not profitable, action needs to be taken, for example, to remove or deemphasize these options or to change their pricing to make a profit. Emcien's dashboards and online, interactive parameter-setting capability permits users to view product lines by a specific characteristic or a combination of specific characteristics; this is done online and completed in a matter of minutes, not days or weeks.

If product characteristics or values are too numerous to categorize, such as inches or centimeters for a product varying from 2 centimeters to 200, there are far too many individual combinations that have meaning. Emcien allows for a quick and easy recategorization of this type of data. Using the previous example, it might be better to combine all measurements into short, medium, long, and extra-long categories. This is easily done in Emcien and makes the data meaningful.

SAP quote and order data is the source of the analytics application analysis of product options (characteristics), and a host of dashboard displays can be reviewed to spot product and regional trends (see Figure 7.33). Historical information is used as the starting point, and current order and quote data is added on an as-needed basis to complete the analysis. The analytics software provides up-to-date analysis with daily updates if that is the frequency chosen and product mix is dynamic.

With easy-to-read online dashboard screens showing complex product trends, the product manager now has a tool to assess what options to emphasize and what options to eliminate. If you want to change the parameters for the analytics, no programmer is needed. This is easily done online with user parameter entries.

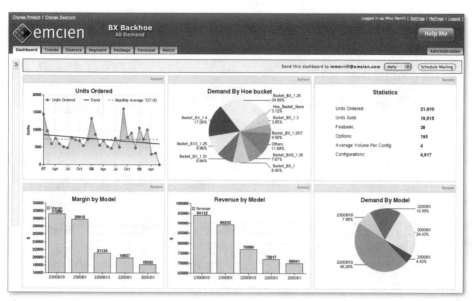

Figure 7.33 Customizable Analytics Dashboard for Configurable Products

Emcien's cluster analysis auto-detects what features and options are being bought together. Unlike BI applications where a user has to query by naming the features, the auto-detection answers the key question of what features are popular. The analytics also reports the margin for the units that have these features, and the sales velocity using the sit-time metric – see Figure 7.34.

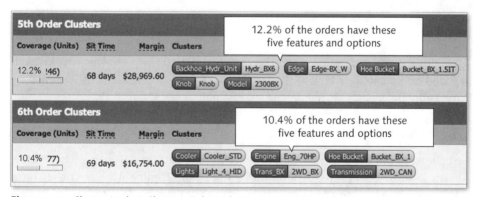

Figure 7.34 Cluster Analysis Showing Sales Velocity Using Sit-Time Metric

Emcien's solution streamlines and automates the long, tedious, manual process of evaluating what options are offered and what options are the best-sellers. The next logical step is to match the LO-VC options to match the best-sellers by region. This does not mean deleting a characteristic or value, but deemphasizing them or hiding them could be a quick change in LO-VC.

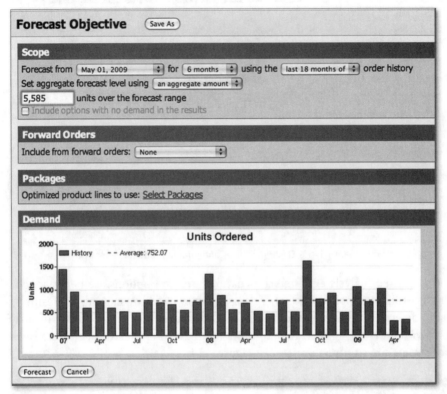

Figure 7.35 Forecast Objective

Emcien also provides trending and forecasts by characteristic and characteristic values and offers a unique look at the forecasting process to supplement SAP APO by forecasting by geographic region and by customer – see Figure 7.35.

7.7.3 Summary

Structured governance of Variant Configuration product model changes is virtually absent in most LO-VC installations, and without the eSpline Avenue Orchestrator application is limited to a time-consuming and seemingly random process of finding all changes and verifying those changes in a review or audit process. In contrast, the Model Compare report is simple and easy to read and to react to. Adding a Model View and a Model Compare report to the suite adds significant value and shortens the change process significantly.

Automated testing for the Variant Configurator including building reusable test cases and ensuring overall integrity of the product model when changes are made is critical to a change process, and supporting the entire testing process can significantly reduce the time to market for changes.

Avenue Orchestrator governs the interaction of third-party analytics and testing and existing SAP modules such as PDR, LO-VC, and IPC.

eSpline's Avenue modules, the analytics solution from Emcien, the LO-VC Testing solution from Fysbee, and the Configit mobile solution for LO-VC models all represent applications that complement LO-VC and IPC. Emcien solves the analytics reporting problem with a solution that translates SAP sales orders and quotes into useful and actionable information. Fysbee ConfigScan fills an automated testing solution gap not offered by SAP. Configit offers a mobile configurator available in scenarios in which SAP Mobile Sales is not applicable.

All of these partner applications are integrated into eSpline Avenue Orchestrator because they support critical process steps in the LO-VC model lifecycle process, or the transactional order fulfillment process.

This section described the first general release of the eSpline Avenue Orchestrator solution to SAP customers. Planned subsequent releases include tighter integration to and from the Fysbee ConfigScan LO-VC Testing solution, improved approval workflow in support of the move-to-production checkpoint, and improved archival capabilities. A tighter integration of Emcien analytics results with SAP APO and Business Warehousing is also planned, as is the ability to influence the LO-VC model lifecycle process by the results of the analytics, for example, removal of rarely sold options, and resequencing of prominent characteristic values to highlight most sold options.

As more requirements surface, eSpline will be adding to its stable of configurable product-related offerings with the goal of continuing to add value to LO-VC and IPC and provide software tools that bring improved productivity to SAP customers.

This section on managing variant configuration has been provided by Don Cochran, Daniel Naus, and David Silverman from eSpline LLC (*www.eSpline.com*). **[+]**

7.8 Summary

In this partner chapter, we introduced you to several partner solutions or add-ons that represent sensible add-ons to the standard SAP system in terms of the Variant Configuration process.

The *Sybit Model Tester* is an instrument used to automate product model testing in SAP ERP by making test cases savable and reproducible. The *Sybit Configuration Visualizer* visualizes the product configured by the user. The 2D visual image is displayed on an additional tab in the IPC and reflects the configuration status at all times. The Configuration Visualizer has a Visualization Modeling Environment (VME) in which you can maintain partial images.

With *CoreVC* and *SmartVC*, AICOMP Consulting introduces two configuration modules that enable you to easily configure processes in SAP ERP. The CoreVC and SmartVC components support a range of configuration prerequisites such as design, material, and machinery process planning. In other words, all of the components are equipped with the same assembly principle.

In the Section 7.4, we showed you how a partner solution makes it possible to fully integrate CAD Automation into SAP Variant Configuration. This solution is a product called *it.cadpilot*, and it was developed by itelligence AG and ACATEC Software GmbH.

By enhancing the ERP-based Variant Configuration (LO-VC), top flow GmbH facilitates optimization potentials in three areas: First, it is possible to optimize the configuration dialog box beyond the standard system. Second, top flow also provides enhanced options for creating and using dependency logic. Third, through the use of an add-on, the top flow Variant Engine can use make-to-stock production despite the use of Variant Configuration.

The ConfigScan Validation Suite from Fysbee SAS addresses common quality and performance-testing issues in the context of product models for variant configuration.

The Avenue Orchestrator from eSpline LCC allows you to control the complete lifecycle of product models — within the SAP business solutions and across third-party configuration tools. It integrates with the product model test environment ConfigScan Validation Suite from Fysbee SAS. For business warehousing functions, an integration with EmcienMix and EmcienMatch from Emcien, Inc. is available.

In this chapter, we let project leads have their say by reporting their experiences in implementing SAP Variant Configuration, and then provide complementary recommendations from the point of view of SAP Consulting.

8   Project Lead Reports on Projects and Project Structures

This chapter will interest all readers who either have a project in mind or can already look back on SAP implementations. You'll learn a lot of new information from this chapter, and it will enable you to compare the procedures for similar projects. In the first section, we'll describe the progress of a project for an SAP customer from the perspective of a project lead. In the second section, we'll discuss the composition of the project team recommended by experienced consultants, and in the third section, we'll introduce a recommended procedure for implementing SAP Variant Configuration.

8.1   "We're Implementing SAP!" — A Project Lead's Experience Report

This report was written by an employee of a large office furniture company that immediately set about implementing Variant Configuration after opting for SAP ERP. The viewpoint here is that of a project lead who is faced with the task of implementing this technology.

> We'd need a separate book if we were to mention all of the points that could actually   [+]
> be significant. Our objective here is more to highlight certain areas and indicate critical
> turning points and success factors.

As you're sitting at your desk, poring over the day's email, a colleague rushes up and breathlessly tells you, "Guess what? We just bought SAP!"

The question that immediately comes to your mind is naturally, "What does this mean to me?" It doesn't mean much of anything until you receive that fateful call from your manager. He tells you something like, "We're going to put in the Variant

Configuration module of SAP, and you're going to run the project!" Sounds great, right? But, what if the only exposure you've had with SAP is seeing those interesting logos on sponsor advertising at Formula One races or your local football stadium? SAP Variant Configuration module? What does that mean?

The reality is that many projects start out in exactly this way. If you're lucky, you're involved in one of the good projects, where you were part of the selection team that chose the software. In this case, you already know what the Variant Configuration module from SAP (also known as VC or LO-VC) is. If not, you'll know it by the time you get done reading this book. Either way, you have some interesting times ahead. In this section, we'll try to show you some of the typical challenges that you'll encounter, and give you some practical advice on overcoming them.

8.1.1 The Marketing Pitch and What Will Follow — Clarify the Prerequisites for Your Work

At some point, very early on, one or many of your executives made a classic startling pronouncement. It probably went something like this: "Hmmm, we keep hearing about all of these really cool things being done with the Internet. I wonder what we could do with it." As that pronouncement went down the management ladder, someone (usually in IT, but not always) made a comment like "SAP can solve that for us!" At that point, all kinds of promises may have been made, both by internal people and by external contacts.

To give SAP credit, they don't usually say things that broad and sweeping. In fact, one of the things we like about SAP is that they usually warn you with statements like, "Be prepared to change your business processes to take full advantage of the software."

At any rate, be prepared for your management team to have some really high expectations of the business results they will get for the money they invest in the software. That's OK; they should expect a good return on investment. After all, that's the whole objective of spending money on technology. Your job is to deliver.

Can you deliver something you don't understand? Typically not. So, your first job is to ensure that you have the expectations of the management team aligned. This is one of the harder jobs you'll encounter as you move forward. Depending on your role in the organization, you can:

- Influence during the selection process
- Influence after the quotation process but before the actual purchase
- Influence after the purchase

In each case, you'll do some of the same things along the way. Remember the objective.

> Align the expectations with the possibilities. Be clear about what's expected from the project and from you yourself.

[+]

First, you need to understand what management is trying to achieve. To do this, you have some interviews to conduct. It's best to start as high as your influence currently reaches within your organization. You're looking for the strategies that the manager is being held to by *his* manager. After all, everyone gets a performance review of some kind, to make sure that your manager's expectations are being met. Also, the fewer stages original ideas and requirements have to go through, the greater the likelihood that you'll implement them correctly.

> **Objective**
>
> Get your manager's strategies, objectives, tactics, and measurable deliverables in writing.

[+]

You're probably thinking it'll be enough to boil the top three to five key points you think are most important down to a one-page summary, but this is rarely enough. Make sure you get the metrics that management uses. Without those metrics, it will be hard to quantify decisions you make later on.

Take this approach to as many different business managers as you can who are impacted by Variant Configuration. This will include just about every functional area in your company that has anything to do with sales, logistics, transportation, or finance. The only area you can probably ignore is human resources. Write down any and all suggestions that you hear. Once you have the one-page summary from each manager, move on to your next task.

> **Finding Common Themes**
>
> Find common themes among the various functional areas.

[+]

These are your leverage points — the things that are important to many functional areas and that you want to focus on. These commonalities are the items that your business finds critical. At the beginning of your work, you might get the impression that multiple items do not align. What's more, sometimes it's not just an impression, but a fact. But don't let that discourage you. Rather, see it as a challenge. This is critical information that you want to keep over the life of the project. You will come back to it later.

8.1.2 Analyze Your Business Processes and Improve Them

A question that always comes up is, "Should I use my existing business processes or something future based?" A lot of debate surrounds this in the consulting community. You need to ask one very basic question.

[+] **What Do You Expect from the SAP System?**

Do you want to achieve the exact same results with SAP that you achieved with whatever software it's replacing?

If the answer to that question is "Yes," then simply draw out your existing business processes and put them in. On the other hand, if you think that investing however much money you spent on an SAP system should result in something better back as a return, then start by understanding your current processes, figuring out how to make them better, and *then* putting in a solution that fits these better processes.

This means you need some kind of business modeling tool. There are many out there to choose from. Many people use traditional process mapping, whereas others use a version originally developed by Toyota called *value stream mapping* or *lean management.*

This methodology is based not solely on the flow of materials, but on the flow of information too. A lot of standard literature is out there for business modeling.

[+] **Communicate Your Strategy to Employees**

Don't forget that effort is needed from all employees for the project to progress successfully. The best ideas are useless if they're not implemented.

Whatever method you use, ensure that you involve the business people who own each process and any business people who are affected by a process. By involving the business people at each step, you'll gain greater alignment and help them understand some of the challenges your business faces.

[+] Develop new processes from your existing processes.

In each case, you typically start with the existing process and then look for places and ways to improve. Once you have the processes drawn out, go over them thoroughly, removing all of the redundancies you can. Look for places where you believe technology will create major advances in the next three to five years. If you see a technology advance, assign one of your IT members to find out everything possible about it.

Why do you need to do that? Think of the explosive Internet growth over the past 10 years. In 1999, business was still thinking about how the Internet might change how business was done. By 2003, major fundamental shifts in business interactions had taken place. Business models changed dramatically, and the needs of the business typically did not keep up. What is there in your business that may benefit by some future thinking in this area?

8.1.3 How Many Instances Would You Like to Have?

At some point, you need to think about how many different physical SAP ERP instances and logical clients you want to run your business on. This covers a couple of different areas.

Criteria for Structuring the System Landscape **[+]**

▶ System architecture and upgrade procedure to new software versions
▶ Strategy for global cooperation in the company

If you plan on using the new Business ByDesign solution SAP offers, which is operated by a hosting company, this discussion is probably not of much interest to you. (For more information, see Chapter 12, Outlook for SAP Business ByDesign.) If you use SAP ERP, you should think about the structure of your system landscape very early on.

Naturally, you'll have an instance of SAP ERP that runs production — the *live system*. What else do you need? Depending on the size of your company, and who is hosting your software, maybe not much or maybe a lot.

A release progression speaks to the number of different systems you'll have to figure things out before putting them into your live system. Many companies follow a three-tier progression that can also have a game system or sandbox system added to it (see Figure 8.1).

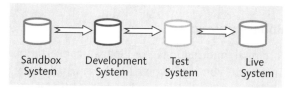

Figure 8.1 System Landscape

At the beginning of any new development, the work is done in the sandbox. This is the "dirtiest" of all of the systems, as everyone is trying new things. The data may not be current and the programs are subject to change without notice. We know of many companies that do not utilize a sandbox, instead simply using a development system.

[+] **Development System versus Sandbox System**

The primary difference between a sandbox and development system is that there is some form of control in development. The release process is beginning to come into play, with significant objects being communicated to all affected parties. The data may still be "dirty," and the programs are being changed, but typically everyone involved knows and gets regular communications over what is changing.

Think of this system as the place where component testing occurs. There won't be much integration testing in the development system, although it may be wise to do integration testing here on larger changes.

The test system is a fairly stable system. It should be a recent copy of the live system, with any changes moved from the development system in place. Data should be almost as clean as in the live system, along with most recent transactional data. This is where all teams perform integration testing to ensure that the solution moving into the live system is correct.

Of course, you have a live system. This is your operational system. This is where you create your purchase orders and sales orders, post your financial data, and prepare your requirements planning. Your information system or business intelligence system is fed from the live system. No one should be changing programs in this system. Data changes may or may not be made in this system, depending on the regulatory requirements of your industry and the ability to move the data you develop from one system to the next.

[+] **Moving Data**

SAP has multiple standard methods (for example, ALE) for moving data between individual systems. The typical problem with all of these methods is that they do *not* move processing information properly, if at all.

8.1.4   The Regional versus Global Approach

Then there's the whole "Are we a regional, multinational, or global?" conversation. This can be a really big deal, so make sure you get involved in it. In essence, this determines how many interdependencies or interfaces you'll end up with and how redundant your data will be. If your business operates autonomously in each geographic region with no cross-region business activity, you would think this makes no difference. However, in all cases, the world is quickly becoming smaller. There is less and less distinction from a customer standpoint over how or where a product is made. If it's on the Internet, you should be able to get it. So, what might this picture look like?

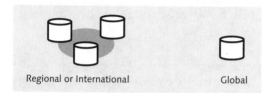

Regional or International   Global

Figure 8.2   Regional, Multinational, or Global

As you can see from Figure 8.2, in either a regional or multinational approach, each individual business entity has its own set of SAP ERP instances. Remember, each instance has its own development, test, and live system. So, if you ever need to share information of any kind about your business, you have a couple of options.

▶ Buy the SAP MDM Master Data Management solution from SAP. They'll like you a lot and be very happy that you continue to spend money with them. Something to remember here is that, as of this writing, the MDM system does not handle configurable products well.

▶ Write your own custom interface between each region's or nation's system. Typically, when you do this you are placing the data into a third-party repository. In any event, you will have a lot of interfaces to manage.

So, because there are many difficulties involved in the regional and multinational approaches, the immediate conclusion is to have one global system. This sounds easy, but can be very difficult to manage. Your business culture may not be ready for this leap. It requires a much higher level of cooperation and integration at the business process level. Be very careful in trying to set this landscape up. It can easily get you into political battles you may not want to fight.

8.1.5   Dealing with Modifications to the Standard System

In almost every case (especially if you put your legacy business processes in) you'll end up modifying the software in some way. SAP has designed a piece of software that is targeted at a wide audience. That means they can't satisfy everyone's exact functionality preferences. It's okay; the world won't end because you have to modify things. Just keep some important points in mind.

What You Should Keep in Mind with Modifications

There are some points you should keep in mind with modifications:

▶ **Selection**
Keep the number of modifications to a minimum. Only do the things that really will give you a competitive edge over the others in your industry. Don't modify something because someone tells you, "That's the way it's always been."

▶ **Release process**
Have a dedicated release process. This ensures that any changes flow through your development and test systems with a series of testing procedures. Only after passing those procedures and meeting the business expectations are they allowed to propagate into your live system.

▶ **Control process**
Govern what you modify. Have a rigorous process of checks and balances that ensure a modification cannot get into your system without review. Track the name of each modified object in a manner that works well for your business. This will also make your life easier when you're upgrading to a more recent software version. If you have a data governance organization, this fits in really well. It's all about ensuring that the design of the system stays consistent and stable.

Now, you certainly don't have to do this three-step program. In fact, many companies don't. The reality is that if you plan on upgrading to a newer release of SAP ERP at any time in the future and have made any modifications to the system, you can plan on additional time, resources, and money being added into your project plan. The key is to minimize the extra hours and additional money you have to spend on the project. In the end, you need to understand:

▶ What you have modified

▶ Where the modifications reside

▶ How to test the modifications (component, integration, and regression test)

▶ What the business impacts of those modifications are

How you accomplish this is up to you. We strongly suggest you give it a lot of thought before you make your first modification.

8.1.6 The Compromises You Can or Cannot Accept

As soon as the executives signed the check to buy SAP, they assumed that life would be grand. You need to bring reality to the situation. This means that you'll be forced to compromise. Get used to it. Implementing SAP is going to be a series of compromises. The key here is making sure you understand when to compromise and when to stand your ground.

> For example, many companies must accept that their tried-and-tested work processes may not be compatible with best practice solutions. Changing this requires a willingness to compromise from every single employee.

[Ex]

The most important thing to remember is that something sets your company apart from the competition. It may be your salespeople, your product, your process, or something else. Someone in your company believes you have a competitive edge. If your business is successful, you probably do have one. Do you know what it is?

> Your mission is to make the compromises that protect your competitive edge yet allow you to "easily" take advantage of new, better features as SAP develops them.

[+]

This is not an easy task. It means you have to be aware of both your business strategies and the SAP strategies. How do you do that? Build your sphere of influence. It's very important that you develop and use your social networking skills to increase the circle of those you influence, because this circle of people will introduce you to others, who will in turn (assuming you do a good job of influencing them) introduce you to still more people. You'll almost certainly find people who are in tune with the real strategies being used in your company.

Company Philosophy

[+]

> Develop negotiating skills and dialogue competence and build up a network. Recognize that your decisions are not only of a professional but also of a political nature. He who works politically must also learn to act politically.

As much as you like to think politics are not part of the game, they are. You cannot afford to forget this fact of business. Part of the journey involves conflicts. You'll almost certainly irritate someone along the way. If that person has better political connections than you do, then your mission is over. So, add negotiation and conflict management to your skill set. Find people and places that can help you improve your social skills. They are a critical piece of making a good compromise. After all, you can't compromise unless you understand both sides of an argument.

This takes care of the internal side, but what about the SAP side? How do you know where SAP is going? You need to understand that before you can make recommendations about compromise, right?

[+]

> **SAP Philosophy**
>
> There are multiple organizations that SAP is promoting and active in. Find them. Many are listed right on the SAP website at *www.sap.com*. Join a user group and start talking to people about what you want to do with the software and why.

In many user groups, the vast majority of the population is passive. They don't contribute. Therefore, they feel that the group is not contributing anything to their success. Wrong. The real problem is that the individuals aren't contributing to their success. The user group cannot help you unless you begin to provide content. You can't get a question answered if you never ask it. For questions relating to Variant Configuration, the one-stop shop user group is the SAP Configuration Workgroup. You'll find more information about this in Chapter 11, Configuration Workgroup, and on the Internet at *www.configuration-workgroup.com*.

The last step is to ensure that any compromise you make meets or exceeds the business expectations. If you don't pay attention to what the business *really* wants (which is sometimes different from what it *says* it wants), you have no chance.

To do this, start with your objective and process maps. Drag the business back into the conversation surrounding the process. Wave the strategies at them. Use your new-found or newly developed communication skills. Remember, in the long run, these skills are as, if not more, important than your technical skills.

[+]   Keep your mind on the process and the end results that you are trying to achieve. Sometimes you have to take smaller steps to get where you want to be.

8.1.7   Finding the Appropriate External Support

By asking yourself one question, you can answer the question about deciding whether to hire an external SAP consulting firm: Do you know everything about SAP software? Probably not, so you'll hire an SAP consultant at some point. In this, you'll be spoilt for choice. Every firm out there will tell you they really understand Variant Configuration, but the reality is there are different levels of understanding. Can you afford to get someone who is learning at your expense?

It's your job to sort the wheat from the chaff. The objective is to find a consulting firm that you can reach a true partnership with. This firm will treat your business as if it is their own. They'll want you to succeed. They'll tell you things you don't want to hear. They'll become part of your team, not just for the short duration of the project, but for years to come.

There are a couple of tactics you can use to ensure that any potential consultants become your partners. There are the standard interview questions as well as some targeted toward Variant Configuration skills.

Questions for Potential Consulting Partners	**[+]**

Naturally, the following questions are not a cure-all for finding the right consultant or consulting firm, but they'll help you make your choice.

- How did you achieve your skill set in Variant Configuration?
- How long have you been implementing Variant Configuration projects?
- How many Variant Configuration projects have you taken end to end?
- Who are your successful Variant Configuration customers, and can I talk to them?
- What professional organizations do you belong to?
  If a company is not a member of the Configuration Workgroup, this is a potential red flag but should not, by itself, preclude you from hiring them.

Although having attended training may certainly be helpful, it's not enough to be able to demonstrate convincing skills. You often only learn the basics in training. It's only in projects that you really learn your stuff. Customer references are essential. Speak with the project leads there, not just with the CIOs. The departments in reference companies also have different value assignment criteria and sometimes have very different answers to the question about whether a project was successful. But a project progressing well doesn't have to mean much; if the project model hasn't been set up imaginatively and consistently, problems will occur in later years. The construction of a house is a good analogy here: A lovely building with an interesting floor plan can also have planned water damage. The Configuration Workgroup is the perfect place to discuss projects (worldwide too) and ask customers questions about their implementations. Your advised partner should be represented there.

Most importantly, have a set of business issues that requires a Variant Configuration solution. Ask the consultant for corresponding solutions, but don't tell them how many solutions you expect. The best consultant will know that you cannot provide them with enough detail in an interview to really solve the problem. However, they'll know enough to give you at least three potential solutions or concepts they would use to resolve it.	**[+]**

Think long and hard before you make your decision. In many cases, the solutions these partners provide will be with you for years to come, so make sure you get a good consulting firm.

8.1.8 Communicate Changes Effectively

You made it through all the easy stuff, so now comes the hard part: establishing and making the changes that will occur in your business public. This is the most over-looked part of any SAP project that we've seen over the years. We continue to forget that people don't *like change*.

You'll encounter resistance to change at all levels in the business — from the manu-facturing worker all the way up to your board of directors.

Common Reasons for Resistance to Change
The two common reasons people resist is that they either: ▸ Don't understand the benefits they can receive, or   ▸ Don't believe the benefits they can receive.

It's time to get that salesperson buried deep within you back out again. To dispel these reservations, you must talk to the employees in your business. Find out what people are expecting, why they are expecting it, and what they believe the benefits are. In addition, find out what they believe the downside is as well. This is all critical data that they'll need to overcome resistance. You must ensure that you understand a person's feelings and emotions about an issue before you can try to deal with it. Take the time to talk to people and make them feel that their issues are being considered in a serious manner, not just to check a box that they were talked to. Make sure to ask clarifying questions and write down the answers to ensure that you understand someone's position before you try to sell them yours.

In the end, you need to focus on what's in it for them. People respond when they know they'll get something back. You'll most likely be asking them to give some-thing up.

8.1.9 Communicate Necessary Compromise Effectively

When making your changes, you'll discover there will be winners and losers in every part of the business. Some managers will see themselves as winners because they do something like reduce headcount, improve processes, save money, and so on. On the other hand, there are managers who need to add staff or have the cycle time of their processes increase. They'll see themselves as losers.

A key point to remember here is that the overall winner needs to be the business as a whole, not one or two functional areas within that business. That means it's your job (again) to convince someone to "take one on the chin" and endure a temporary setback.

To do this, start with yourself: How do you respond when someone wants you to do something you don't want to? Imagine this feeling and convey it. This will make it easier when you're dealing with people who feel like they're getting the worst part of the deal.

> **You Need To Do Two Primary Things**
>
> ▶ Listen to what others have to say.
>
> ▶ Acknowledge their feelings about the issues they bring up.

If you can do these two seemingly simple things, you'll go a long way toward defusing the situation. After you get true understanding of the issue, and only then, (this also means the different perspectives of the issue) you can go about finding ways to overcome it.

Look back to the value stream maps and the business engagement sessions you had early on. In many cases, you'll find something there that allows you to focus on an improvement in an upstream or a downstream part of the business that, in turn, has longer-term benefits to the area feeling like a loser.

8.1.10 Train Your Employees

Part of this exercise will be to train the employees affected by changes. This will naturally include training the project team members. Remember, there are many ways to learn and many ways to receive training. Not everyone learns in the same way or at the same pace. It will always take longer than you think. Ensure that you sell this starting way back with that first conversation with your boss when he told you, "We are going to put in the Variant Configuration module of SAP, and you are going to run the project."

The Variant Configuration training will be one of the longest, if not the longest, lead time items you have. It also requires some specific things out of your project team members. Ensure that your project team members:

▶ **Will be with your company for at least five years**
It will take a year or so (at minimum) before these people are really proficient. You want to ensure that you get at least three or four years of productive work out of them.

▶ **Think abstractly**
SAP Variant Configuration is an exercise in abstract thought. The concepts of object-oriented programming come into play on a regular basis. The people you have in this role need to be able to grasp those types of concepts quickly.

> ► **Are proactive in seeking out learning**
> These people will have to supplement their knowledge every day. There is no place you can go that will teach a person how to deal with every possible configuration scenario. The people you have in this role need to be able to seek out and find their own solutions.

When it comes to Variant Configuration work, we can't stress enough how differently each person will react. We do believe that anyone who *wants* to do this work can. Whatever you do, don't just send a person off to SAP training and expect him to come back a fully trained, functioning professional capable of performing miracles. SAP training can be good and it can be bad. It all depends on the trainer *and* the trainee's attitude.

We cannot stress enough just how much attitude plays in this game. Some of the concepts are abstract, and a person who believes he can't grasp abstract concepts, won't. The individuals need to have the attitude and motivation to *want* to learn abstract concepts. Some will be better than others. That happens in every type of learning. You need to be ready for it and be able to adjust.

You'll also meet people who you're confident do very good work, but who have decided for whatever reason not to progress in their careers. This is more an attitude problem than anything else.

[+] **Choose the Right Employees**

It's crucial to develop objective measures that allow you to set criteria and measure employees against them. When you've found these criteria, use them.

Be prepared to let employees go or move them to different parts of the company. The Variant Configuration area is a critical resource that, done right, doesn't have to be a big team. Just a few of the right people can make a huge difference. Your job is to find the right ones. One method is to ask some different questions of people.

[Ex] **Questions for Your Employees**

> ► Do you prefer to play chess or checkers?
> ► Do you prefer to play role-playing games or first-person shooters?
> ► Do you want a map or will you just find your way?

This type of questioning is simply a method of determining what a person enjoys — abstractions or absolutes. Go for the abstractions every time.

8.1.11 Problems After Going Live

You made it. You did all the right things, navigated all the rough roads, and the system is live. Your boss is happy, and you're thinking you get to go back to your old job and read more emails ... until a problem occurs.

There's a problem with the SAP system. Once you go live, every issue suddenly becomes critical and time-sensitive. The reality is that not every issue will shut you down. Look for the things that disrupt your customers first and get them solved. After all, they're the people paying the bills.

One of the main things we see people doing wrong is trying to solve problems without a process to do so. You need to define a sequence of events that shows people the places to look and items to look for. Not every problem is the same. Most of the time, the data items involved are, because there is a finite number of data items, involved in Variant Configuration. Although they are used in multiple ways, it typically comes down to the same set. If it's not in that set, then you have probably customized something somewhere.

It would be great to be able to tell you, "Look here, then here, then here," but we have no idea how you implemented your model, and the range of possibilities is just too large to make a "one size fits all" problem-solving process work. It's up to you. But figure it out for your model, and then document it and teach it. It'd be best to do this when implementing your product model, not once the damage has been done.

The thing to remember is that people learn at different rates. You need to be patient and speak in ways that get your message across. Find analogies to everyday problems in people's lives and try to explain the SAP system problem in those terms.

Communicating Problem-Solving Processes [Ex]

For example, let's say you have a problem with a set of characteristics not appearing in the proper place at the proper time. The person you're working with loves cars and works on them all the time. In this case, you would ask, "What thinking do you do to diagnose the car not starting?" The person will go through a litany of things like check the gas, then the spark, and so on. Your job is to then explain that checking the gas is like making sure all the characteristics are in the class. Checking the spark is like making sure the characteristics have object dependencies assigned to them.

The point of this little exercise is to explain that it's just another small problem. They solve problems every day using a process. Showing them the relationship between their existing process and the different data items used in the SAP problem-solving process will help them make the connections between objects much faster.

Sometimes this little trick needs to be applied in the other direction as well. You may have an operations manager who is irate because his demands are not exploding in the way he thinks they should.

8.1.12 Changing Mass Data

Sometimes the issues that come up are problems with the master data. No matter how good you or your people are, you're only human and will make mistakes. Sometimes, those mistakes cover a lot of territory in SAP. For example, you have a part in a lot of different BOMs, and the quantity in each of those BOMs is wrong. Maybe it is 1, but it should have been 2. You need to fix them in a hurry. It will take too long to key them in one at a time. There are multiple ways to do this in the standard system to speed things up. In this example, you can use either Transaction CS20 (Change BOM Data) or Transaction CEWB (Engineering Workbench).

Many mass maintenance methods are available to you. It's a veritable alphabet soup of transaction codes that would require a separate book to describe. Here's a list of some important transactions you can use here:

- ▶ CA75 (Replace Production Resources and Tools)
- ▶ CA85 (Replace Work Centers)
- ▶ CA95 (Change Reference Operation Sets)
- ▶ CS20 (Change BOM Data)
- ▶ C223 (Change Production Versions)
- ▶ CEWB (Engineering Workbench)
- ▶ CLMM (Change Classification)
- ▶ MM17 (Change Material Master Data)
- ▶ MM50 (Extend Materials)

Then there are other tools such as the Legacy System Migration Workbench (LSMW), SAP GUI Scripting, and plenty of third-party add-ons.

Just remember: The speed with which you can change things is the same speed you can mess them up. You need to understand a few things about the system before you start using these tools. You need to understand:

- ▶ How each object is used — conceptually, logically and physically in a single thread.
- ▶ If you are using Engineering Change Management (ECM), how that factors into the object state. Engineering Change Management is described in Chapter 10, Challenges in Variant Configuration.

▸ How each object you are changing impacts related documents (such as sales orders) in the system.

▸ What will happen to those related documents when you make your change.

▸ How you will manage those results.

▸ Who you should communicate the changes to.

Think long and hard before you click that Execute button. We strongly advocate taking each change into a development or test system and trying it there before you actually do it in the live system. Have a back-up plan just in case things go wrong. Know what steps you'll need to take and how you'll execute them.

8.1.13 Changing Business Models

In this day and age, business models change rapidly. The business that cannot adapt to change will die. That means you need to be thinking down the road and deciding how you can adapt your product model to the changing business needs.

If you're lucky, or your bosses all think about the future, you've been thinking about how to build flexibility into your model from day one. There's another factor to consider, though. What about SAP? Doesn't their business model change too? And when it does, what will you do? You change the software.

You need to stay on top of those changes. SAP is pretty good about telling you what to expect in the coming years. Yes, sometimes you really need to go out on a limb and try to read between the lines. The hardest part is convincing your business that you've read the tea leaves correctly.

The burning question is, how do you even begin to find the tea leaves to read, much less interpret them?

How To Find Out Information on SAP's Direction	**[+]**

Get informed and go where the knowledge you need flows.

▸ Join the Configuration Workgroup.

▸ Ask your account rep.

▸ Better still, find someone who deals with the topic on a multidisciplinary basis at SAP and looks for alliances between sales, product management, and development.

▸ Attend one of the many conferences that SAP puts on — Sapphire, Tech Ed, and so on.

▸ Join a user group such as ASUG or DSAG.

Once you see and hear all of the information, you have to consider what it means for your business. Once you get that done, run it by someone you trust to make sure your thinking is right. Then begin holding that thinking up against the strategies you have from the business. Remember, though, those strategies may be out of date. Make sure you are thinking three to five years down the road, not using just today's processes. Then you get to sell your ideas.

In many cases, this means you have to find the primary influencer within your company, and then influence him over a very long time. This can be very hard, but the payoffs can be huge. The reality of the payoffs here is that you get the best things for your company plus you learn a lot about yourself. Those of us who choose to continue growing will continue to prosper, regardless of the circumstances.

This looking for and collecting information and promoting your ideas actually brings us to the upshot of this section again because we told you at the beginning that you'll go through these processes many times. The key thing here is that you emerge from your journey stronger.

8.2   Roles in a Variant Configuration Team

In this section, we'll discuss the different consultancy team roles that ideally ought to be filled when implementing SAP Variant Configuration. Consultants are frequently asked what differentiates a Variant Configuration project from any other SAP software implementation project. The answer is the use of master data.

Creating, maintaining, and transporting complex product models puts every Variant Configuration project to the test. This raises the question of putting together the right project team. It's not enough to have a specialist for the Sales and Distribution (SD) module in the project team. Nor can a specialist for production planning manage a variant project alone. The solution's complexity and integrity are reflected in exactly the same way in the project team.

We want to point out these differences on the next few pages to help you identify pitfalls early on and counteract them in good time. We'll also discuss IPC-specific scenarios and special features.

8.2.1   Expertise and Experts

The question that's raised at the beginning of every project is how best to staff it. Important roles include:

▸ Solution architect
▸ Variant Configuration expert (VC expert)
▸ Modeler
▸ IPC expert

- Pricing expert
- Master data expert
- Project lead

Solely owing to Variant Configuration, very few companies run SAP ERP. In other words, such a project is always based on a broader application context. Like every rule, there are exceptions: (1) the implementation of the *SAP Vehicle Management System* (VMS) and (2) the implementation of the *SAP Apparel and Footwear Solution* (AFS). Variant Configuration is the focus of both these solutions. So, exceptions aside, a Variant Configuration project is about changing an existing process (with nonconfigurable standard products) to a new process with configurable products.

Solution Architect

Because Variant Configuration is about changing an existing process, the solution architect needs to invest particular attention in analyzing the existing solution to identify possible changes to this process in good time and schedule these changes as the project proceeds. As the name suggests, a requirement of the role is to identify and design the architecture of the business process solution.

Solution Architect Requirements	[+]
- Analytical capabilities to record actual processes - Conceptual capabilities to describe target processes - Many years of project experience with Variant Configuration	

VC Expert

As we've mentioned, material master data is especially important. This is the area where the VC expert ought to invest special attention. By that we mean a VC expert always needs experience in using material master data. He should be able to understand existing products and product structures and identify possibilities for efficiently reorganizing the product structure. Because of this, it's very important that he have a good understanding of the *SAP Product Lifecycle Management* (SAP PLM) solution. He also has to be able to recognize logical relationships and describe them.

VC Expert Requirements	[+]
- Knowledge of SAP PLM solution - Capable of understanding configuration models - Marked ability to solve logical problems	

Modeler

The modeler is the most important specialization of the VC expert. He knows how to translate logical relationships into a language the system understands. He needs to be able to read requirements on a configuration model from business process requirements and product requirements and turn them into knowledge about the product. The modeler also needs to be able to work in a team, an essential requirement because of the multidisciplinary association. He should be able to understand and implement technological requirements and those from sales and marketing. The modeler is very much like a programmer in that he needs to translate knowledge into a configuration language. Even if it's "only" a formal language we're talking about here, a very abstract imagination is needed for this. He also needs to be able to program simple program modules in ABAP. This is absolutely essential for creating variant functions or user exits, in other words, optional project-specific program enhancements already provided in the standard SAP solution.

[+] **Modeler Requirements**

A modeler is a VC expert who has to make the grade in terms of:

▶ Great abstract imagination
▶ Experience in creating configuration models
▶ Good ABAP knowledge
▶ Marked ability to work as part of a team

IPC Expert

The IPC expert is a special kind of VC expert. The IPC expert can show the differences between the LO-VC variant configurator and IPC and notice them when using the IPC. He can analyze these differences and modify the models if need be so that they'll be able to run within LO-VC and in the IPC. He's also very familiar with the technical architecture and the options for using the IPC. He should know all about the different scenarios used in the IPC and thoroughly understand the data transfer flows needed for this. Because user exits are implemented in the IPC differently than SAP ERP, it's essential he knows Java.

[+] **IPC Expert Requirements**

An IPC expert is a VC expert, who also:

▶ Knows the different options that can be used in the IPC
▶ Can understand and modify a configuration model
▶ Knows the differences between LO-VC and the IPC
▶ Is familiar with the IPC architecture

> ► Knows about data retention and data transfer flows for the IPC
>
> ► Knows ABAP and Java well

Pricing Expert

The pricing expert should be able to analyze the existing pricing and adjust it based on the VC expert's or IPC expert's specifications. He needs to be proficient in SAP pricing to do this. His profile also needs to include knowledge about IPC architecture and technical pricing requirements based on that. He needs to know ABAP and Java so he can create missing conditions and formulas and variant pricing in SAP ERP and for pricing with the IPC in SAP CRM, for example.

Pricing Expert Requirements
► Is very knowledgeable about pricing in SAP ERP and the IPC
► Is familiar with the IPC architecture
► Knows ABAP and is also very knowledgeable about Java
► Knows about the area of user exits in pricing

Master Data Expert

Large amounts of data need to be created and tested, particularly when many configurable products are introduced in a relatively short amount of time. That's why with these types of projects you need to start training experts early on to maintain configuration master data. It's their job to create and test the data that's needed based on the VC expert's, modeler's, and possibly the IPC expert's specifications. They generally use a sample model as the basis for creating other models.

Sample Model	[+]
A sample model is a product model that's been created with the knowledge of experienced configuration experts and contains all modeling aspects for ways of looking at specific project problems.	
Several thousand sample models may be created in a project because each product family requires different things from modeling that may have to be dealt with using different modeling approaches.	

Master Data Expert Requirements	[+]
► Has strong logical thinking skills	
► Can grasp concepts quickly	
► Is familiar with the SAP PLM solution	

Project Lead

In Section 8.1, "We're Implementing SAP!" — A Project Lead's Experience Report, we already detailed the important jobs a project lead has to do. It's the project lead's job to identify and quantify the necessary roles and priority areas at the beginning of the project with the assistance of the solution architect. He also has to ensure that any knowledge that's missing when he's filling the roles in the project team can be provided through training. It also helps a lot if the project lead himself has experience with SAP PLM and Variant Configuration. In many past projects, this has resulted in better results.

Project Lead Requirements

Besides the general requirements, a project lead also needs:

▸ Experience in SAP PLM and Variant Configuration

▸ Project experience in Variant Configuration

8.2.2 Putting Together and Structuring the Project Team

Knowing the roles we described in the previous sections for a Variant Configuration project doesn't always make it easy to put together a specific project team. Roles that are perfectly tailored in theory usually can't be filled in exactly that way in practice. You'd also barely be able to limit the specific roles in the everyday running of the project.

We'd need another book to describe the method for building up relevant role knowledge exactly, so we'll only give you the basics here.

The solution architect and project lead need project experience in Variant Configuration, something they can only increase by participating in implementation projects. The VC expert can build up his knowledge by attending different SAP PLM training courses in the area of logistics. VC experts can end up becoming modelers or IPC experts as a result of their experience using LO-VC or the IPC. In the process, they need to take advantage of the training courses on offer. Typically, the pricing expert has already obtained experience with the SD module in SAP ERP, but it's essential that he gets relevant training in this area. Based on that, he'll then gain experience with the IPC too.

When all's said and done, a balanced mix of the described roles in a team is a crucial success factor for a Variant Configuration project.

8.3 ASAP for Variant Configuration Projects

The *ASAP Implementation Roadmap* provides the methodology for implementing or enhancing SAP software. ASAP is an abbreviation for AcceleratedSAP or simply for "as soon as possible." The ASAP Implementation Roadmap is a project procedure developed by SAP that's been tried and tested over many years. The methodology, also known as ASAP Roadmap, is divided into five phases:

1. Project preparation

2. Business blueprint

3. Realization

4. Final preparation

5. Go-live and support

It's an obvious choice to use this methodology for implementing SAP Variant Configuration too. Figure 8.3 illustrates the ASAP method for a Variant Configuration project.

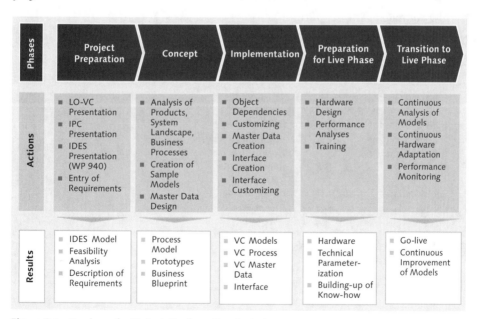

Figure 8.3 Roadmap for Variant Configuration Projects

[+] The following pages don't represent the complete ASAP Implementation Roadmap. We'd need a separate chapter for that, so we're only going to deal with special features of Variant Configuration projects here.

8.3.1 Project Preparation

Usually, lots of workshops take place during the project preparation phase (see Figure 8.4) to teach about the different SAP configurations. You compare the product requirements with the configurations and corresponding scenarios here. This will help you to choose the right configuration option and perform a feasibility analysis.

You should ask yourself the following questions here:

► What scenario do I want to use?

► Should I use configure to order?

► Should I use engineering to order?

► Do I need to implement slotting?

► Do I need to implement racking?

► Do I need to think about cabling in the set of rules?

► Does a BOM explosion take place?

► How complex is the set of rules?

► Do I need user exits to process invoices?

► Do I need to consider interfaces?

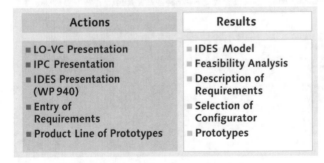

Actions	Results
▪ LO-VC Presentation ▪ IPC Presentation ▪ IDES Presentation (WP 940) ▪ Entry of Requirements ▪ Product Line of Prototypes	▪ IDES Model ▪ Feasibility Analysis ▪ Description of Requirements ▪ Selection of Configurator ▪ Prototypes

Figure 8.4 Project Preparation

Once you've picked the right option, you should choose a representative product line to create a prototype. You need to create this prototype on a separate sandbox system. If you want to avoid the effort of creating your own model in this phase, you can use an existing IDES model.

8.3.2 Business Blueprint

You're going to have to put a lot of effort into designing the master data in the business blueprint phase (see Figure 8.5). You should check every product line separately. Any misjudgment in designing the master data in this phase could lead to an enormous amount of work in a later phase. Even so, don't succumb to the temptation of keeping to every aspect of the master data design in the business blueprint, because a certain amount of flexibility is essential for modeling. You'll get a healthy mix if you define the structure of the classification clearly. At this point, it's critical that you have a working prototype for the product line to be implemented because you'll only be able to demonstrate the complex relationships of Variant Configuration and the classification using a prototype.

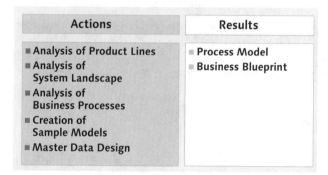

Actions	Results
■ Analysis of Product Lines ■ Analysis of System Landscape ■ Analysis of Business Processes ■ Creation of Sample Models ■ Master Data Design	■ Process Model ■ Business Blueprint

Figure 8.5 Business Blueprint

In this phase you should choose either a *golden client* or Engineering Change Management for the implementation. Decision-making support is available in the description of the golden client in Section 8.3.6.

8.3.3 Realization

In the realization phase (see Figure 8.6) you implement the processes described in the business blueprint into the system. The modeler creates a representative product model for every available product line in this phase. The person maintaining the master data copies this model and completes it with additional products.

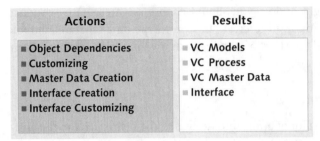

Figure 8.6 Realization

The best way to implement this method is through scheduled weekly workshops between the modeler, people responsible for the products or product managers, and product developers. For IPC projects in particular, where you want external customers to be able to access the configuration, you need to put in a lot of time to modify the interface. This modification can range from standard customizing (XCM) through to redesigning the client completely. The following examples illustrate some of the efforts required in customizing IPC interfaces. You can find more information about the IPC in Chapter 5, Special Features of Product Configuration in SAP CRM.

[Ex]

IPC Interface Customizing

▶ **XCM Customizing**
Customizing using Extended Customizing Management for the IPC can take a few hours or days.

▶ **CSS Customizing**
You can change Cascading Style Sheets in a few days.

▶ **JSP Customizing**
It can take you a few days up to a few months to customize IPC Java Server Pages, and you need corresponding knowledge of the IPC client architecture.

▶ **IPC Client Customizing**
If you need to create a new client, you'll have to factor in several months or years of development effort.

8.3.4 Final Preparation

When you've completed the developments, you move into the final preparation (see Figure 8.7). This phase deals with setting up the live system, which involves installing the hardware and setting the technical parameters. You have to prepare and implement training for basic users, experienced users, and those responsible for maintaining the system. This is the latest phase during which the setup or training can be done for a separate modeler.

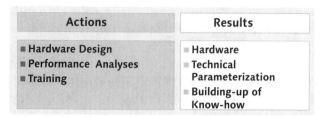

Figure 8.7 Final Preparation

| Technical Parameterization | [+] |

You ought to attach great importance to the technical parameterization for IPC, particularly because the behavior of the IPC and the whole application depends very much on sound parameterization. In this phase you should especially set the parameters of the *Virtual Machine Container* (VMC).

8.3.5 Go-Live and Support

Once you've successfully finished the previous phases, you complete the project procedure in the go-live and support (see Figure 8.8) phase. But don't make the mistake of thinking that go-live means that the project's over and done with. It's live and you've just fired the starting shot.

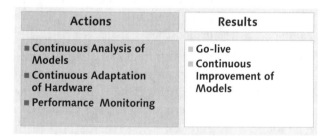

Figure 8.8 Go-Live and Support

8.3.6 Golden Client Approach

As already described in the previous section, creating and maintaining material master data assumes a particular role in a Variant Configuration project. Distributing an SAP system and the way you do this is normally clearly defined.

You generally distribute customizing and customer enhancements between a development system, test system, or consolidation system and a live system.

However, this is a particular difficulty, especially for Variant Configuration. Materials are extended by knowledge that can very easily exceed the complexity of customer enhancements. This is why you need to check materials that have been extended this way to ensure that they're correct. You can't resort to the usual transportation and testing methods here, though. (Note: Refer to the Sybit Model Tester partner solution in Chapter 7, Enhancements and Add-Ons in the SAP Partner Environment.) This means you need to look at new concepts to guarantee the accuracy of a live system despite this obstacle. The golden client concept provides one of the answers to this question.

The best way to define a *golden client* is as a client where you can save data untouched, application programs can work stably, and you can't carry out transactions.

Many customers use a golden client in their test systems.

[!] The main difference between the golden client and a live system or test system client is that you're absolutely forbidden to perform any transactions.

You're not allowed to create sales orders, manufacturing orders, or stock orders in this client. However, you're supposed to create and maintain Variant Configuration data in this kind of client and then copy it to other clients so that you can use it in transactions.

The main reason transactions are strictly forbidden in this client is the goal to guarantee that you can maintain and change Variant Configuration objects easily. Figure 8.9 shows an example of a recommended system landscape.

As soon as an SAP document has been created, whether it's from a sales order or quotation for a configurable product, you need to know that you're no longer allowed to make certain changes to Variant Configuration objects without using *Engineering Change Management* (ECM) and archiving (for more information see Chapter 10, Challenges in Variant Configuration).

[Ex] One painful example is where, without Engineering Change Management, the system refuses to delete a characteristic that's been assigned to a configurable product that's already had an order created for it. Creating, deleting, and assigning characteristics are key requirements for efficient modeling.

Instead of using a golden client for maintenance, some customers create and maintain their Variant Configuration models directly in a live environment. The main problem with this strategy is that end customers have immediate availability to these models and products without the modeler getting the chance to test them beforehand. This increases the likelihood of you being able to create and save an order that contains an inconsistent or incomplete product. From a business point of view, purely technical consistency often doesn't make sense or isn't satisfactory either. For more information, see Chapter 2, Creating a Product Model for SAP Variant Configuration.

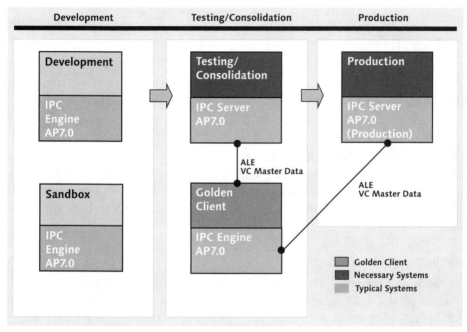

Figure 8.9 System Landscape When Using LO-VC and IPC

Models created in a golden client should be tested in a test system and then released in a live system.

8.3.7 Specific Features of IPC Scenarios

Unlike LO-VC Variant Configuration, the IPC is designed for use in many different scenarios. That's why we're going to highlight the special features of IPC scenarios and explain how they impact the project directly.

CRM Internet Sales

Using the IPC in the Internet sales scenario is a specific requirement for presenting the configurable product. You need to present the configurable product to customers quickly and clearly. The more the Internet shop is aimed at an end user, the greater this requirement increases.

A trader or technical purchaser will appreciate that a highly complex product needs a few seconds to process a complex calculation. An end user will unfortunately not have the same appreciation. As a result, there's a lot demanded for performance and availability in this type of project.

This is why you also need to schedule dedicated work packages in the project plan to optimize the configuration models and performance.

In this kind of project, modifications to the standard JSP interface in the IPC based on company specifications and new technologies usually result in large work packages.

SAP Internet Sales in ERP

There's a similar requirement for Internet Sales in ERP or e-commerce as there is for CRM Internet Sales, but with the difference that, owing to the absence of CRM as a separate system layer, there's greater demand on the back-end system for ERP Internet Sales. As a result, this scenario is more suitable for B2B than B2C. We'd only recommend B2C if you're expecting a manageable number of users.

Mobile Sales

Mobile Sales represents a particular architectural challenge in conjunction with configurable products. By using web technology to display the configuration as of IPC 5.0 (AP 7.0), you need to operate the IPC with a specific architecture in Mobile Sales. You need to have in-depth knowledge of this architecture and be familiar with data distribution. You should also have sound knowledge of Visual Basic und Tomcat for this type of project.

CRM Telesales — Customer Interaction Center (CIC)

In CRM Telesales you use the same models available in CRM. The interface can and should be customized based on what Telesales employees need. The configuration interface is only available to internal employees, which means it can be designed more easily and differently than in an Internet Sales scenario. But if you combine a CIC with Internet Sales, we recommend you use an identical or only slightly customized version of the interface. A customer typically indicates what he wants on the provider's website or webshop. This helps the Telesales employee understand what the customer wants when the quotation is subsequently processed in-house.

Besides the display of price margins, particular requirements in Telesales include up- and downselling functions, specific availability information and opportunities to sell accessories.

SAP for Automotive: Vehicle Management System

The *SAP for Automotive* industry solution provides several SAP standard transactions combined in an interface tailored for the automotive industry and includes a portal that's been customized according to a dealer's preferences. These standard transactions also include the IPC because automotive manufacturers have to provide vehicles that allow individual modifications for customers. It's absolutely essential in this context that the IPC expert has basic knowledge of VMS. This is the only way to guarantee that the IPC can be integrated properly into VMS and the dealer portal. In addition to this specialist requirement, the IPC expert must also be an expert in Unix operating systems and Oracle databases. There's no other industry where Unix and Oracle are used so widely.

Apparel and Footwear Solution

The SAP AFS solution is a customized CRM solution for the apparel and footwear industry. What's distinctive about this solution is that you use configurable products to be able to display the apparel and footwear sizes. Modeling and manual data replication take a back seat in this solution because the system generates them using *grids* (sizes or matrixes). Particular demands are placed on data replication in this solution because it basically differs from other scenarios.

Specific Integration into Non-SAP Systems

Possibly one of the most difficult scenarios when using the IPC is specific integration into non-SAP systems (also known as a stand-alone scenario) because you have to pay special attention to the integration between an SAP system and external system in this type of scenario. You should only undertake these kinds of projects in very close cooperation with SAP because they require comprehensive knowledge of the IPC API. SAP Custom Development can provide additional help here. You'll particularly find what you need here if a standard SAP solution can't help you, but you don't want to miss out on the reliability of the SAP solution.

8.4 Summary

In this chapter, we outlined the customer viewpoint of a project lead who's entrusted with implementing a Variant Configuration project. The first section was deliberately different from the rest of the sections in this chapter, because it was the only way to keep the report authentic.

In the second section, we dealt with the different consultant roles from a project lead's point of view and, in the third section, we described the ASAP implementation method that has been used successfully for years in "traditional" SAP ERP implementations.

The aim of the section about IPC scenarios was to explain that you have a lot of options to respond to industry-specific (SAP for Automotive, Apparel and Footwear Solution, and so on) and technical (online and offline use) conditions. This differs significantly from the LO-VC approach in SAP ERP. The fact that you can advance to SAP Custom Development as a development partner in the IPC area in particular is seen as a solution for individual approaches.

Hopefully we achieved our goal with this chapter of providing you with a transition from pure theory in the previous chapters to practice in the next chapter, using examples of customer implementations.

Before the implementation of such an integrated solution as Variant Configuration, a reference situation is particularly interesting for customers and partners. In this chapter, we will present examples of companies that use SAP Variant Configuration in many different ways.

9 Customer Reports on the Introduction of SAP Variant Configuration

This chapter will interest all readers who either have a project in mind or want to look back on completed implementations. You'll learn a considerable amount of new information from this chapter, and it will enable you to compare other companies by contrasting quantity structures or scopes. Our selection of companies is not from any particular industry; however, we focus on manufacturing companies because this is where the most experience exists. We also concentrate on complex and very comprehensive applications to show you large quantity structures. You can, of course, also set up and logically integrate Variant Configuration with only one material master and two characteristics, but the reading material for such a project would consequently be relatively short.

▶ **Getriebebau NORD**
In the first section, we discuss the project at Getriebebau NORD, a manufacturer of configurable gears and motors. This customer installation involves challenges such as integration into the overall logistics, an enormously high number of variants (35 million) and approximately 1,500 order items per day.

▶ **Krones AG**
Krones manufactures filling and packaging systems and represents highly complex Variant Configuration, for which several thousand materials and hundreds of configurable assemblies are processed through a six-step BOM structure.

▶ **Hauni Maschinenbau AG**
Hauni produces machinery for tobacco processing and is distinguished by a highly complex structure in conjunction with dummy assemblies and an extremely large quantity structure. The end product here therefore consists of a subset of up to 50,000 components from 3,000 configurable BOMs that cover 12 levels in some places.

▶ **Felix Schoeller Group**
We use the Felix Schoeller Group, a manufacturer of specialty papers, to document an example of smaller product models, for which we have focused strongly on storing object dependencies easily in the form of tables. An IPC application has also been implemented here in the SAP NetWeaver Portal.

▶ **Hüls Corporate Group and Hülsta**
The example of Hülsta, a manufacturer of high-quality furniture, illustrates the integration of SAP Variant Configuration into a graphical application. Examples include header configurations, processing object dependencies asynchronously, and agreeing Variant Configuration with predefined sales types.

▶ **Baldor Electric**
Baldor Electric is an American manufacturing company producing electric motors, power transmission products, drives, and generators. Since 1997 Baldor has operated a live SAP ERP implementation. For the first 10 years they ran SAP ERP in combination with legacy tools for product configuration. To streamline business processes Baldor finally started displacing the legacy configuration tools by SAP Variant Configuration in 2007.

9.1 Progress of the Project at Getriebebau NORD

Getriebebau NORD develops, manufactures, and distributes solutions for drive technology. The product selection consists of gears, geared motors, electric motors, and frequency inverters with a torque range of up to 200,000 Nm. The company headquarters are based in Bargteheide, near Hamburg, Germany.

Getriebebau NORD operates internationally with 35 plants and commercial agencies in 32 countries. By December 31, 2007, Getriebebau NORD employed a staff of 2,500. The number of employees has increased since then. Sales for the NORD Group in 2007 amounted to approximately EUR 350 million.

Getriebebau NORD currently uses SAP R/3 4.7. The system landscape consists of a classic three-tier landscape:

▶ A live system with one client
▶ A consolidation system with one client
▶ A test system with two clients

Presently, 652 users actively work with the system. In the final expansion phase, approximately 1,200 users at NORD will work with SAP software worldwide.

The company has been actively using Variant Configuration since the beginning of 2007. A two-and-a-half year long implementation project preceded the implementation, the duration of which resulted from the huge number of variants. This variety currently consists of 35 million different variants.

9.1.1 Initial Situation

Getriebebau NORD uses many components of the LO-VC configuration. Some of the areas they are used in are as follows:

▶ **Configuration in company departments**
This includes configuration in sales and in production (assembly) and the automatic processing of the production BOM with the relevant routings. *Service materials* are configured as part of the service and repairs processing. Configuration is also used within procurement processes when motors are ordered.

▶ **Configuration across all areas**
Configuration is also used across all areas at Getriebebau NORD, for example, when BOMs and routings are used in shipping processing. Configurable materials and, to a large extent, material variants are used for this purpose.

Essentially, up to 100% of the production BOMs in inquiries, quotations, or sales orders is determined by the configuration when they are entered. However, because special parts are also frequently used (in particular with motors and special motors), the BOM determined is often subsequently processed using Transaction CU51E. Parts are added to and removed from the order BOM during this processing. When this processing is completed, the BOM is fixed, which means it is separated from the original BOM so changes cannot be introduced during the processing of master data.

BOM structures are very broad and very deep owing to the enormous variance of products. A structure of up to seven levels is used within the configuration. This achieves very broad variability. The actual configuration here always occurs at the highest level. The lower levels are mainly defined by inheriting characteristics or characteristic values.

Objective

One objective of the implementation was to process all NORD product groups using the configuration. In addition, all relevant processes should, where feasible, be supported by the configuration. This applied in particular to sales, logistics, and production and assembly.

Project Scope

Different departments and areas with a total of 14 employees actively took part in the implementation project. The internal project team consisted of employees from sales (two employees), development (two employees), IT (six employees), engineering (two employees), controlling (one employee), and purchasing (one employee).

At the start of the project, only one employee in the company had any knowledge about LO-VC Variant Configuration from SAP. At the end of the project, every member on the project team had a level of knowledge that can be considered expert knowledge.

SAP Consulting provided external support during the implementation project. The scale of external consultation amounted to approximately 100 person days. The project itself was designed as a *coaching project*, because the intention was for the participating employees to obtain relevant knowledge within the framework of the implementation. This meant they would not have to depend on external consultants for implementing changes and further developments.

9.1.2 Measures

Because entering a sales order should take as little time as possible (approximately 1,500 order items are currently entered per day), it was necessary to make as much information as possible available for users quickly and compactly. The interface design option was actively used here (see Figure 9.1).

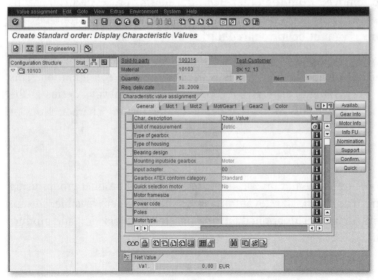

Figure 9.1 Interface Design in Sales Order

All kinds of important information was not contained in the standard system within Release SAP R/3 4.7, for example, the availability check up to component level and detailed information about motors and gears. To compensate for this shortfall, a corresponding custom development was used here to integrate a separate availability check into the system (see Figure 9.2).

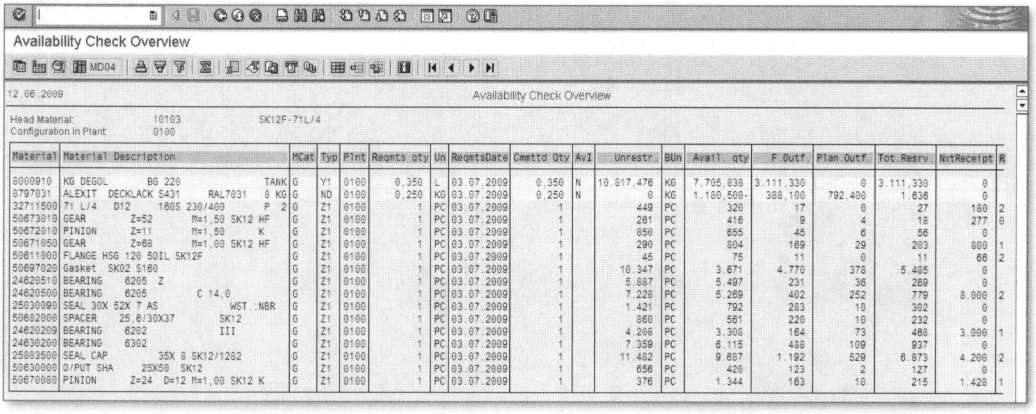

Availability Check Overview

MD04

12.06.2009 Availability Check Overview

Head Material: 10103 SK12F-71L/4
Configuration in Plant: 0100

Material	Material Description	MCat	Typ	Plnt	Reqmts qty	Un	ReqmtsDate	Cmmttd Qty	AvI	Unrestr.	BUn	Avail. qty	F.Outf.	Plan.Outf.	Tot.Resrv.	NxtReceipt	R
8000910	KG DEGOL BG 220 TANK	G	Y1	0100	0,350	L	03.07.2009	0,350	N	10.817,476	KG	7.705,838	3.111,330	0	3.111,330	0	
8797031	ALEXIT DECKLACK S431 RAL7031 8 KG	G	ND	0100	0,250	KG	03.07.2009	0,250	N	0	KG	1.180,500-	388,100	792,400	1.636	0	
32711500	71 L/4 D12 160S 230/400 P 2	G	Z1	0100	1	PC	03.07.2009	1		449	PC	320	17	0	27	180	2
50673010	GEAR Z=52 M=1,50 SK12 HF	G	Z1	0100	1	PC	03.07.2009	1		261	PC	416	9	4	18	277	0
50672010	PINION Z=11 M=1,50 K	G	Z1	0100	1	PC	03.07.2009	1		850	PC	655	45	6	56	0	
50671050	GEAR Z=68 M=1,00 SK12 HF	G	Z1	0100	1	PC	03.07.2009	1		290	PC	804	169	29	203	800	1
50611000	FLANGE HSG 120 50IL SK12F	G	Z1	0100	1	PC	03.07.2009	1		45	PC	75	11	0	11	66	2
50697020	Gasket SK02 S160	G	Z1	0100	1	PC	03.07.2009	1		10.347	PC	3.671	4.770	378	5.485	0	
24620510	BEARING 6205 Z	G	Z1	0100	1	PC	03.07.2009	1		5.887	PC	5.497	231	36	269	0	
24620500	BEARING 6205 C 14,0	G	Z1	0100	1	PC	03.07.2009	1		7.228	PC	5.269	402	252	779	8.000	2
25030090	SEAL 30X 52X 7 AS WST.:NBR	G	Z1	0100	1	PC	03.07.2009	1		1.421	PC	792	283	10	302	0	
50682000	SPACER 25,6/30X37 SK12	G	Z1	0100	1	PC	03.07.2009	1		860	PC	561	220	10	232	0	
24620209	BEARING 6202 III	G	Z1	0100	1	PC	03.07.2009	1		4.208	PC	3.308	164	73	468	3.000	1
24630200	BEARING 6302	G	Z1	0100	1	PC	03.07.2009	1		7.359	PC	6.115	488	109	937	0	
25803500	SEAL CAP 35X 8 SK12/1282	G	Z1	0100	1	PC	03.07.2009	1		11.482	PC	9.687	1.192	529	6.873	4.200	2
50630000	O/PUT SHA 25X50 SK12	G	Z1	0100	1	PC	03.07.2009	1		656	PC	420	123	2	127	0	
50670080	PINION Z=24 D=12 M=1,00 SK12 K	G	Z1	0100	1	PC	03.07.2009	1		376	PC	1.344	163	10	215	1.428	1

Figure 9.2 Availability Check at Component Level

Separate buttons (which were developed using the options for designing interfaces in the configuration editor that are contained in the standard system) were used to make the necessary detailed information for gears and motors available (see Figure 9.3). The characteristics and values that should not necessarily be ranked first in the sales view were stored here.

A custom-developed user status controls the release of sales order items. This user status also controls the transfer of requirements (see Figure 9.4).

Finally, we should also mention the option of storing the characteristic value assignment. This can only be done temporarily in the standard system (the interim result is lost after the sales order is exited) and was developed by SAP based on Getriebebau NORD's requirements. This is a particularly interesting function for sales because it means duplicate entries can be avoided for inquires or when sales orders relating to the inquiry are created (see Figure 9.5).

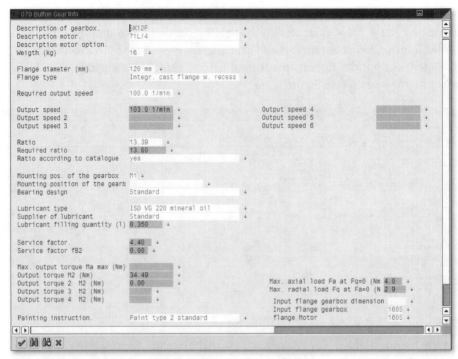

Figure 9.3 Detailed Information About Gears

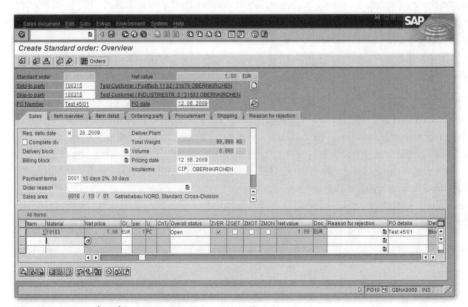

Figure 9.4 Transfer of Requirements Using a User Status

Figure 9.5 Permanent Storage of Characteristic Value Assignment

9.1.3 Results

The current quantity structure of the product model looks as follows:

Model Part	Quantity Structure
Configurable materials (KMATs)	281 header materials
Material variants (MATV)	627
Configurable BOMs	Approximately 25,000
Configurable routings	Approximately 3,400
Classes	Approximately 100
Interface profiles	Approximately 20
Characteristics (internal/to be value assigned)	Approximately 700
Object dependencies	1,300 constraints 722 preconditions 3,500 selection conditions 1,475 procedures 3,749 variant conditions (Scales ignored) 527 variant tables 20 database tables
Configuration profiles	Approximately 280

Table 9.1 Quantity Structure of Getriebebau NORD Product Models

In addition to the mentioned model parts for Variant Configuration, Getriebebau NORD also uses the following data for the product model:

Object characteristics for the supplying plant:

▶ Display and hide characteristics or values

▶ Select BOM category

- ► Default values for capacity planning
- ► Quantities (oil, color)
- ► Process texts in routing
- ► Control keys

Function modules for determining motors:

- ► Determining gear translations
- ► Material availability

Changes and enhancements, referred to as *modifications*, to the standard program were only made when absolutely necessary. The required releasability is the background for this, in other words, the desire is to keep the process of upgrading the standard SAP system as free of complications as possible. The following changes were specifically implemented:

- ► **Modified program parts**
  The following program parts were enhanced or modified:
  - ► Availability check
  - ► Plant change when creating orders
  - ► Configuration storage when entering orders or requests for quotation
  - ► Uploading and downloading configuration data
  - ► Maintenance views
  - ► Selecting characteristics within a "sales order monitor"
  - ► Order costing (KALCVARCOND)
- ► **Specifications**
  A specific team essentially implements new specifications. The specifications themselves come from areas or departments affected by the specifications:
  - ► The technical office and work scheduling/engineering issue technical constraints.
  - ► The IT department specifies system-based constraints.
  - ► Sales determines price-relevant constraints.

Object dependencies are generally maintained in cooperation with the technical office, engineering, and IT.

Problem Areas

Problems occurred repeatedly during the implementation owing to the varied range of products. The problems differed in nature, so the search for solutions differed too. This affected the following areas in particular:

▶ Quality of master data

▶ Correcting the decision about which processes should be dealt with in the configuration and what extent they would have

▶ Changing the assembly or supplying plants in the sales order

▶ Availability check (type and scope)

▶ Using the configuration internally or externally

▶ Displaying logistical processes within the configuration

9.1.4 Conclusion

If the project were to be set up again under the same initial conditions, the following points would have to be clarified and decided on beforehand:

▶ Legacy systems should be removed completely.

▶ Everything to be processed in the future using configuration would have to be specifically defined in advance.

▶ Target requirements would have to be clarified exactly.

▶ The processes that configuration is to be used for would have to be clarified.

▶ The topics to be dealt with would have to be planned specifically in advance.

The project achieved the following improvements:

▶ **Cost savings**
Implementing configuration has resulted in savings in many areas. In particular, this includes a reduction in the error rate in BOMs and associated incorrect stock removals.

▶ **Improved runtime behavior**
Using Variant Configuration in the SAP ERP system has also improved the runtime behavior considerably. Attention basically needs to be paid to what is being configured and what the scope of the products to be configured looks like. This is crucially important when choosing the necessary hardware.

Besides the just mentioned prerequisites and points that should be clarified in advance, the following factors also need to be considered in future.

Future Tasks

Various organizational rules have to be taken into account for using the configuration. Against the background of maintaining configuration data in the live system in particular, using engineering change management is a mandatory requirement. Because distributing data via ALE is only useful in limited cases, a long-term valid solution should always be found here before the project starts.

Functions are only checked or controlled sporadically to see if they are correct. However, a special program within the planning run establishes whether the required components are created in the assembly plant.

Variant Configuration is also included in the worldwide rollout project. Against this background, consideration has already been given during the implementation phase to the configuration being used internationally. This includes special units of measures in the U.S. and necessary language requirements.

Recommendations for Similar Projects

We can recommend the following for implementing the solution in a similar project.

A crucial decision to be made before an implementation is whether the configuration will only be used by internal employees or also by external and/or suppliers. This will have a major effect on the effort required for designing the interface.

Another important task is always to check the level of materials or processes to be determined automatically by object dependencies because otherwise the number of characteristics to be value assigned will increase the amount of time needed for entering orders.

You should never map logistical processes within the configuration logic because this can cause difficulties in different places in downstream processes. If cross-company sales, plant changes, fixed BOMs within cross-company stock transport orders, and quantity contracts are used in the company in particular, this can lead to unexpected effects that can be corrected in a live operation only with difficulty. This is why a detailed analysis of business processes supported by the use of Variant Configuration is essential.

9.2 Configurable Materials at Krones AG

Krones AG, located in Neutraubling, Germany, plans, develops, manufactures, and installs machines and complete systems for filling and packaging technology. With approximately 10,000 employees, the total cash flow into the company for the group in 2007 was a solid EUR 2,156 billion.

9.2.1 Project

Relevant expertise is needed to use SAP Variation Configuration, and this first had to be developed in many departments.

External consultants were involved in the implementation phase, and the relevant employees attended standard SAP training courses. Internal IT employees very soon took over tasks, and internal training was developed in the company.

The configuration structure of a complicated machine with thousands of components requires you to know the machine as well as the configuration options. For this reason, the configuration was developed by design engineers who were generally unable to demonstrate any IT skills and were trained as "configurers," as they are called at Krones, only through internal training and learning by doing.

To this day, almost 100 employees are still involved in setting up and maintaining the product structure and associated Variant Configuration.

Of course, before you can benefit from a configuration, you have to maintain large amounts of master data. A product model at Krones would roughly include the following:

▶ Several hundred classes
▶ Several thousand characteristics
▶ Several hundred variant tables
▶ Several hundred pages of object dependencies
▶ Multilevel BOM structures with several thousand materials, of which several hundred are configurable

9.2.2 Results

Krones AG currently uses SAP ERP, release version ECC 6.0, with approximately 4,000 users and operates one common system for all locations.

They have been using Variant Configuration since the SAP implementation in 1998. It is currently used for their full product range and contains several hundred configuration models. The product structure is very complex. The sales configuration (product and price configuration) with up to 2,000 characteristics and a very complex object dependency provides a starting point.

The sales configuration of the header material is also included in the same configuration model for developing the sales order BOM, network and individual routings. The company is currently working on using the configurable purchased material.

BOMs

The BOM structure of the relevant machine consists of configurable assemblies. It can be configured at multiple levels and is distributed over several *branches*. The usual configuration tools are used to develop the sales order BOM:

► Preconditions

► Selection conditions

► Procedures

► Class nodes

► Object characteristics

► Variant tables

These are enhanced by ABAP function modules that are used to read data into the configuration (for example, from a CAD system) or implement the syntax of the object dependency for nonsupported functions (for example, existing design programs).

Engineering-to-order) is needed to process the product range of the relevant customer with the machines. The BOM is multilevel and deep configuration, and the originally configured sales order BOM is still processed manually by engineering after it has been created.

Continuous engineering change management ensures that production for an order whose details have been received relatively late runs smoothly. The workflow system integrated in SAP ERP is used for controlling the process.

Order Processing

Standard Transactions such as CU51 and CS62 and custom developments based on these standard transactions are used for processing orders. A change number is automatically generated and a multilevel BOM is released to support the employees processing the orders. They can also navigate easily in the multilevel BOM and access different status reports – also see Figures 9.6 and 9.7. The advanced process is controlled by networks, and scheduling is performed through an interface for I2.

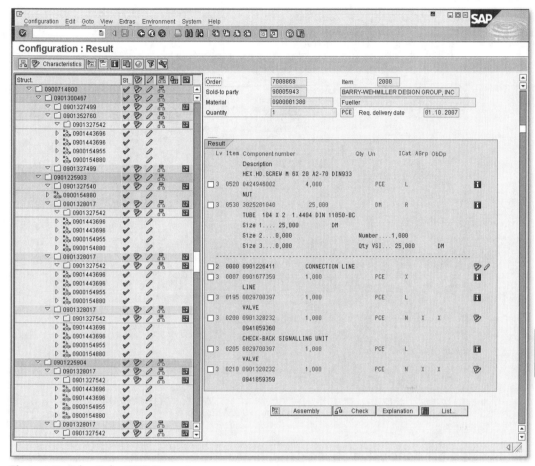

Figure 9.6 Sales Order BOM Processing

Maintaining Master Data

Master data is maintained using worklists and engineering change management for master data. To prevent the advanced technical development of machines and the associated changes to the configuration from interrupting current production, the master data is provided with new series production statuses every quarter: The master data for the configuration and the BOMs is maintained with the time stamp and released for use with the series production status.

Configuration is used as a strategic tool and affects every department along the value chain — from sales to product development, order processing, scheduling, production, assembly, and delivery to creating the documentation.

Figure 9.7 Characteristic Value Assignment in BOM Structure

9.2.3 Conclusion

The SAP LO-VC engine handles this amount of data very well and mostly works without any problems. Because Krones has not allowed the engine to be modified, it can take advantage of SAP maintenance services at any time. This gives the company a certain amount of security in working with the quite sensitive Variant Configuration tool.

It is almost impossible to quantify the savings made from implementing the configuration. However, a decreasing number of employees in order processing is a good indication that using this process has been more than worth it. In this context, we also have to mention the standardization of the product range and reproducibility of products despite make-to-order development.

The company would not do much differently in a new implementation. At the start, the configurers had the job of configuring the machines as fully as possible. To do this, of course, they had to delve deeply into capabilities of the configuration tool — each doing as well as he could. This is why the process has become overly complicated today. Maybe less would be more here, in particular, specifications, allowed methods, reuse, not using nondocumented tool functions, employing best practices when creating object dependencies. In other words, standardizing the structure of the product model is beneficial.

9.3   Progress of the Project at Hauni Maschinenbau AG

Hauni Maschinenbau AG develops, produces and distributes machinery for tobacco processing and filter and cigarette production. Its headquarters are based in Hamburg, Germany. The company also has a number of production subsidiaries in Germany, France, Switzerland, the U.S., Hungary, and Malaysia. It has numerous sales offices around the world too. Hauni has 3,600 employees and turned over EUR 629 million in 2007.

The company currently uses SAP R/3 Release 4.7, and an upgrade to ERP 6.0 is planned for 2009. The system landscape is based on a three-tier architecture:

- A live system with one client
- A consolidation system with one client
- A test and development system with two clients (one development and customer client each)

Approximately 2,000 users work with this SAP system landscape worldwide.

Variant Configuration has been used actively since R/3 Release 3.0F went live in 1998. The project for upgrading from R/2 to R/3 started in 1997.

In the BOM area, a custom development for automatically determining BOM components was already being used in the R/2 system. This custom development should be removed and replaced with BOM-based Variant Configuration.

Objective

Based on this "component description" for the custom development, a large amount of the data needed for Variant Configuration (such as class items or object dependencies) was created automatically, meaning the company could use a certain structure from the legacy system. To enable this, rules and regulations for naming and using object dependencies, characteristics, and classes had to be created. This included only being able to use selection conditions and procedures for changing component quantities and class items in the BOM area. The point to emphasize here is that tables and functions are not used in this area. In some cases these rules may lead to more complex selection conditions and sometimes to necessary additional BOM structures, but overall they made it easier to maintain data and read BOMs.

Figure 9.8 shows an example of additional BOM levels for selecting components.

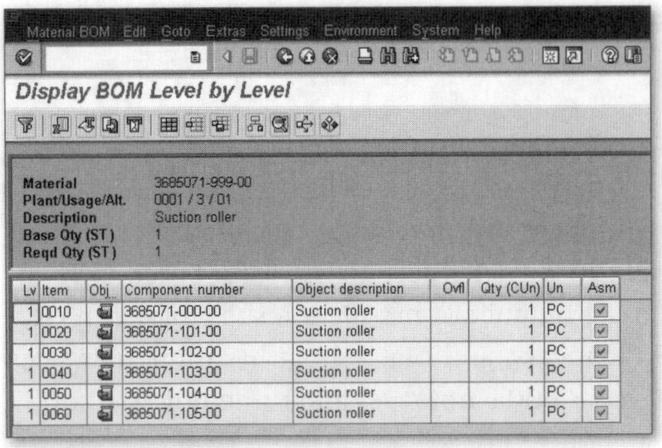

Figure 9.8 BOM Structure for a Suction Roller

Every variant for a component is summarized in this dummy structure. The relevant object dependency for selecting the component is available on each item.

The focus during the implementation was therefore on Variant Configuration in the BOM area.

9.3.1 Personnel Resources

Based on the very large-scale and highly integrated approach, the entire duration of the project was eighteen months. At the beginning, the Variant Configuration part of the project was started with two employees from a main area (which is today integrated into the IT organization). These two employees organized the scope for using Variant Configuration. They received the following support:

▶ Ten employees were trained throughout the project to maintain object dependency within BOMs.

▶ At the same time, three sales employees were trained to create variant classes. This ensured that employees could work with the same knowledge on a multidisciplinary basis.

About 40 SAP data administrators currently maintain the master data for Variant Configuration in the area of BOMs. They are also responsible for neutral master data maintenance in the whole system and have other duties as well.

In the SD area, a group of three people currently create and maintain approximately 200 marketable products and 300 related classes including the entire pricing.

9.3.2 Result

After the project was completed, the use of the configure-to-order or engineer-to-order process was integrated into almost every area of the company. The sales configuration here is mainly used for pricing. The production BOM is exploded by using the low-level configuration, in other words, using object dependencies on BOM items. The corresponding routing for selecting the operations needed is derived to the same extent. They also use the network from the project system to select processes.

No BOMs are processed yet in the quotation process. Only the configuration is filled in the sales document at this point, which results in a binding price. This configuration is subsequently completed in the order. All configurable BOMs for result-oriented order BOMs are then fixed. They may also be manually revised at a later stage, so the company uses the *engineer-to-order scenario* by using the order BOM. The scope of this manual revision can vary greatly. Some machines that are almost 100% configured (CTO process), but some are only 60% configured (ETO process). Because the quantity of machines is not high, it makes no sense to modify the object dependency for every special case. The employees who oversee the order details have the necessary knowledge to compile machine components manually in these individual cases.

Transactions CU51 and CS62 and some custom-developed programs based on the functions of the CAVC* function group are used here for a manual revision.

Reports that enable order BOMs to be checked logically were also custom developed. In the example of all the variants for a component being integrated into a dummy BOM, the system expects exactly one component in one order BOM. The result after the object dependency has been processed is shown in Figure 9.9. Exactly one suction roller was selected from the superstructure.

Every other situation would automatically lead to errors being issued.

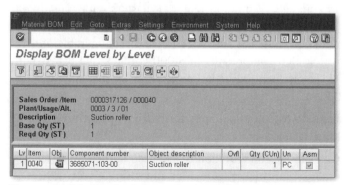

Figure 9.9 Selecting the Relevant Suction Roller

These checking mechanisms are necessary because values are already permitted in the order configuration that (although they are structurally allowed) have not yet been implemented. The employees who oversee the order details and the design engineers respond to these error messages and create the necessary components.

During the project phase it was already clear that the large BOM structures would cause runtime problems. This is why an order BOM is fixed exclusively in the batch. This was done through custom development. This ensures that the system load and locking problems are reduced when orders are being processed because Transaction CS62 can be used for manually reprocessing order BOMs. This transaction only locks the specific single-level BOM and not the complete order item, like Transaction CU51.

[+] A special feature of the whole process is that the sales characteristics are strictly separated from the technical characteristics. A machine is described completely by sales characteristics. In many cases, technical characteristics are simply a copy of the corresponding sales characteristics. Only technical characteristics are used in object dependency in BOMs, networks, and routings.

The background for this is as follows: Engineering change management is used to change class assignments and characteristic values in the area of sales characteristics. This means that a characteristic value may be omitted from an option as of a certain date because this option is then a series. As a result, however, the object dependencies in the downstream processes do not have to be changed because the corresponding technical characteristic is queried here. All the company has to ensure is that this characteristic is now always set accordingly.

Engineering and sales are therefore independent of each other and consequently more flexible.

Quantity Structure

There are currently approximately 1.5 million material masters with roughly 4.6 million plant segments in the system. These are distributed across about 1 million neutral material BOMs, 250,000 of which are configurable. A configurable product neutrally consists of up to 50,000 components in 3,000 configurable BOMs with approximately 2,000 200 classes and 1,000 relationship rules or dependencies distributed across up to 12 BOM levels.

A configured BOM contains up to 20,000 components, whereby a product can be described with up to 200 characteristics.

Besides its BOM-based use, Variant Configuration is also used in the area of sales — in the SD module specifically — for pricing quotations and orders.

The conversion to R/3 and the Variant Configuration implementation meant the entire logistical process was simultaneously changed over to SAP software.

9.3.3 Conclusion

If the project were to be set up again, the following points in particular would have to be taken into account:

▶ The BOM structure would have to be modified correctly

▶ Certain functions would have to be standardized or restricted

If the focus is on optimizing the system runtimes, a flatter BOM structure would be more favorable, although in this case alternatives to the current master data structure that arrange the BOM in a clearer way would have to be found.

Almost all Variant Configuration options are used in configuration or in developing 300 classes. The only reason this is manageable is that very few people maintain the data here.

In the rest of the application, a check would be needed to see whether it would make sense to standardize and, in doing so, restrict certain functions. Although this might increase the amount of maintenance effort involved, the reward would be more clarity and simpler applications. It also appears in this application that, as is the case with the entire R/3 system, standardizing processes and applications is more beneficial than sustaining special programs and isolated solutions.

9.4 Variant Configuration at the Felix Schoeller Group

The Felix Schoeller Group has been producing high-quality specialty, photographic, digital-imaging, decorative, and technical specialty paper for a wide range of applications for over 110 years. The company employs 2,400 people in eight locations in Germany, the U.S., Canada, and Russia. In 2007, it had total sales of 364,000 tons and an overall turnover of EUR 714 million.

The Felix Schoeller Group is currently using SAP ERP 5.0, and the ERP system landscape is divided into the classic three stages of a test, a consolidation, and a live system. Approximately 900 users work with the ERP system worldwide. SAP Supply Chain Management (SCM), SAP NetWeaver Business Warehouse, and the SAP portal solution are also incorporated into this landscape. The IPC server responsible for integrating Variant Configuration into the portal is a stand-alone installation (without CRM) in Release 4.0. The portal solution is SAP Enterprise Portal 5.0.

Variant Configuration has been in use in the decorative business area since 1997, although it was only used there with one configurable characteristic. In 2006, it was introduced into the digital imaging (DI) business area with a greater range of functions. Digital imaging was mapped to converted paper here. Converted paper is paper cut-to-size in format and converted paper rolls (as used in plotters, for example), and each needs to be packaged in different customer-specific packaging.

9.4.1 Project

The implementation project took about 14 months, and SAP Warehouse Management (WM) was implemented at the same time as Variant Configuration, so these projects were closely integrated.

Analysis and Statement of Objectives

The objective of implementing Variant Configuration in the DI business area is to enable make-to-order processes (MTO) for the first time and map them in the system in an integrated way. An analysis performed here before the project showed that the huge number of variants is only accrued in the last production or packaging stage. Allowances should be made for this when using Variant Configuration in a CTO scenario. Variant Configuration should also be used in upstream stages to meet standards in a make-to-stock scenario (MTS). Figure 9.10 illustrates the balance between make-to-stock production and make-to-order production.

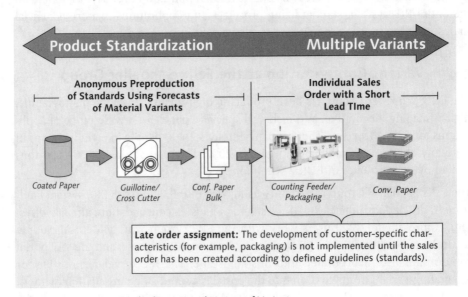

Figure 9.10 Product Standardization and Variety of Variants

MTO scenarios should reduce the master data effort involved in material masters, so that variants produced and sold as one-offs do not have to be manually created with a lot of effort like before and be deleted after they have been used once. By being able to create BOMs and routings automatically for created material variants, the effort involved in master data should also be reduced for make-to-stock scenarios (MTS) as well. The information contained in the material masters should also be enhanced by configuring and specifically defining products using characteristics.

Project Scope

At the beginning of the project, the IT employees in this company only had basic knowledge of Variant Configuration, so external consulting from SAP Consulting, which mainly provided support in product modeling, was engaged early on. This consulting was of a strong coaching nature, so that IT and user departments could quickly speed up the concepts and project independently. The external effort only occupied about 25 person days in total.

Only one employee from the user department and one employee from IT were involved in the product modeling and in implementing the configuration structures including classes, object dependencies, variant tables, super BOMs, and super routings. They were able to use the knowledge they had gained so far in the project to maintain and extend the product models without any external support, so that current changes to products and processes could be modified in Variant Configuration. The relevant IT and user department employees responsible carried out the integration into the different modules in the SAP ERP system (such as SD, CO, and QM) and into the shop floor systems. The internal effort involved about 300 person days in total.

9.4.2 Results

Relatively simple variation configuration product models could be developed based on the requirements. The characteristics to be defined manually are limited to a minimum of five characteristics because many characteristics are preallocated with default values. A maximum of 30 characteristics can be defined manually. The remaining characteristics are derived through object dependencies and variant tables. As a result, there was greater approval of the configuration in sales in particular, because there was only a limited amount of effort involved in the characteristic value assignment for standard configuration here. Complex configurations were nevertheless possible for exceptional cases.

The absolutely necessary characteristics are available on the Entry Area tab in the configuration (see Figure 9.11).

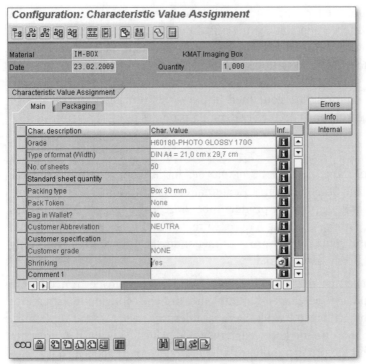

Figure 9.11 Initial Screen for Configuration

The configuration is also used in the material master system, where material variants are created for MTS scenarios. This means they are created in relation to a configurable material and reference this header material. BOMs or routings do not have to be created manually.

A *single-level BOM explosion* occurs in the configuration (see Figure 9.12). The super BOM is evaluated here based on variant tables. This immediately allows the sales employee or person creating the material master to see whether all of the necessary BOM components are maintained in the system. This leads to greater transparency in sales in particular about whether all of the necessary data and components for the product are already created. An availability check was not performed at the component level in this case.

If the configuration is consistent, a planned order is automatically dispatched and can be processed further in sales order processing. We therefore use *assembly processing*.

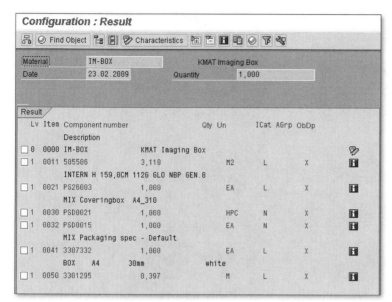

Figure 9.12 BOM Configuration Result

Something else to note here is that characteristic-based variant pricing is not carried out in order configurations.

Another advantage of the model is that you do not have to have any knowledge about the object dependency for maintaining BOMs because BOM components are only maintained through variant tables. On the basis of Variant Configuration, materials can also be used in BOMs again, but this is only possible in the form of material variants. The development to sales order is only supposed to occur for the last production phase; in other words, there is no multilevel configuration in the sales order.

Because class nodes are used to integrate components into BOMs, recursion problems occurred from time to time when material variants were used as possible initial components. The system normally recognizes this and prevents or corrects it, but mixing the different product models means there may be combinations that could block new material variants from being assigned to super BOMs. This is why it is essential to consider clearly defined component classes and assign material variants to them.

Quantity Structure

The following quantity structure applies for the current product model (see Table 9.2).

Model Part	Quantity Structure
Configurable materials (KMATs)	8 header materials
Material variants (MATV)	4,100
Configurable BOMs	Approximately 70
Configurable routings	Approximately 110
Classes	21
Interface profiles	8
Characteristics (internal/to be value assigned)	Approximately 100
Object dependencies	25 constraints 1 precondition 50 selection conditions 198 procedures 48 variant tables 9 variant functions
Configuration profiles	8

Table 9.2 Quantity Structure for Product Model

The following object characteristics and functions were also used:

Object characteristics that are used to determine the following data, for example:

▸ Displaying or hiding characteristics or values

▸ Controlling the IPC

▸ Managing errors

▸ Preselecting components

▸ Text items in BOMs

Functions:

▸ Displaying errors or information in the user interface

▸ Determining specific assemblies

▸ Setting text items

IT maintains and modifies product models based on what the user department needs. However, the modifications here are kept to a minimum because the highly fluctuating parts of the configuration were deliberately stored in variant tables and not directly in the object dependency. The maintenance for variant tables is split into tables for high-level and low-level configuration.

High-level tables define the product properties allowed and are used to set product standards. Two employees from global product management were trained to maintain these tables.

Low-level tables are evaluated using object dependencies and consequently form super BOMs and super routings. Two employees from sales order processing were trained to maintain these tables.

9.4.3 Extending Variant Configuration Using the IPC

In October 2008, an order entry application was added to Variant Configuration in the SAP web portal.

A standalone installation of the *Internet Pricing Configurator* (IPC) was used to do this. Product models are transferred to the IPC using the standard function for creating a knowledge base in ERP. The only problems here were caused by characteristics with hyphens in their names, which necessitated a customer-specific patch in the IPC. SAP provided the patch.

The configuration logic for product models was also extended and modified in the portal based on the requirements of a configuration. The integration into the existing portal functions was done by calling the IPC through a URL. This means a user can go directly from the order entry area in the portal to the IPC (see Figure 9.13).

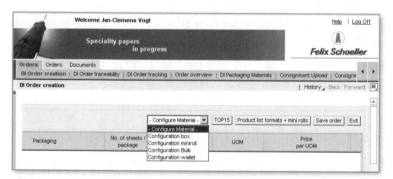

Figure 9.13 Going from the Order Entry Area to the Configuration

The standard interface was then used in the IPC, whereby modifications were only made using the *Extended Configuration Management* (XCM) of IPC Customizing. The object dependency in this case has severely limited the configuration interface, which means this interface fundamentally differs from the interface for internal use in the ERP system (see Figure 9.14).

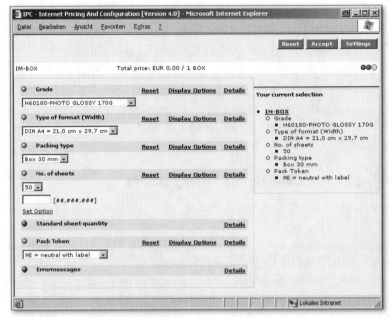

Figure 9.14 IPC Configuration Interface

The value assignment of characteristics can then performed in the IPC. By clicking on the Accept button, the configuration is transferred to the order entry area. The characteristic value assignments are then transferred by calling a URL, and the configured item is created in the order entry area.

This function was able to be carried out with external involvement of approximately 15 person days and barely more internal involvement.

This additional function guarantees the continuity of the MTO scenario by all sales processes (including portal functions), and the objectives stated at the beginning of the project were achieved in full.

9.4.4 Conclusion

We can establish that Variant Configuration was the only thing that could prevent the master data from increasing uncontrollably. It would be very difficult to integrate the MTS or CTO process, and a wide range of products would have to be offered to the customer or user of the Customer Relationship Management system (CRM). The implementation made everything easier. In Logistics, make-to-order production partly involves a change in the process flow. We recommend using tables when maintaining the product model because stored knowledge is much easier to read than source code.

9.5 SAP at Hülsta and in the Hüls Corporate Group

The Hüls Corporate Group employs over 4,000 people and is an association of mid-sized German furniture manufacturers which, as well as the Hülsta, Parador and Rolf Benz brands, includes companies such as Loddenkemper, Ruf, and Arte M. Its headquarters are based in Stadtlohn, in the Westmünsterland region of Germany. Two production facilities for Hülsta and some other plants belonging to the corporate group are also located there and in the surrounding area. There are also several sales offices worldwide and the production locations.

The decision to use SAP software as the standard software was made in the company headquarters back in the 1980s. The group is currently using SAP ERP 6.0. The system landscape has a classic three-tier structure:

- A live system with one client
- A consolidation system with one client
- A development system

More than 1,500 users (and counting) now actively work with the system.

Besides the ERP system landscape, the group also uses a separate SAP R/3 system for Human Resources and an SAP R/3 Logistics system. Other SAP systems such as SAP NetWeaver Business Warehouse, SAP Supplier Relationship Management (SRM), a Customer Relationship Management system (CRM), and SAP NetWeaver Portal extend the application portfolio.

9.5.1 Initial Situation

Hülsta has been using Variant Configuration since changing over and migrating from SAP R/2 to SAP R/3 in 2000. Starting with Release 4.0, two release upgrades have been successfully completed to date, the first one to Release 4.6, and the next one to the current Release ERP 6.0.

As well as Hülsta, other companies within the Hüls Corporate Group have been using Variant Configuration. For example, to manufacture mattresses, a selection of several thousand end products with a configurable material and therefore a single super BOM is mapped. If inventory management is needed here, the material variants provided for this are used.

The following descriptions all refer to Hülsta, but the other companies in the group have similar data models.

In the initial situation there were dedicated items and BOMs that were created semi-automatically in the SAP R/2 legacy system on the basis of numerous custom-developed tools.

9.5.2 Preparation

As part of the preparations for the higher-level migration from SAP R/2 to SAP R/3, two IT employees were trained in Variant Configuration at the SAP training center in Walldorf, Germany. Together with an external consultant and employees from work scheduling, an initial draft of a data model was prepared in a few weeks. Additional intensive workshops were held with employees from sales and production before the final decision to use such a data model was made.

The benefits developed in the first concept and the scope for using Variant Configuration have been proven and are still valid today. The following are a few examples:

▸ Higher quality of master data

▸ Quick and efficient maintenance of master data

▸ Uniform configuration profiles

▸ Naming conventions for all objects involved

▸ Database tables and function modules that can be used within Variant Configuration

To reduce risks with SAP R/3 going live, however, only 1 of the 50 furniture programs was initially converted to Variant Configuration.

After this project was completed successfully, all of the other programs were successively converted to Variant Configuration over a period of three years.

Eight employees from the relevant user departments were mainly trained internally to do this. The aim of this knowledge transfer was to enable user departments to create and maintain super BOMs plus object dependencies and class nodes independently.

9.5.3 Project Objectives and Results

One challenge of the project was to build a bridge from the old world to the new world. Marketing documentation in the furniture industry is pretty much standardized.

Developing the Current Quantity Structure

Even if individual furniture assemblies are multivariant, marketing documentation details a multitude of frequently descriptive item numbers for more or less distinct variants that are subsequently often referred to as types and for which only the color is variable.

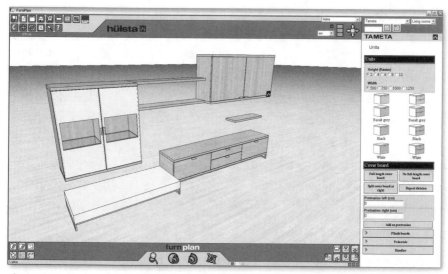

Figure 9.15 Screenshot of a Graphical Configurator

Because no electronic furniture configurators (see Figure 9.15) were being used yet across the board at implementation time (and are still not to date), the paper-based type list (see Figure 9.16) is an essential communication tool for sales.

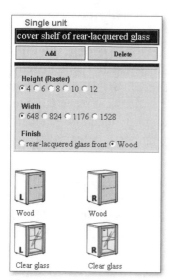

Figure 9.16 Example of a Type List for the Furniture Industry

However, there was a data model based on a low number of configurable items that was actually no longer needed in the configured types. This contrasts with the current quantity structure at Hülsta (see Table 9.3).

Quantity Structure	
Order items per month	150,000
Value-assigned characteristics of a multilevel sales order item (instance)	10–60
Number of procedures	14,000
Number of selection conditions	13,000
Number of characteristics	3,000
Number of furniture programs	50
Number of KMATs per furniture program	50–200
Number of class nodes	3,000

Table 9.3 Quantity Structure of Master Data

The figures clearly show how static documentation is based on sales types. These types appear to be very multivariant owing to their dimensions, fronts, and surface finishes, but are offered as predefined variants in favor of unique purchase orders, so the assembly that is four grids high and 648 mm wide also has a unique PO number (corresponds to the sales type) as the assembly that is six grids high and 824 mm wide.

Header Configuration

However, based on the material matching within sales order processing, all of the same kinds of sales types result in a KMAT (configurable material) and consequently in a BOM. Because a sales order is not made up of one configurable product, but can contain more than 40 items that are all configurable, it does not make sense to configure every item manually in order processing. The company therefore had to be able to enter a sales type as a sales order item without the program actually branching to the configuration dialog box (see Figure 9.17).

At the same time, however, a number of characteristics are identical for all or at least many items, so a higher-level instance that inherits the characteristic value assignment on sales order items was needed. A consulting firm was used here to implement the header configuration into the sales order as an add-on (see Figure 9.18).

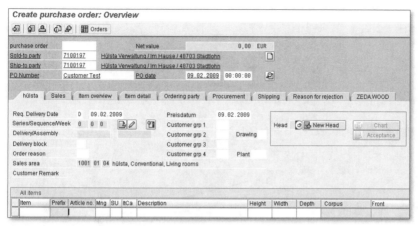

Figure 9.17 Branching to the Header Configuration from the Purchase Order

Sold-to party	7100197				
Material	90000999				
Quantity	1,000.000	STK	Item	999901	
Req. deliv.date	09.02.2009				

Characteristic Value Assignment

Char. description	Char. Value	Inf...
H0000_S_PROGRAMM	xelo	ℹ
H0000_S_FRONT	maple natural	ℹ
H0000_S_KORPUS	maple natural	ℹ
H0000_S_ABDE	maple natural	ℹ
H0000_S_SEITE	white lacquer	ℹ
H0000_S_RAHMEN	white lacquer	↵
H0000_S_PANEEL	maple natural	ℹ

Figure 9.18 Detailed View of Header Configuration

As a result, the characteristic value assignment of the header characteristics (for example, the surface finish or handle versions together with the value assignments from types such as measurements) forms the overall value assignment of the configurable material. This means the time-consuming manual characteristic value assignment can frequently be avoided in sales processing, and the established type numbers can be used on order confirmations or invoices in communications with the customer.

The data for the automatic value assignment (for example, the partial configurations relating to dimensions that are already known for sales types from the sales documentation) is maintained in a separate database table. Reference characteristics are needed for using the information for the configuration. A reference characteristic on the MATERIAL field is used to enter the key information for the sales type automatically into the configuration (see Figure 9.19).

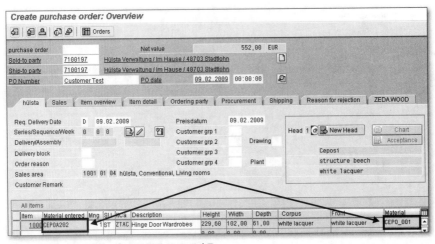

Figure 9.19 Matching the KMAT Material Type

By using a procedure to call a function module for determining item data, all of the relevant information is determined from the database table and written into the corresponding characteristics. The configuration of sales order items is consistent owing to the combination of header configuration and use of function modules and database tables. Because of the header configuration add-on, the configuration can now also be run in the background, so the configuration dialog box remains hidden from the user.

Planning Software

However, the majority of Hülsta's purchase orders are not entered into an SAP dialog application by manually entering sales type numbers (see Figures 9.19 to 9.21), but are instead created using a graphical configurator. (You were introduced to the graphical configurator in Figure 9.15.)

[+]   You can download a consumer version of this *Furn Plan* configurator by going to *http:// www.now-by-huelsta.de/index.php/en/en/home* and selecting the PLANNING SOFTWARE menu option.

You can use this planning software to graphically plan and display customer commissions, but you can also control the SAP software through a bidirectional *communications port* (COM) interface so you can create, change, or delete order items and control some characteristics of Variant Configuration directly through the planning software.

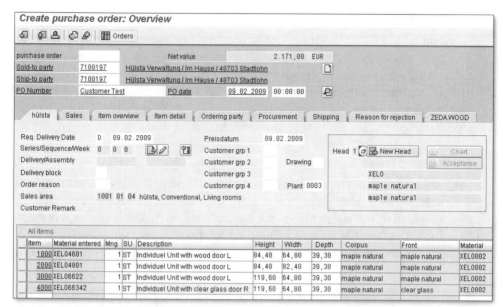

Figure 9.20   Hülsta Sales Order (SD View of Graphical Configuration)

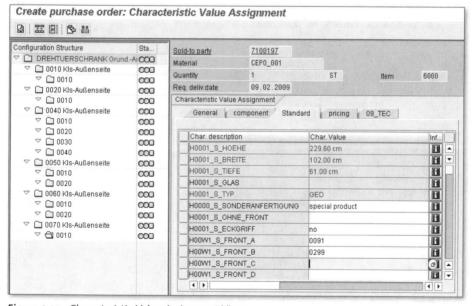

Figure 9.21   Characteristic Value Assignment View

Performance

The main critical factor in Variant Configuration is performance. Owing to the high number of sales order items configured at multiple levels, the dialog response times for entering sales orders are a constant cause for concern, but it only took a few days of development effort to find a solution. To improve response times online, two identical configuration profiles were created and assigned for each KMAT, one single level (see Figure 9.22), the other multilevel (see Figure 9.23).

Customizing Tables

You can use separate Customizing tables to specify that the single-level profile is automatically used when sales orders are being entered. As soon as you save the sales order, all of the sales order items that are configured at a single level but also have a multilevel profile are reconfigured asynchronously on a multilevel basis. This technique means the user can avoid the long runtimes for multilevel configuration.

The company could keep the creation of master data in the SD area calculable by standardizing configuration profiles on the KMATs, so that copies can be made frequently. To a large extent, creating master data is then actually based on maintaining fewer separate master data tables.

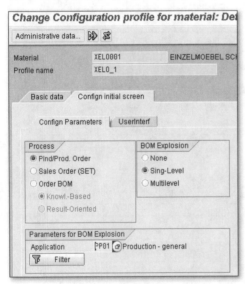

Figure 9.22 Single-Level Configuration Profile Setting (in Online Operation)

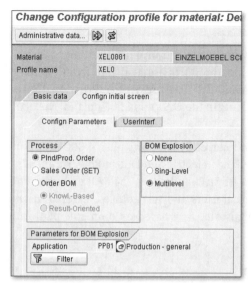

Figure 9.23 Multilevel Configuration Profile Setting (in Batch Operation)

9.5.4 Conclusion

The extremely rapid growth of individualization in furniture in recent years by mixing colors and materials and through special sizes for customers would barely have been efficiently possible without using a Variant Configuration option that is fully integrated into production. Variant Configuration enables us to implement the needs of our customers much more securely and flexibly. Even if we do not have any key figures for the volume of savings from BOM changes, we can clearly see that some changes that previously took several days can now be completed in a few hours or even minutes.

In the end, though, it has been proven that using Variant Configuration significantly improved the entire master data quality. This is largely due to the integrated data model that forces all areas using SAP software to control their business processes to coordinate actions across departments. The quality of the whole process improves as a result.

9.6 Product Configuration at Baldor Electric

Baldor Electric is an American manufacturing company producing electric motors, power transmission products, drives, and generators. The Baldor mission statement is, *"To be the best (as determined by our customers) marketers, designers and manufac-*

turers of industrial electric motors, mechanical power transmission products, drives and generators."

Baldor products are sold worldwide to distributors and original equipment manufacturers in more than 70 countries. Baldor products are available from 50 sales offices and warehouses in North America and 26 offices serving international markets. These products are produced at 26 plants in the U.S., Canada, England, Mexico, and China.

Baldor implemented SAP ERP in 1996 and currently supports over 4,000 users on SAP ECC 6.0. In 2007 Baldor acquired Reliance Electric and Dodge Power transmission, more than doubling the demand on its SAP landscape. The SAP environment is made available via SAP GUI to all internal users and sales offices. External users have e-Commerce capabilities through a custom, Java based e-Commerce platform that calls SAP BAPIs when required.

9.6.1 Starting Point of the Project

A significant portion of Baldor's business comes from custom and configured products. These custom orders were either configured manually through communication between the buyers, Baldor Sales, and Baldor Engineering, or the products were configured using legacy, non-SAP systems. These systems were often disjointed, making them difficult to maintain and difficult to provide the customer with a complete solution without consulting engineering.

The power generator product line was especially lacking in configuration tools to help facilitate the accurate configuration of custom generators for Baldor's customers and resellers. The lack of configuration tools coupled with the sales force's limited knowledge of the generator products resulted in engineering having to spend too much time making repeated configurations instead of working on new products and in a lack of standardization between custom generators.

The goals of Baldor's Variant Configuration project were to:

▶ Give Baldor a competitive advantage in the generator market by offering better and more accurate configuration, quoting, and documentation tools

▶ Reduce the time engineering has to spend reviewing and designing custom generators

▶ Create a solution that does not disturb existing business practices and allows Variant Configuration to be expanded to the custom motor and custom power transmission businesses in the future.

One of Baldor's key information services strategies is to have a small group of individuals representing the information services needs of the various sections of the business. This team is referred to as the "Core Team." It is made up of business specialists representing all of the key sections of Baldor. Each team member reports to the management of the business section he represents. The Core Team is responsible for setting priorities and direction for all information services projects. For Baldor's Variant Configuration project, the Core Team representative for configured products was designated as the project coordinator, supported by the Core Team members representing sales, engineering, and manufacturing. This Core Team representative was responsible for the overall order processing of configurable products and any integration decisions.

In addition to the Core Team, a group of specialist from the generator business was assembled to determine the specific requirements of a Variant Configuration solution for the generator business. This group was made up of representatives from engineering, marketing, and sales.

One information services resource was designated as the variant model lead responsible for all product modeling activities. This information services resource was supported by one off-site Variant Configuration consultant from eSpline Consulting. Other ABAP development resources were tasked as necessary during the project.

9.6.2　Key Characteristics of the Project

In the subsequent sections we'll highlight the basic findings and key elements of the project.

Development Landscape and Product Data Replication

The Baldor SAP system landscape already consisted of separate development, test, and production systems. However, after a few months of preliminary development, it became clear that the development of Variant Configuration needed to be isolated from other ABAP development and testing. The existing development clients had too much potential of quotes and orders being created using the KMAT, thus locking the variant model from changes. During this time, the project team also realized changes needed to be made to the overall naming convention of the variant model. To accomplish both goals, they decided to create a small development client with no master data, and "move" the variant model to this client.

With the help of eSpline's export service, the variant model was exported from one development client in XML format. Once exported, mass changes were made to the variant model, correcting naming conventions, characteristic values, and even characteristic data types. Once these changes were complete, the model was imported to the new development client using the XML ALE IDOC. The result was a complete success.

Baldor now operates with a dedicated gold VC client. The only master data contained in the client are the elements of the variant model and specific materials needed for the super BOM. Additional materials are moved from production to the development client using standard ALE MATMAS IDOCs.

Baldor relies heavily on the SAP Product Data Replication (PDR) services to transport the variant model between systems. PDR is a fast, reliable, and efficient solution for updating the variant model across all of the systems in Baldor's system landscape. It is no longer necessary to keep track of what changes have been made in which system. It is Baldor's strict practice to only change the variant model in the gold VC client, and then create PDR baselines to move those changes to the test systems. Once testing is complete, the same PDR baseline is moved to production. The PDR replication table ensures that only new or changed elements of the variant model are moved to the new systems. Transaction UPS03 also provides a detailed view for troubleshooting. Although it was a difficult process to get all of the necessary notes, patches, and IMG settings correct for PDR to work properly, Baldor estimates it now saves 3 hours per move. This savings will increase as the variant model becomes more complex.

Baldor's Best Practices for Variant Configuration

Baldor has established best practices for the variant model that focus on the use of variant tables and constraints whenever feasible. Procedures are only used to control preset values (set_default) and hide characteristics (is INVISIBLE). Most other logic is housed in table constraints. The business and engineering logic experts put this logic into matrix format – typically in Microsoft Excel spreadsheets. In many cases, the Variant Configuration modeling expert is able load these spreadsheets directly into variant tables using Transaction CU60e. This approach allows the nontechnical business personnel to own the majority of the logic built into the variant model. It also makes changing the variant model easier. Often, only variant table changes are required to update the model.

A combination of selection conditions and class items are used in the super BOM logic. Class items are used for core components of a generator (engine, alternator, controller, etc.), and selection conditions are used for all other materials.

The Product Modeling Environment for Variant Configuration (PME VC – Transaction PMEVC) is an integral part of the variant model development process. All constraints and procedures are written using the PMEVC, and all knowledge bases for the IPC are managed from the PMEVC. The all-in-one capabilities granted by the PMEVC not only make it easier to develop the variant model, but also simplify sharing the variant model with nontechnical personal. Interested parties are able to see how all of the elements of the variant model work together within one transaction.

Integration with Sales Processing

In the discovery phase of the variant project, Baldor determined that there are two types of configured products: make-to-order (MTO) and engineer-to-order (ETO). MTO configurations result in a 100% accurate BOM and require no review by engineering personnel. ETO configurations result in a partial BOM and usually contain special requests from the customer requiring a review by engineering. The project team decided to focus on the ETO process in the first phases of the project.

Several long-standing business processes make it difficult for Baldor to implement variant configuration as an end-to-end system, and one of the goals of the project was to keep from disturbing the existing flow of business. As a result, they decided to integrate variant configuration with Baldor's existing ETO process. This gave the opportunity to introduce Variant Configuration into the SAP system landscape without disturbing downstream processes. By forcing all Variant Configuration orders down the ETO process, it also ensured that the generator sales and engineering personnel had a chance to review every configured order and thus provided a real-world hardening of the variant model without impacting production.

To the end-user, Variant Configuration behaves normally in the quote and order transactions. However, once the quote or order is saved, Baldor has created a custom solution taking the information rendered from the VC Engine and creating a unique engineering special report (see Figure 9.24). The characteristics and values, variant pricing, and resulting BOM are all copied into the engineering special report, which then kicks off a review process for the ETO order. First, Baldor sales support validates the pricing and any custom requirements with the customer, and then the order is released to engineering. Engineering uses the characteristic values from Variant Configuration, the resulting BOM coming from Variant Configuration, and any custom requirements entered by the customer or notes provided by sales to design the custom generator.

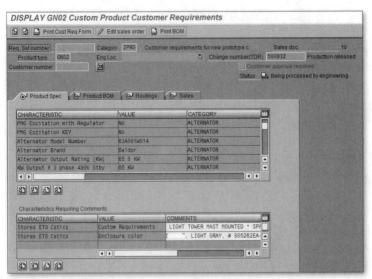

Figure 9.24 Engineering Special Report with Configuration Information and Notes

The end-user benefits from a standard order entry solution for configuring a custom generator that prevents them from making invalid configurations, but allows for customizations. Baldor sales benefits because all custom generators orders come in a standard format that helps in communication with engineering and the customer. Engineering benefits because Variant Configuration has most of the logic required to build the generator. Also, the engineering special report now comes in a standard, more efficient format. An added benefit of this revamped ETO process is the ability to use the information coming from the VC Engine to create various Adobe Forms documents for the customer, sales support, engineering, and manufacturing.

Product Configuration User Interface

The seamless integration of Variant Configuration and IPC provided Baldor with the opportunity to provide different user interfaces to its two categories of users:

1. Variant Configuration through the SAP GUI sales quote and sales order screens for the Baldor sales offices
2. IPC integrated with the Baldor e-Commerce site for external customers, distributors, and resellers

The Baldor sales office users are very comfortable with the SAP GUI client, and need the ability to review the variant pricing, as well as other aspects of the quote and order, as a part of the custom order. The VC Engine within the standard sales quote and order transactions gives the Baldor sales office personnel the full capability to manipulate configuration and pricing in order to get the sale.

Baldor has an established custom Java e-Commerce site to support all outward-facing users (see Figure 9.25). Baldor integrated AP 7.0 with the latest IPC 7.0 user interface to provide these users with the ability to create custom generator quotes in the SAP ERP system. Using a standard punch-out approach, users are redirected from the e-Commerce site to IPC when a custom generator is requested (see Figure 9.26). When a successful configuration is made, the users are directed back to the e-Commerce site, where they are asked to enter explanations for any custom requirements that came up during their configuration and save the quote.

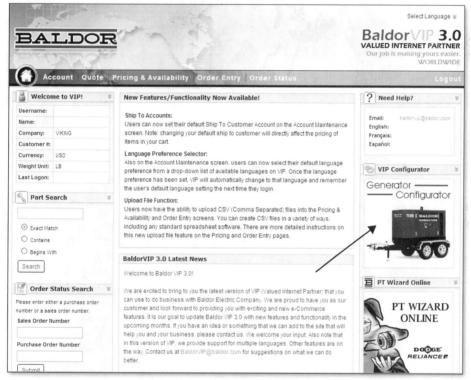

Figure 9.25 Initial screen of e-Commerce site with IPC launch

Once the quote is saved, the user has the ability to print several types of documents related to the variant order:

▶ A standard quote document detailing the selections made in IPC and pricing for that configuration.

445

▶ A submittal package for generator resellers that prints the selections made in IPC plus any relevant documents to the IPC configuration. This is accomplished by a custom BAPI using the resulting variant BOM to determine all relevant documents attached to materials in the BOM, or documents loaded directly on the super BOM. All of this information is delivered in a report the reseller can modify to include their pricing and services.

▶ A specification package for resellers or customers that provides detailed application and performance data about the configured generators. This report is built using the same technology as the submittal package, but with different documents.

[+] Baldor has also prototyped, and is reviewing, the ability to render the IPC UI inside the SAP GUI client. This gives all users the same UI experience without any sacrifices. The results have been very promising.

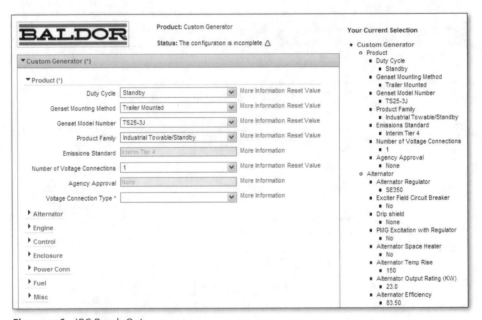

Figure 9.26 IPC Punch Out

9.6.3 Basics of the Variant Model

For the generator product lines, the variant model is composed of the following elements:

▶ One KMAT for all generator product lines

▶ One primary variant class (300)

- 250 characteristics in the primary variant class
- 150 constraints
- 100 variant tables
- 5 procedures
- 395 selection conditions
- 400 materials in the super BOM
- 4 classes items in super BOM
- 320 pricing conditions of type VA00

9.6.4 Conclusion

During the course of the Variant Configuration project at Baldor, a great deal of experience has been gained.

Lessons Learned   [+]

Get good help: Baldor's partnership with eSpline significantly increased their confidence in running a successful project. eSpline provided excellent hands-on help when required, but more importantly, they provided experience and best practices. It was never Baldor's intention for eSpline to do any of the Variant Configuration modeling or IPC development. Baldor wanted to partner with a provider that could ramp up the internal expertise of Variant Configuration.

Get business buy-in: The success of most information services projects usually depends on how much the business really wants it to get done. However, in Baldor's Variant Configuration project, it was not enough that the business believed in the efforts. The project did not start to take off until the engineering, marketing, and sales personnel started owning the logic required to build the variant model.

Since Baldor's Variant Configuration project was led by the Baldor Core Team, all of the solutions and process implemented were built with all products in mind. Efforts are already underway to start converting legacy configuration tools to SAP Variant Configuration for the (much larger) electric motor segment of the business. In addition to these new variant models, the more complex make-to-order business process will be developed to bypass engineering completely. Baldor will continue to invest in the use of SAP Variant Configuration and IPC for the next several years.

With over 10 years of customizations to standard SAP ERP, the fear at Baldor was that Variant Configuration would not be able to coexist with the proven, mission-critical business process built into Baldor's SAP ERP solution. However, Baldor was able to use a standard Variant Configuration modeling approach and standard IPC implementation coupled with custom business processes and programs to achieve a successful end-to-end process for custom generators. The solution has improved every phase of the custom order experience, but all of the success is derived from a good configuration engine.

9.7 Summary

This chapter dealt with real-world SAP implementations of Variant Configuration that illustrate you can use several different methods to get the same result. In other words, there's no specific method or predefined quantity structure for the implementation.

The more individual each customer and material is, the more individual the need for the solution, project, and achieved success after the installation is. It has become clear that even very different requirements — from a configuration for a complex multilevel plant construction to a graphically supported furniture configuration, or a relatively simple example of a paper configuration developed using a smaller product model — can be mapped in the standard SAP system.

If you've followed our line of thought up to this point, you'll agree that product configuration is a complex and multilayered topic. This chapter examines some of the challenges you may encounter in the course of product configuration.

10   Challenges in Variant Configuration

Variant Configuration, like any other interesting field of activity, is associated with a range of challenges. This chapter singles out three specific issues for particular attention that frequently arise and quickly present a challenge in connection with product configuration. These are:

▶ **Performance optimization**
The fulfillment of a complex configuration task requires a correspondingly intense use of resources. In particular, the computing time required to process the configuration steps involved may have a significant impact on interactive processing flow. Early identification of potential trouble spots is essential to the elimination of bottlenecks.

▶ **Change management**
Both the options of a configurable product and the requirements of customers are subject to change. If you want to ensure that changes to the product model in your application are incorporated into the current logistics process in a way that can be both monitored and reproduced or if you want to be able to take account of your customers' requests for change as part of the current production process, you may benefit from using the change management functions for master data and orders.

▶ **Complex system configurations**
Interdependencies arise when configurable components are grouped together in complex system configurations. These dependencies are sometimes difficult to define and control. This chapter explains how you can distinguish between three types of system configuration and which modeling tools to use for each.

Each of these three issues is discussed in a separate section. In practice, they frequently occur side by side. For example, if you have a complex system configuration, you will soon be confronted by runtime problems. In such cases, the scale of the challenge increases accordingly.

10.1 Performance Optimization

This section discusses runtime behavior during the execution of a configuration task. Everyone wants a software solution that offers the shortest possible response times to minimize the impact on workflows. As soon as we feel that the system is keeping us waiting, we feel that our workflow is being disrupted. We then speak of performance problems and try to come up with possible improvements. The natural approach to performance optimization is to identify bottlenecks and seek to eliminate these where possible. One key factor in determining the quality of any performance improvement is how early the potential bottlenecks can be detected. If a significant investment has already been made in the setup and operation of configuration models, any major changes to these are going to be difficult to accept.

As stated in Chapter 1, Basic Principles of Variant Configuration, variant diversity is the central issue here and is the very reason for the existence of variant configuration. The number of possible configurations increases exponentially in tandem with the number of characteristics involved. Algorithms with exponential growth are difficult to control in IT. In dealing with variant configuration, you can expect to encounter potential performance problems, and you should therefore take the runtime behavior of the software into account when devising and testing any solution.

10.1.1 Performance Bottlenecks — Occurrence and Influencing Factors

Long runtimes may occur in both high-level configuration, where a configuration task is processed interactively, and low-level configuration, where the specific component structure is determined automatically on the basis of the configuration result and one or more super BOMs. Long runtimes in high-level configuration are a more critical issue because interactive processing is disrupted in this case. We'll therefore examine this problem in more detail here.

High-level configuration is normally part of sales order processing and can be broken down into the following processing steps:

1. **Opening the configuration**
   The configuration model and the configuration (if one already exists) may need to be loaded from the database for this step. If configuration tasks of a similar type have previously been processed, the model may still be stored in a system-internal buffer, in which case there is no need to reload it from the database. A newly created configuration needs to be initialized when it is opened.

2. **Value help for characteristics**

   The possible values that are available for a characteristic can be determined using the value help function. In LO-VC Variant Configuration the values can be displayed permanently or when you press the $\boxed{\text{F4}}$ key.

3. **Selecting values**

   When a characteristic value is selected interactively, the value is set in the configuration and, if the relevant setting is made in LO-VC Variant Configuration, a check is always run against the configuration to evaluate the relevant object dependencies.

4. **Checking the configuration**

   If the system is configured so that a check is not automatically run against the configuration model each time a characteristic value is selected, this check can instead be executed on request. The configuration structure is also recalculated as part of this process. The configuration is always checked whenever the focus shifts from one component to another in a configuration with a BOM.

Runtime behavior is influenced by several factors during the execution of configuration tasks. The main factors are as follows:

▶ **Configuration model**

   The size and structure of the configuration model have a significant influence on runtime behavior. There is no definitive best way to model a certain configuration task. SAP Variant Configuration offers an extensive range of modeling tools. You can therefore optimize a configuration model using very different approaches. However, your objective should always be to achieve efficient runtime behavior.

▶ **Application context**

   The context determines the way in which the configuration task is processed. The processing sequence and user behavior may be very different depending on the specific configuration task at hand. As a result, runtime behavior can only be assessed within a specific context.

Different Contexts	[Ex]

If the configuration is running in sales Transaction VA01 in SAP ERP, the classic scenario is one where LO-VC Variant Configuration is used directly by a sales employee in dialog mode. However, the same configuration model can also be accessed simultaneously by multiple end customers in an Internet Sales scenario. Different software modules are used in this case. Specifically, the IPC or the AP Configuration Engine serve as the configuration tool in place of LO-VC Variant Configuration.

▶ **Software**

In terms of the software used, the database system and the operating system in particular may differ within the same application context. This may have a noticeable effect on runtime behavior in certain cases.

▶ **Hardware**

The design of the hardware layout is of key importance in the case of frequently used applications. When it comes to the processing of configuration tasks, account must be taken both of the number of transactions that may be running in parallel for each time unit and of the precise content of the configuration task when designing hardware resources.

The application context and software used are not examined here as part of a possible optimization strategy. We view the design of the hardware layout and, in particular, the server architecture employed to ensure correct scaling, as forming a very definite sub-task of an implementation project.

An analysis of the exact content of the configuration task within the specific application context is required to provide some background to the specific set of problems at hand. We won't examine the topic of hardware design in more detail here. When setting up the configuration model, you make fundamental decisions that have a lasting effect on runtime behavior during the processing of a configuration task. This specific influencing factor is discussed in greater depth in the following.

10.1.2 Reasons for Performance Bottlenecks

If you use a small, manageable configuration model, you run little risk of encountering the kind of performance bottlenecks that plague large configuration models. Unfortunately, however, this is not a simple matter of size. After all, configuration models can grow, and experience shows that they may indeed branch out in very different directions. Here, we provide some pointers to help you detect potential performance bottlenecks in a configuration model. Models that do not stand out based on these criteria may still contribute to long runtimes. Conversely, models that fit one or more of the criteria listed below may still result in perfectly adequate runtimes. Nevertheless, it is advisable to scrutinize the modeling in such cases and to focus on runtime behavior.

Number of Characteristics

The number of characteristics required to provide a complete description of an entire configuration or to describe a configurable instance, for example, a configurable component, is one very simple indicator of the complexity of a configuration task. Often, a model contains auxiliary characteristics in addition to the product options that are visible to the user. These are used to determine intermediate results of calculations. Naturally, this depends to a large extent on the processing steps that are performed for the characteristics. For models containing 100 or more characteristics, you should pinpoint which characteristics are essential and why.

Number of Characteristic Values

The number of possible characteristic values provides another indication of the configuration model's complexity, particularly if single values are defined. If the number of single values defined for a characteristic is in the region of 100, it is always worth probing a little deeper.

Structure of the BOM

For BOMs, the number of components is less significant than the specific BOM structure, the number of BOM explosions executed, and the times at which these occur. The use of BOMs should be kept to a minimum in the high-level configuration. Simple selection conditions have short processing times. If you have far in excess of 10 alternative components in a BOM, modeling based on a class node may prove faster in certain cases. Using a class node always makes a model easier to manage.

Object Dependencies

In terms of object dependencies, it is important to note in particular the interaction between dependencies of different types and the interaction between these and the BOM explosion process, as well as of the quantity structure. These interactions should always be subjected to rigorous testing.

> Object dependencies of the Action type can be regarded as obsolete, not least from the point of view of performance. You should therefore avoid using configuration rules of the Action dependency type as far as possible and use the Procedure dependency type instead. **[+]**
>
> Some configuration rules that are formulated as dependencies of the Procedure dependency type are occasionally very long, extending over several pages. As a rule, it is worth scrutinizing procedures like these.

At each configuration step, inferences from procedural object dependencies are reset and execution is repeated. The benefit of using declarative object dependencies in constraints is that they are only executed when a change to the configuration (for example, setting a characteristic value) makes their execution necessary. However, if characteristic values derived from procedures trigger the execution of constraints, this benefit becomes a drawback because constraints are also associated with a many, largely unnecessary, repeated evaluations.

Constraints

Constraints are a very efficient way of formulating cross-component interdependencies. However, you should be aware of the fact that a small number of concisely formulated constraints may sometimes result in a very large number of occasionally unnecessary evaluations. If these involve the transfer of values from lower-level BOM components to higher-level BOM components, this may result in additional loops in the BOM explosion.

Variant Tables

We have already recommended several times that you exploit the full potential of variant tables where possible. The layout of variant tables may play an important role in runtime behavior. Very large variant tables represent a performance bottleneck and are to be avoided if possible. If you come across variant tables containing hundreds of thousands of cells, you should investigate alternative modeling solutions. You can link a variant table to a database table in Transaction CU61 (Create Variant Table) or CU62 (Change Variant Table). However, you should only do so if the variant table contains approximately 100,000 cells or more (for example, tables with more than 10 columns and 10,000 rows). Linking smaller variant tables to database tables has a negative impact on runtime performance. In general, variant tables should be structured so that each cell contains exactly one single value.

Knowledge Bases and Runtime Versions

If you use the IPC, you'll also need to use knowledge bases and their runtime versions. The runtime version of a knowledge base is generated as a snapshot of the configuration model that is valid on a certain key date. The key date is interpreted as the VALID FROM validity date in the runtime version. If you make frequent changes to your model and generate new runtime versions each time you do so, the large number of runtime versions may result in performance bottlenecks during loading of the correct runtime version in each case. Frequently, it may be sufficient to overwrite existing runtime versions instead of creating new ones. The size of a runtime version is occasionally determined by the number of texts that exist for characteristics and characteristic values. If you have maintained many long texts in multiple languages in your model, it may be useful to generate a runtime version for one specific target language only. When creating runtime versions, BOMs should only be included if they are actually needed to process the configuration task.

If a large number of configuration results are stored in SAP ERP, long runtimes may also occur when these results are accessed. This situation may arise owing to the creation of a large number of sales orders for configurable products over a long period of time. A very large number of characteristic value assignments are stored in the system in such scenarios.

SAP Note 917987 (General Performance in Variant Configuration) contains additional **[+]** recommendations for creating high-performance configuration models for LO-VC. SAP Note 1081650 (Modeling Tips for the User of the IPC) provides additional information about using the IPC or AP Configuration Engine.

10.1.3 Performance Analysis

You should conduct rigorous ongoing testing of your configuration models during their creation. This testing should focus on both functional aspects and runtime behavior. You should seek to simulate as accurately as possible the real-life conditions in which the models are to be used and the typical interactive behavior of the users who will subsequently process the configuration task on a regular basis. You can then optimize the model. For example, the sequence of value assignments may impact runtime behavior significantly. In a model that makes considerable use of value restrictions, it is useful to begin by assigning values to characteristics that are highly selective and therefore significantly restrict the number of variants.

If long runtimes occur during stress testing, a more detailed performance analysis is recommended. However, such an analysis should also be undertaken if runtimes are classified as being "still acceptable". This is because further enhancements of the model quickly lead to unacceptable response times in these cases.

The simplest tool you can use to analyze a processing sequence triggered during an interactive configuration in LO-VC Variant Configuration is the *trace function*. You can access the trace function during interactive configuration, for example, during sales order processing, by following the menu path EXTRAS • TRACE. The trace functions controlled by the individual menu items are specified below:

▶ **Settings**
Define the trace parameters.

▶ **Activate**
Start logging.

▶ **Display**
Display the log.

▶ **Deactivate**
Deactivate logging.

▶ **Delete**
Delete the log.

Selected trace settings are provided as an example on the left of Figure 10.1. The DYNAMIC DATABASE checkbox is set under TRACE AREAS. The *dynamic data base* (DDB) is the part of the configurator where the characteristics that occur in a configuration are managed, together with their values and, where relevant, with their restricted value ranges.

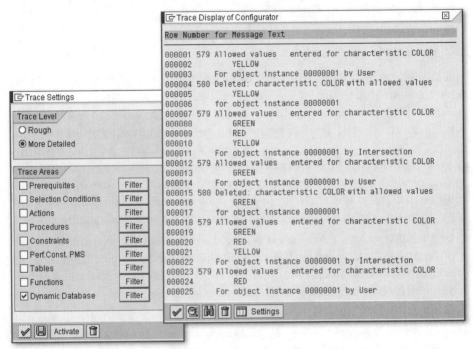

Figure 10.1 Trace Function in LO-VC: Settings and Log

The trace log shown on the right of Figure 10.1 describes a sequence of configuration steps for the Hello World example presented in Chapter 1, Basic Principles of Variant Configuration. Here you can see that the user has assigned various values in sequence to the COLOR characteristic. The value YELLOW was set first and then deleted, followed by the value GREEN and then, finally, the value was set to RED.

The FILTER function in the trace settings allows you to restrict the information displayed in the log. For example, you can restrict logging to certain characteristics when using a dynamic database. Note also the PERF.CONST. PMS checkbox, which you can set in the trace settings. PMS stands for *Pattern Matching System*, which is the part of the configurator that controls the execution of constraints.

For information about performing more in-depth analyses with LO-VC, refer to SAP **[+]**
Note 1121318 (Analysis of the Performance of Dependency Knowledge in VC). This
note discusses an advanced tool for performance analysis in Variant Configuration. A
key feature of this tool is its ability to calculate the specific runtimes for individual processing steps.

Trace functions and additional tools for performance analysis are also available for
the IPC or SCE and the AP Configuration Engine.

For more information, see Transactions STAD (Business Transaction Analysis) and ST05 **[+]**
(Performance Analysis) and SAP Notes 997111 (JRAM Instrumentation of the Configuration) and 901689 (Engine Trace for Configuration).

10.2 Change Management

Change Management represents a core logistical function. It allows you to plan, manage, document, and monitor changes during all phases. In the SAP environment, the
Engineering Change Management (ECM) component allows you to manage changes
to master data. SAP also has an *Order Change Management* (OCM) component, which
allows you to manage changes to production orders. Both forms of change management are relevant for Variant Configuration.

10.2.1 Engineering Change Management (ECM)

In Chapter 2, Creating a Product Model for SAP Variant Configuration, we examined
the constituent parts that make up a model for Variant Configuration and explained
how these are created. We'll assume, at this point, that your model is fully maintained. In other words, all required master data has been created, and the model
has been tested to an adequate degree and corrected and enhanced as necessary.
The model has been released for the processes in the logical chain and is already in
operational use.

According to the approach we have taken in this chapter, this type of model in Variant Configuration is not rigid. Rather, it is dynamic and constantly evolving. The
planning, documentation, and monitoring requirements are much more demanding
when it comes to enhancing the model than was the case during the development
phase. ECM allows you to plan changes to the model in a controlled manner and to
effect these changes at the desired point in time.

ECM offers the following benefits:

- All changes are monitored and clearly documented.
- Changes are planned and implemented within a defined validity period (usually on a specific validity date).
- Each model object has a history because each development status is stored.
- All essential processes automatically recognize the valid development status.
- Several related master data changes are grouped together under a shared change number.

The example below explains how to use master data change management.

A New Processor for a Configurable Computer

Several processors are offered to customers as alternative components for a configurable computer. After a trial run of three months, the lowest-performing processor is to be removed from the selection. A new high-performance processor is to be offered simultaneously so that variability remains constant for the customer.

So, which items of the configurable computer's model are affected by these planned changes?

- **Material**
  A new material is created for the processor in the material master.

- **Documents**
  Changes may be made to the documents, for example, drawings or specifications. In this case, new versions are created for the relevant document info records.

- **BOM**
  The new processor is added to the BOM, and the lowest-performing processor is simultaneously deleted.

- **Routing**
  The routing may need to be adjusted. This may affect new or changed operations, changed times, changed PRT assignments, or component assignments.

- **Characteristic**
  The characteristic for querying the preferred processor must be changed. The value for the lowest-performing processor must be deleted and a new value added for the new processor. It is also necessary to check where this characteristic is used, in particular in connection with the value that is to be deleted. The where-used list for characteristics and characteristic values in Transaction CT12 is ideally suited to this purpose.

▶ **Variant class**

Changes are generally made to the variant class. However, this is not the case in our example because no changes need to be made to the assignment of characteristics to the class.

▶ **Configuration profile**

Changes to the configuration profile are often necessary. This is the case, for example, if changes to the assignment of object dependencies or other settings are required.

▶ **Object dependencies**

It is necessary to check the object dependencies, which are detected for characteristics and characteristic values using the where-used list. Often, no changes need to be made to the object dependencies for the high-level configuration. Meanwhile, a selection condition for the BOM item of the new processor is required for the low-level configuration in the BOM and routing.

▶ **Variant table**

The checks conducted in the high-level configuration to establish completeness and consistency and determine the assigned values are often based on variant tables. The considerable advantage of this approach is that you only need to change the contents of the relevant variant table, whereas the object dependencies themselves can remain untouched.

In the following sections, we'll discuss in more detail how changes to the model for the configurable computer are scheduled with ECM.

A *change master record*, which is identified by a *change number*, is created to make the changes discussed above using change management.

> The change master record is often simply referred to as the change number. It contains the essential settings for the planned changes.   **[+]**

Creating Change Numbers: Initial Screen

When you start Transaction CC01 to create a new change number, the initial screen shown in Figure 10.2 is displayed.

Here you select a type for the change number. The following types are available:

▶ Change Master

▶ ECR (engineering change request)

The difference between these two types lies in the support provided during the change process rather than the actual result of this process. The result in terms of changes with a corresponding change number is the same in both cases.

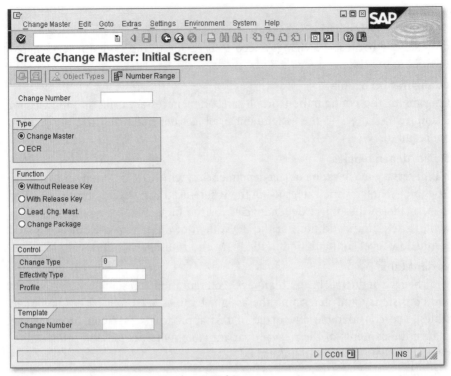

Figure 10.2 Creating a Change Number: Initial Screen

In contrast to a *change process* with the Change Master type, the ECR type allows you to carry out additional steps defined in the system status network using change numbers:

1. The relevant objects are determined after the start of the request phase.

2. The relevant departments are tasked with checking the planned changes for relevance and potential problems.

3. All check results are logged during the check phase.

4. All changes are confirmed during the approval phase.

5. If the engineering change request is converted into an engineering change order, all departments affected are tasked with implementing the changes.

6. The changes that are actually implemented are logged.

7. When the engineering change request is released, it is *frozen* and can no longer be changed. Any necessary corrections require a new change number.

The function of a change number determines whether a *release key* is to be used.

If a release key is not used, all changes are immediately visible and therefore also active for all areas along the supply chain, provided that these areas use validities from the change number.

Using release keys means you can also restrict the validity of the changes. In this case, the changes are only available for the relevant areas when the changes have been released using the release key.

Changing the Processor With and Without a Release Key   [Ex]

The processor is to be changed after a trial run of three months. If a release key is not used, the changes made to the BOM are immediately available for all areas after these changes are saved. This means, for example, that if material requirements planning creates planned independent requirements for the next six months for the configurable computer immediately after the changed BOM is saved, these planned independent requirements will use the old BOM for the first three months and then use the changed BOM for the three months after that. If, on the other hand, a release key is used, material requirements planning will use the old BOM for all six months in this scenario. The changed BOM would only be effective for the last three months in this period if a corresponding release key were to be set in the change number.

Note that the use of a release key is also necessary if changes to master data are to be taken into account as part of Order Change Management. This is discussed in more detail in Section 10.2.2, Order Change Management (OCM).

When you create change numbers, you can also use *leading change master records and change packages*. These allow you to create change hierarchies and structure change numbers accordingly. The leading change master record represents the root node of one of these hierarchies, whereas the change packages represent the leaves or end nodes.

The *change type* is a control element in change management, similar to class types in classification and material types for the material master. The change type is only available for engineering change requests and is not available for change master records. In the relevant Customizing settings, you can, for example, assign the function of a user status profile to the change type to enhance the previously mentioned system status network.

A change number can also be assigned a *validity type*. If a validity type is not assigned, the development status of a change is determined by the Valid From key date. In our example of the processor replacement, a specific validity type is not required. You can define different procedures for validity types in Customizing. For example, you can specify that the changes created with a certain validity are only to apply for a very specific interval of time.

[+]

Using Templates

When you create change numbers, you can also use a profile of default values or a copy template.

Creating a Change Master: Change Header

A description must be entered in the *change header* (see Figure 10.3). You must also enter a Valid From key date here. This is the date as of which the changes are to take effect. This validity date applies if special validity types are not used. Entries are required for other validity types, such as the relevant range of serial numbers. In addition, you must specify a status for the validity number. This status determines whether master data can be changed with reference to the change number, whether the Valid From date can subsequently be changed, and whether a distribution lock applies to changed master data.

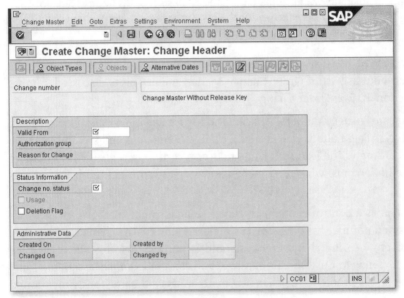

Figure 10.3 Change Header in the Change Master

Creating a Change Master: Object Types

Additional entries are required in relation to object types. A list of relevant object types is shown in Figure 10.4.

In this list, you must select the object types you want to change with this change number as Active. As you can see, the list includes all of the essential object types that make up a model in Variant Configuration. However, the following configuration model items are not supported as object types for a change master:

▶ Variant function

▶ Structure of a variant table (the table contents can be changed with change management)

▶ Class (the characteristic assignment can similarly be changed with change management)

▶ Configuration of a material variant

The checkbox in the Actv. column (see Figure 10.4) must be selected to ensure that objects of the relevant object type can be changed with a change number. If only this checkbox is selected in this column, each object of the object type can be changed with the change number, as shown in Figure 10.4 for task lists, characteristics, object dependencies, and several other rows.

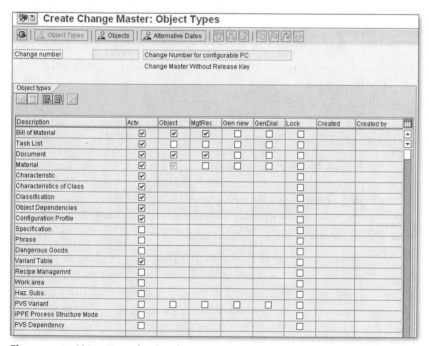

Figure 10.4 Object Types for the Change Master

If not all objects of an object type can be changed with the change number, the Object checkbox must also be selected. In this case, objects with the relevant object type can only be changed with reference to the change number if they have been explicitly entered in the object list, which you access by clicking on the Objects button. The checkbox for the Material object type is selected by default. This means that, in our example, we must enter the material for the new processor in the object list for the Material object type. Doing this generates an *object management record*.

You can also select the following checkboxes in the list of object types:

- **New object management record is generated**
  If the MgtRec (new object management record is generated) checkbox is selected, the object list is automatically extended for this object type if an object of the relevant object type is created or changed with reference to the change number.

- **Generation with new creation only**
  The Gen new (generation with new creation only) checkbox triggers the automatic extension of the object list when objects are created for the first time but not when they are changed.

- **Generation with dialog box**
  If the GenDial (generation with dialog box) checkbox is selected, a dialog box opens each time the object list is extended.

Let's examine this last point for a moment. Why are automatic extensions of the object list permitted? These don't necessarily go against the principle of an explicit object list. If a problem occurs with explicit object lists, it can only be resolved by creating master data with internal number assignment. In this case, the object key cannot be manually assigned to the list beforehand.

Figure 10.4 also shows a button labeled Alternative Dates. This button is linked to a function that allows you to use alternative Valid From key dates within a change number. In our example, we want the material we're creating to be immediately available for the new processor. We therefore create an alternative date in the present and assign the object management record of the material for the processor to this alternative date.

Once we've created the change number as a change master record as described, we can start to implement the changes themselves.

Making Changes to a Material

First, we create the material for the new processor. The change takes immediate effect based on the alternative date we just created. Only active materials can be assigned to the BOM. This also applies to BOM changes that are merely planned. In addition, this is the only way to trigger the procurement of the material, for example, by the purchasing department. We can then change the BOM. Here, we replace the lowest-performing processor with the new processor. Note that documents can also be changed with reference to the change number.

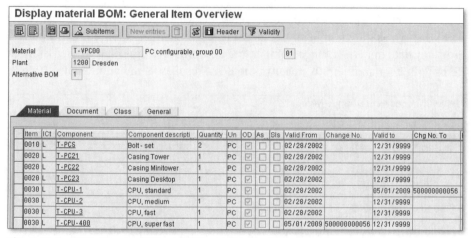

Figure 10.5 BOM Changes With and Without Change Management

Making Changes to the BOM

The BOM illustrates perfectly the difference between changes made with and without change management, and therefore also between changes with and without a history. Two changes were made to the BOM shown in Figure 10.5. First, the component quantity in item 10 was increased from 1 to 2 without the use of change management. In item 30, processor T-CPU-1 was then overwritten with processor T-CPU-4 with reference to change number 500000000056.

Once an object is changed for the first time with reference to a change number, that object is then subject to the *ECM history requirement*. This means that all subsequent changes made to that object can only be made with ECM, in other words, with reference to a change number. Otherwise, it would be impossible to ensure a consistent change history for the object.

The change to item 10, which was made without change management, isn't discernible in the BOM display. If the BOM is exploded on any date on which the BOM is valid, the changed quantity (2) is always displayed as item 10.

In contrast, the change of processor in item 30 has a history. Technically speaking, if an item is changed using change management, the item is copied rather than changed. New information is added to the old item regarding the Valid From key date. The relevant date from the change number is assigned to the copied item as its Valid From key date. The changes are then made to the copied item. If the BOM is then exploded on a date prior to the Valid From key date from the change number, the old processor is used, and the first three items with item number 30 appear in the BOM display, as shown in Figure 10.5. If the BOM is exploded on or after the Valid From key date from the change number, the new processor is used, and the last three items with item number 30 appear in the BOM display, as shown in Figure 10.5.

BOM Changes for a Change Number						
Hierarchy	Chg. Order	Item	Item Categ	Descriptio	Quantity	Unit of Me
▽ 📝 Change Number: 500000000056				Change Number for configurable PC		
▽ 🔩 Material BOM 00001708				Usage 1		
▽ 🔩 Alternative 01		1200 / T-VPC00		PC configurable, group 00		
▷ 🔩 Item:0030	OLD	T-CPU-1	L	CPU, standard	1	PC
▷ 🔩 Item:0030	NEW	T-CPU-4	L	CPU, super fast	1	PC

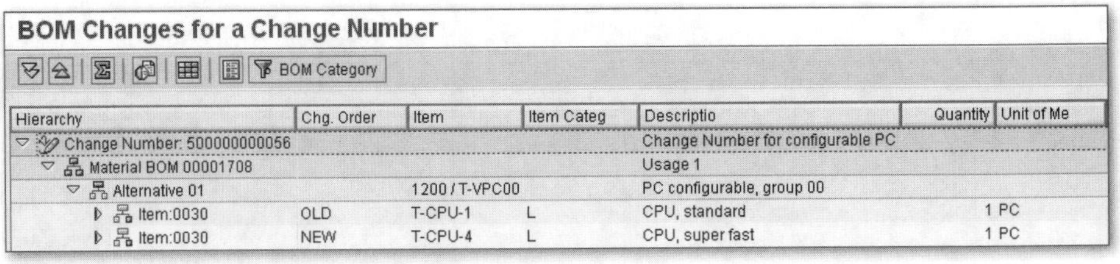

Figure 10.6 Documentation of BOM Changes

In Transaction CC03, you can select the menu option ENVIRONMENT • REPORTING • BILL OF MATERIAL to view details of the changes made to the BOM with reference to the change number. An overview of the evaluation for our example is shown in Figure 10.6.

You can make the preferred change to the processor in one of the following two ways:

▶ **Overwrite**
In our example, we have simply overwritten the material entry in the item containing the material that represents the old processor. This change can be detected on the screen shown in Figure 10.6, where the Chg. Order column contains entries with the Old and New change categories.

▶ **Delete and create again**
Alternatively, we could have deleted the item with the old processor and created a new item for the new processor. The corresponding change categories are DEL and NEW.

These two options produce very similar results. The only major difference relates to the assignment of object dependencies. With the first option, the object dependencies are copied and need to be adjusted. With the second option, object dependencies are newly assigned. These are either new object dependencies or the old object dependencies modified as required.

Changes to the Routing

After the BOM is changed, you can make any necessary changes to the routing. You need to change the BOM before changing the routing if, for example, you want to make changes to component assignments.

Changes to Characteristics

In characteristic maintenance, we need to add a characteristic value for the new processor to the relevant characteristic and delete the value for the old processor using change management. As with the BOM, we need to take account of the degree to which changes affect the object dependencies of characteristic values. We must also check the assignment of variant conditions to the characteristic values.

Changes to Variant Class and Configuration Profile

Changes may then need to be made to the variant class and configuration profile. However, this is not necessary in our example. The variant class needs to be changed if characteristics are to be deleted or added. Changes to the configuration profile are necessitated by various settings or the assignment of object dependencies.

Changes to Object Dependencies and Variant Tables

In terms of object dependencies, the selection condition for the changed BOM item must be adjusted in our example. Because we've already defined the new characteristic value, we can make this change now. It's also necessary to check the degree to which the characteristic and, in particular, the old characteristic value for the old processor, is used in object dependencies, variant tables, and other objects of Variant Configuration. The where-used list in Transaction CT12 (see Figure 10.7) is ideally suited to this purpose. Once you locate the relevant object dependencies, you must check these from the point of view of the planned changes. This check must incorporate both the syntax and the where-used list for object dependencies in Transaction CU05. Increased use of variant tables is of benefit in this instance, because it means that the actual object dependencies do not need to be changed. Instead, you simply modify the contents of the variant table, which should be a much simpler task.

Result of the Changes

Finally, you need to check the uses in planning tables and material variants in the change result (see Figure 10.7).

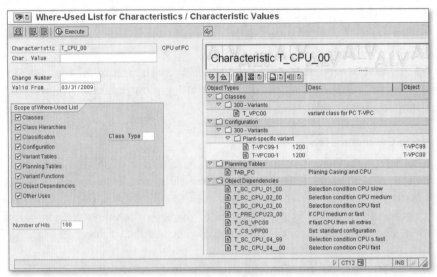

Figure 10.7 Where-Used List for Characteristics/Characteristic Values

The preferred model changes are then available and active in the system. You can now use the changed model without further delay. But what exactly does this mean? If you now create a sales order that's to be delivered within a short space of time, the old development status is still used. You can choose between the three older values for the processor characteristic on the value assignment screen of the configuration. If the configuration profile is set accordingly, the super BOM will be displayed in the sales order. It will only contain the three BOM items with the three old processors.

The new development status comes into play if you create a sales order with a delivery date that falls on a date further into the future. In this case, the new high-performance processor is shown on the value assignment screen. The old low-performance processor, on the other hand, is no longer included in the offering. The creation of sales orders is not essential to test the enhanced model. You can also test the model in configuration simulation in Transaction CU50.

10.2.2 Order Change Management (OCM)

Production orders that are already in processing (which means material consumption or initial confirmations have been posted) cannot usually be changed automatically.

In this case, you can use *Order Change Management* (OCM) as a tool to make controlled and at least partly automated changes to production orders. Let's begin by looking at the basic data:

- **Reason to use OCM**
  OCM tools are useful if your products have long lead times or if sales-order-specific make-to-order production is required, as in Variant Configuration.

- **When to use OCM**
  You use OCM when you want to make changes to sales orders, in particular, changes to the configuration, even though the production order is already in processing. It also makes sense to use OCM if structural or technological changes are to be incorporated into existing production orders.

Let's assume the scenario is similar to that shown at the top of Figure 10.8. Here, a sales order is created for a configuration material. This sales order generates a corresponding requirement. A planned order is created for this in the planning run (MRP). The planned order is then converted into a production order and released. Production starts and is documented by the posting of goods movements or confirmations. A change is then subsequently made to the sales order. Or, to be more precise, the configuration is changed in the sales order.

These changes to the configuration result in an alternative explosion of the BOM and routing. However, these changes cannot normally be copied automatically to the production order. Because goods movements or confirmations already exist for the production order, the Read PP Master Data function can no longer be used.

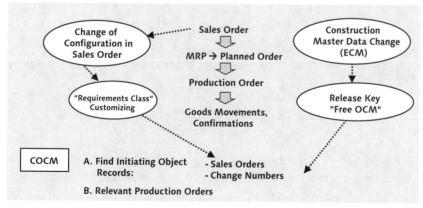

Figure 10.8 Searching for Initiating Object Records in the OCM Process

Let's take another scenario. In some situations, similar to the one we described where a change is made to an existing sales order, it may be necessary to make changes to the master data in production orders. Certain key questions need to be answered in relation to this scenario:

▶ **For which materials can production orders be taken into account in OCM?**
To activate a material for OCM, you must assign an overall profile for Order Change Management on the Work scheduling tab in the material master, for example, PP0001 — Overall Change Profile (OCM selected) (see Figure 10.9).

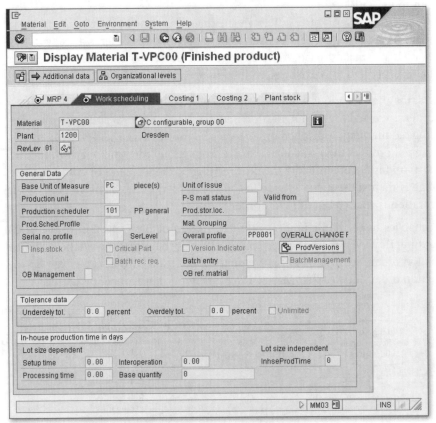

Figure 10.9 The Work Scheduling Tab in the Material Master

▶ **For which sales orders can subsequent changes be made with OCM?**
A sales order only writes an initiating object record for OCM if a *requirements class* in which OCM was activated is found for the sales order item. The Customizing settings for requirement classes include an EC Management (trigger order change management) checkbox, which determines whether an initiating object record for OCM is to be written for production orders. This checkbox must be selected.

The requirements class is determined by the *planning strategy set in the material master.* For detailed information about the settings for requirements classes, refer to Chapter 4, Customizing SAP ERP for Variant Configuration.

▶ **Which changes to master data are taken into account in OCM?**

Changes to master data such as the BOM and routing must be made with change management (ECM) to ensure that these changes can be taken into account in OCM. However, a change number only writes an initiating object record for OCM if the change number has been assigned a release key that also releases the changes for OCM. Release keys can be configured in Customizing. When you assign a release key to a change number, the input help (F4) indicates whether the changes are released for OCM.

We can distinguish between the following three OCM processes:

▶ Sales order change process

▶ Master data change process

▶ Change process for assembly orders

We'll discuss the most important OCM process, namely, the sales order change process. This process is triggered by a change to a sales order. In the following example of this change process, we assume that a relevant requirements class is used, which means that initiating object records are written for OCM.

Changing a Sales Order for a Computer [Ex]

A sales order is created as the basis for an ongoing procurement process. The customer then calls and requests certain changes to the configuration, specifically, a minitower (instead of a tower), a medium-speed (instead of a fast-speed) processor, in blue with a gloss finish (rather than without) and including the MS Office suite (rather than Word alone).

As shown in Figure 10.8, the actual OCM process starts with the search for initiating object records in Transaction COCM. To access this transaction, follow the menu path LOGISTICS • PRODUCTION • SHOP FLOOR CONTROL • CONTROL • ORDER CHANGE MANAGEMENT • CHANGE PROCESS • INITIATING OBJECT RECORDS from the SAP menu. The initiating object records can be sales orders and change numbers.

Figure 10.10 Initiating Object Records

The results of the search in our example are shown in Figure 10.10. Initially, only the initiating object records are displayed (in this case, sales order item SalOrd 12159/10). If you then click on the execute icon, the system searches for the procurement elements affected by the changes (in this case, production order PrOrd 60000083). Up to this point, the OCM process can be used as a selection tool, after which the relevant production orders can be edited manually. However, you also have the option of having these edited automatically, and this is the option described here.

Once you have this information, you can start editing the procurement elements in Transaction COCM1 (see Figure 10.11).

Order change process - Procurement elements

	Procurement element	Material	Plant	SimOrder	FindC	ChckC	Chng	Closed	Initiating objt	Reason	Change No.
	PrOrd 000060000083	T-VPC00	1200						Sales order 12159/10	SlsOrdCh	

Figure 10.11 Procurement Elements

Editing consists of the following steps, which are illustrated in Figure 10.12.

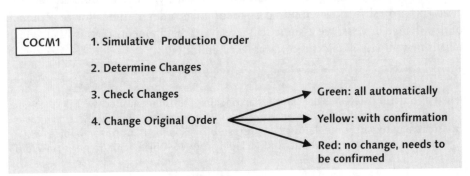

COCM1

1. Simulative Production Order

2. Determine Changes

3. Check Changes → Green: all automatically

4. Change Original Order → Yellow: with confirmation

→ Red: no change, needs to be confirmed

Figure 10.12 Initiating Object Records for Procurement Elements in the OCM Process

You can access these steps by clicking on the execute icon and then execute each separately in sequence or together in a single run.

The first step, Create Simulation Order, generates a simulated production order, which you can display by clicking on the display simulation order icon.

The second step, Determine Changes, compares the simulated production order with the original production order. To view the results of the comparison, click on the changes overview icon. An additional status, NOCH, is assigned to the original production order to prevent manual changes to the production order.

The third step, Check Changes, results in the assignment of another status, CGFB, to the original production order. This prevents the posting of additional confirmations and goods movements. It is assumed that production has stopped when this status is assigned. The purpose of this check is to determine which changes can be made with OCM. The overall profile assigned in the material master plays a key role here. You'll find detailed information about overall profile settings in Chapter 4, Customizing SAP ERP for Variant Configuration. In general, we can distinguish between noncritical, critical and prohibited changes.

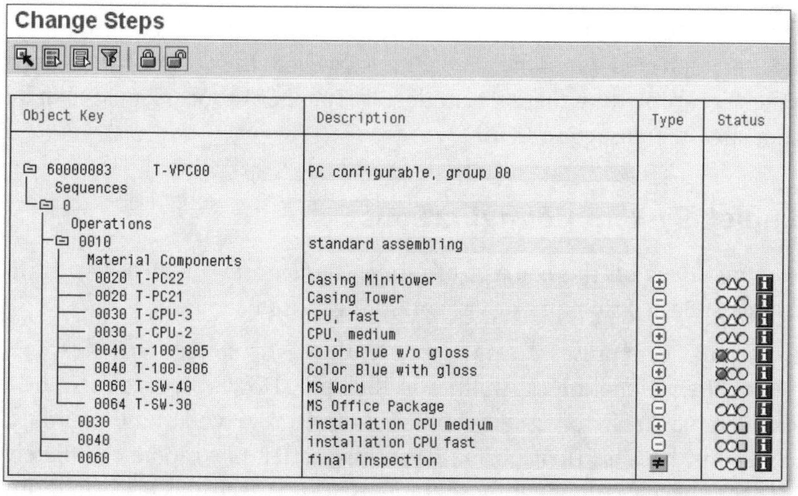

Figure 10.13 Checking the Change Steps

Figure 10.13 shows the change steps determined for our example, together with the corresponding change indicator in the Status column. The noncritical changes to operations 0030, 0040, und 0050 refer to the replacement of the processor and its casing. These changes can be made without any problem. The prohibited changes refer to the replacement of the material component Color Blue without Gloss with the material component Color Blue with Gloss. Prohibited changes cannot be accepted automatically using OCM. The other changes to material components are classified as critical. In these cases, the component changes require manual confirmation.

In the fourth step, Change Original Order, you change the original production order. In this step, the critical changes are either confirmed or rejected. The two status indicators NOCH and CGFB are then deleted.

The second OCM process, that is, the master data change process, follows the same steps as the sales order change process described here. The only difference in this case is that a simulated production order is not required in the master data change process. Instead, the changes can be determined immediately and directly from the change number.

The third OCM process, the change process for assembly orders, is used, as its name indicates, if *assembly processing* is used in sales and distribution. This is determined by the requirements class used. In assembly processing, the sales order generates the procurement element (in this case, the production order) directly by bypassing MRP. In this case, the system tries to transfer changes made to the sales order directly to the production order. If OCM is activated for the material and sales order, OCM also tries to change the current production order automatically without user intervention. All steps in the OCM process are executed in the background. If critical or prohibited changes are detected when the changes are checked, the OCM process stops and issues a message, providing the relevant details. The OCM process must then be continued manually in Transaction COCM1.

10.3 Complex System Configurations

In this section, we focus on the concept of *system configuration*, which was introduced in Chapter 1, Basic Principles of Variant Configuration.

This topic is one that gets pulses racing among enthusiastic configuration experts because it pushes their modeling capabilities to the limit. However, it can also be a source of considerable stress for users whose daily tasks revolve around complex, configurable systems, because their business processes often need to be completely adjusted to suit the specific configuration solution used. The decisions made in terms of the software solution employed in this area of conflicting requirements are often very wide-ranging and can only be changed at a high cost.

In this section we'll provide a simple definition of system configuration, and then present three examples, which may appear to be a little academic. However, these have been deliberately selected because they can be easily understood and will help you distinguish between various levels of complexity.

10.3.1 System Configuration — Definition

Our first example of a system configuration is an open-plan office with numerous matching and, in some cases, adjoining desks, cubicles, shelves, and partitions (see Chapter 1, Basic Principles of Variant Configuration). The technical configuration of mobile communications networks, control cabinet installations for shop floors, emergency power supplies for hospitals or underground railways are other examples of system configurations. We could provide many more examples of systems where a range of configurable components interact with one another.

The following definition seeks to capture the essence of system configurations, which distinguishes these from the individual product configurations with which we are already familiar as an integral part of the standard business process solution in the SAP Business Suite.

> **Definition of System Configuration**
>
> A system configuration is a configuration task involving a number of configurable components, where the interaction between these components in the configuration cannot adequately be described in a static BOM hierarchy.

Essentially, system configuration concerns the properties of the *configuration structure*. As we have seen, a super BOM can be used as an instrument for describing any configurable components when configuring a specific, clearly defined product. In accordance with our definition, a system configuration always refers to configurations where a super BOM cannot adequately describe the structure of the components.

The wide range of issues that necessitate the use of system configurations can be divided into the following three categories:

▶ **Dynamic modification of the BOM structure**
Here, a specific configuration requires the enhancement of the predefined super BOM to include additional, possibly configurable components.

▶ **Interlinked configuration structures**
The configurable components are interlinked in a way that cannot be expressed in the BOM. This is the case, for example, if connection conditions apply to neighboring components. For example, the adjoining edges of two desks in an open-plan office should always have identical dimensions.

▶ **Composition problems**
A system is gradually built and expanded from a set of configurable and combinable components, based on the ongoing addition of clearly defined component instances.

We'll examine these aspects in the following sections using three typical examples of system configurations. Of course, all of these aspects may also occur in combination within a single system configuration.

10.3.2 Dynamic Modification of the BOM Structure

In this section we illustrate the problems associated with enhancing the BOM structure with configurable components along with a practical example.

Wood-Framed Window Example

A configurable wood-framed window can easily be described using parameters such as width, height, interior paint color, exterior paint color, glass type, and direction of opening (to the left or right). In this case, it should be easy to itemize any components required in a super BOM. The double-glazed glass pane is normally manufactured or sold at a separate manufacturing level.

Windows may also have various types of sash bars. These bars can be attached to the window with a sealant or may be used as part of the frame to divide the window pane. The configuration of the bars may be based on predefined geometric patterns or the customer's individual requirements. A sample configuration is shown in Figure 10.14.

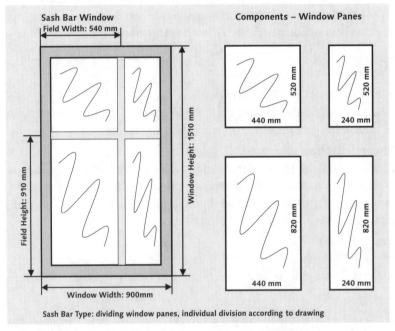

Figure 10.14 Sash Bar Window with Panes of Varying Dimensions

The number of glass panes with various dimensions is based on the specific configuration selected. The predefined super BOM is clearly pushed to its limits in this case. The window pane component must be included in the BOM several times based on the details of the specific configuration to ensure that the various dimensions are represented. This multiple use of configurable components, which is also known as *dynamic instantiation*, is in conflict with a statically designed super BOM.

If the possible arrangement of the sash bars is restricted to a small number of geometric patterns, we can take a pragmatic approach and add several different panes to the super BOM as components. In this case, the number of components corresponds to the maximum possible number of panes with different dimensions. The effort involved in formalization is therefore relatively large. This approach is also inadequate if, as in the example of the wood-framed window with a customer-specific sash bar configuration, the number of component variants cannot be determined accurately in advance.

Sales Order and Order BOM Configuration Processes

LO-VC Variant Configuration allows you to manually enhance static BOMs in one of the following two ways:

▶ Using the *sales order* configuration process

▶ Using the *order BOM* configuration process (ETO production scenario)

Figure 10.15 shows some of the settings that can be made for the configuration profile.

The Sales order process, which can be selected on the CONFIGN PARAMETERS tab in the configuration profile, allows you to manually add configurable (and nonconfigurable) components to the sales order as subitems — provided that the MANUAL CHANGES ALLOWED checkbox is selected in the SALES ORDER tab of the configuration profile. The configurable components can each have different values. The manually added components are visible in the sales order. Only in the sales order is the overall structure of the configuration known. A production level based on the overall system cannot be supported. Because windows are normally fully assembled when sold (rather than as construction kits of frames, hinges, and panes), the *sales order* configuration process is not ideally suited to our wood-framed window.

If you select the Order BOM process on the CONFIGN PARAMETERS tab, the BOM is completed as part of technical postprocessing of the sales order. Select the following path in the SAP menu to access Transaction CU51 for technical postprocessing of the order-specific BOM: LOGISTICS • PRODUCTION • MASTER DATA • BILLS OF MATERIAL • BILL OF MATERIAL • ORDER BOM • MAINTAIN (MULTI-LEVEL). After you create and release a sales order, you can manually configure and change a sales-order-specific BOM in technical postprocessing. A configurable component can be added as an additional BOM item with different values assigned during configuration. The logistics process for our wood-framed window example can be represented very well using this approach.

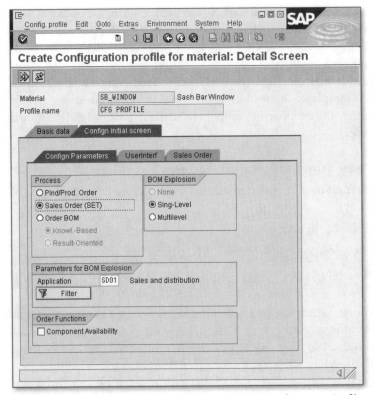

Figure 10.15 The Configuration Parameters Tab in the Configuration Profile

Order BOMs for Components

If the Result-Oriented radio button is selected for the sales order configuration process in the configuration profile, you can similarly create order-specific BOMs for various configured components (with reference to the same component material), and these BOMs can then be postprocessed manually.

An order BOM is identified by the three IDs for sales order, order item, and material. If you also want to adjust the BOMs of configurable components for a specific order, a problem will occur with identification, at least when you have several identical component materials with different configurations. To distinguish between order BOMs for components with different configurations, a material variant is created for each of the different configurations of the component. In other words, a new entry is created in the material master for each.

This process is referred to on the user interface as the *instantiation* of the configured component. As a result, the order BOMs for the components can also be identified by the three IDs for sales order, order item and material (or material variant).

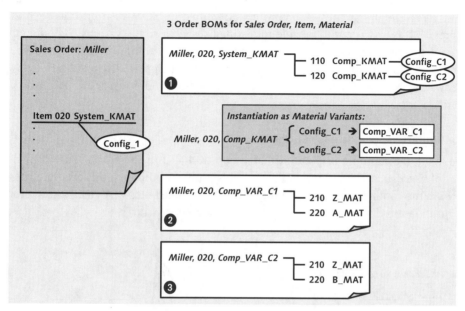

Figure 10.16 Order BOMs with Instantiated Components

Figure 10.16 provides a schematic representation of an example of the order BOM configuration process with instantiated components. Here, item 020 of the Miller sales order contains a System_KMAT material with the assigned configuration result, Config_1. In the corresponding order BOM (Miller, 020, System_KMAT), the configurable material Comp_KMAT appears in items 110 and 120 with the assigned configuration results, Config_C1 and Config_C2 (❶). To allow order BOMs to also be maintained for these two components, the configured components Comp_KMAT, Config_C1 and Comp_KMAT, Config_C2 are instantiated as material variants (Comp_VAR_C1 and Comp_VAR_C2). As a result, we can identify and distinguish between the relevant order-specific component BOMs (❷ and ❸).

However, if extensive use is made of instantiated material variants, the material master needs to be extended to a significant degree. In this case, the generic LO-VC solution for describing variants using parameterization reaches its limits.

The configuration-specific enhancement (of the super BOM defined as a master data record) that we have described here is supported in the SAP Business Suite and is fully integrated into the logistics processes.

10.3.3 Interlinked Configuration Structures in LO-VC

In many cases, dependencies between components are easily formulated if you can distinguish between the relevant components based on their hierarchical nesting or their type. The hierarchical structure is determined by BOM relationships such as "part of". It is important to be able to distinguish between the various components in a relationship. You can differentiate between components of different types based on their BOM item, their component material, or the assigned variant class. In many cases, however, these criteria prove insufficient because additional relations between components exist.

Configuration Structure for Sudoku

We'll use the example of Sudoku that was introduced in Chapter 1, Basic Principles of Variant Configuration, to illustrate this problem. Although this example goes beyond the confines of product configuration, it demonstrates very clearly how a lot of information can be contained in static component structures.

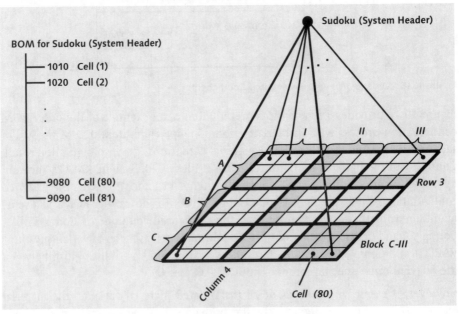

Figure 10.17 BOM and Configuration Structure for Sudoku

The number fields contained in a Sudoku grid of 9×9 cells are arranged in nine rows, nine columns, and nine blocks of 3×3 cells. If we want to represent the Sudoku structure in LO-VC Variant Configuration, the most obvious approach is to model the cell as a variant class (or a material) called CELL and to assign it the number field as the VALUE characteristic. The Sudoku system is represented by the variant class (and material) SUDOKU and is assigned a BOM with 81 items to represent the 81 cells (see Figure 10.17). The first cell in BOM item 1010 is in row 1 and column 1, the second cell in BOM item 1020 is in row 1 and column 2, and so on. However, this model does not adequately map the complex structure of rows, columns, and blocks and the individual position of a cell in the overall structure. To describe the position of a cell, we must also assign the characteristics ROW, COLUMN, BLOCKROW, and BLOCKCOL-UMN. This allows us to describe the multidimensional structure and apply declarative configuration rules of the constraint dependency type.

The following constraint uses the BOM item to determine all cells in column 4 and assigns the relevant values to the COLUMN and BLOCKCOLUMN characteristics. Note that only three of the nine cells are listed in the condition part.

```
Rule: Identification of All Cells in Column 4 and Block Column II
OBJECTS:
  SD is_a (300) SUDOKU, C is_a (300) CELL.

CONDITION:
     PART_OF (C, SD, '1040')
  OR PART_OF (C, SD, '2040')
... and so on for all further cells in column 4 up to ...
  OR PART_OF (C, SD, '9040').

RESTRICTIONS:
  C.COLUMN = '4', C.BLOCKCOLUMN = 'II'.

INFERENCES:
  C.COLUMN, C.BLOCKCOLUMN.
```

With one rule for each column and row, we can construct the structure of the cells in a total of 18 rules. The structure of the Sudoku puzzle, with its rows, columns, and blocks, which we were unable to describe with a BOM, can now be fully represented using the characteristics of the cells.

Configuration Rules in an Interlinked Structure

With this approach, the Sudoku rule, "Each column contains each digit exactly once," can easily be formulated in a constraint:

[+]

Rule: Each Column Contains Each Digit Exactly Once

```
OBJECTS:
  SD is_a (300) SUDOKU,
  C1 is_a (300) CELL, C2 is_a (300) CELL.

CONDITION:
  C1.COLUMN = C2.COLUMN AND C1.ROW <> C2.ROW.

RESTRICTION:
    C2.VALUE  IN ('2', '3', '4', '5', '6', '7', '8', '9' )
            IF C1.VALUE  =  '1'
  , C2.VALUE  IN ('1', '3', '4', '5', '6', '7', '8', '9' )
            IF C1.VALUE  =  '2'
```
... and so on for values 3 to 9 in cell C1

```
  .


INFERENCES:
  C2.VALUE.
```

What exactly is described in this constraint? The cells (C1 and C2) in each column are compared with one another in pairs. A cell (C1) that contains a value restricts the possible value range for the other cells (C2) in the same column because none of the other cells (C2) can contain this same value. The two rules for rows and blocks can be expressed in a similar way with constraints.

For reasons of completeness we want to mention here that the following, self-evident short form of the RESTRICTION part is not supported:

C2.VALUE <> C1.VALUE.

This example illustrates the powerful modeling capabilities of LO-VC Variant Configuration. It enables the precise and compact formulation of declarative rules for complex configuration structures. However, the fact that it is relatively easy to formulate these rules should not lead you to underestimate the inherent complexity of the problems associated with variant diversity. The set of rules described above can be applied in practice but will result in long computing times owing to the large numbers of combinations involved.

[+]

Our Sudoku example also highlights another issue. Configuration tasks with a complex configuration structure usually call for a graphical user interface that can adequately represent the properties of the configuration structure. After all, nobody wants to have to grapple with a list of 81 entries when attempting to solve a Sudoku puzzle.

10.3.4 Composition Problems in SCE Advanced Mode

Product configuration with the SAP Business Suite primarily supports a hierarchical configuration structure that can be formulated in BOMs. The configuration structure, which is based on an overall system (such as a window or a Sudoku puzzle), is described as a decomposition problem in a top-down approach. However, there are also systems in which the configuration task requires the constructive merging of configurable items into a larger whole structure. These structures are composition problems based on a bottom-up approach. Product configuration tasks such as furnishing an open-plan office space, assembling a conveyor-belt system, or setting up control cabinet installations for factories and power stations all fall into this category. Such configuration tasks normally consist of a range of configurable items and, owing to their complexity, require a specially tailored high-end configuration solution based on a best-of-breed configuration engine.

The *Sales Configuration Engine* (SCE) offers an *advanced mode* that is specially designed to handle complex composition problems. This section describes how SCE advanced mode works. We'll draw on the example of a configurable bookshelf system. This easily understood example of configuring a shelving system for the orderly arrangement of a large number of books is a simplified version of the real-life configuration task mentioned of configuring a control cabinet installation.

Based on predefined rules, new books can continue to be placed on the shelves until these are full. The dimensions of the shelf units are determined by the books they are to accommodate. New shelf units can be added if required. Figure 10.18 shows an example of a shelving system and its basic components with two BOMs.

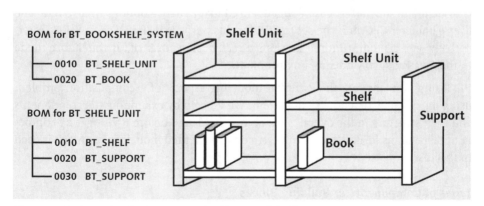

Figure 10.18 Example of a Composition Problem: Bookshelf System

Embedment in the Logistics Processing

From a business perspective, logistics processes such as production and staging are usually based on components. In large-scale composition tasks, these can normally be handled as "simple" configurable products or, from a configuration point of view, as decomposition problems. Components are not usually assembled into an overall system by the component manufacturer. Instead, this step is normally executed in the location where the system is to be operated.

The integration of the configuration tasks described here into a business process solution is always problem-specific and therefore cannot be standardized. Consider, for example, the Sudoku example discussed earlier, where a problem-specific user interface is required for interactive system configuration.

In practice, the number of composition problems that occur is relatively small when compared with the number of decomposition problems that arise in product configuration.

In view of all of these considerations, the best approach is to implement the integration of the SCE into a business process solution for a complex composition problem as part of a custom development project on a problem-specific basis.

Product Model Set-up for the SCE Advanced Mode

From a purely technical perspective, the SCE uses master data, in other words, an entry in the runtime version of a knowledge base to detect a configuration problem modeled in advanced mode. An advanced mode model or corresponding knowledge base with runtime versions can be created in two different ways. The first option is to maintain the master data with advanced maintenance options in SAP ERP. The user parameter CFG must be set to SCE for this purpose in Transaction SU03. Alternatively, you can edit runtime versions directly using advanced maintenance mode in the Java Product Modeling Environment.

We'll now demonstrate the essential modeling aspects of a composition problem using the example of a bookshelf system. We'll begin by examining the components and their structure in the overall system in more detail, before investigating component behavior as formulated in rules. An object-oriented modeling approach is used in this case.

Types of Components or Building Blocks

The components of the composition problem are described as *types of building blocks*. A type of building block consists of a configurable material and a variant class (usually with the same name) with the characteristics that distinguish the specific type of building block. In the following example, the types of building blocks are identified by names starting with BT_.

Building Block Types Defined for the Bookshelf System

▸ **BT_Bookself_System**
This type of building block is our starting point and represents the overall shelving system for the books.

▸ **BT_Book**
This type of building block can represent an individual book or a multivolume edition consisting of books of the same type. Properties such as the no_volumes (number of volumes) or the height, width, and thickness of the individual volumes are maintained as characteristics.

▸ **BT_Shelf_Unit**
This type of building block represents an individual shelf unit within the shelving system, consisting of horizontal shelves and lateral supports. A new shelf unit can simply be added to the existing units with the use of an additional support. The shelf unit is described using characteristics such as left_support and right_support with the values yes and no.

▸ **BT_Shelf**
Characteristics such as length, depth, thickness, and occupied_length are assigned to this type of building block.

▸ **BT_Support**
Lateral supports enclose the shelves to the left and right. Characteristics such as support_direction with the values left and right are therefore assigned to this type of building block.

Static and Dynamic Component Structures

The hierarchical aspect of the component structure is described in a static BOM using types of building blocks respectively the configurable materials of the types of building blocks. In contrast to the super BOM or maximum BOM, which describes the maximum scope of the components for a decomposition problem, composition problems use a *minimum BOM*. This merely defines the minimum and maximum number of building blocks of a specific type in a certain function, in other words, in a certain item in the BOM.

The compositional aspect of the problem is expressed during configuration by virtue of the fact that the components of a certain type of building block can be instantiated into concrete building blocks as required. Every building block (or, to put it another way, every instance) is assigned a building block ID or instance ID that is unique within the configuration. A building block is defined by a specific type of building block, which defines the properties of that building block. The building blocks themselves contain the characteristic values.

Figure 10.19 shows the static component structure of the types of building blocks, which is defined using BOMs. The arrows labeled with BOM and the respective item number represent BOM relationships. Each of these arrows leads to a target component that forms part of the source component. In other words, these arrows represent the has_part one-to-many relationship that exists between the components. The minimum and maximum number of components that can be referenced are indicated at the tip of each arrow. We can also go in the opposite direction of the arrows, from the target component to the source component to track the part_of one-to-one relationships between components. Figure 10.19 also shows the characteristics of the types of building blocks, including their data types. These will be important later. Here, num stands for numeric, and char stands for character format.

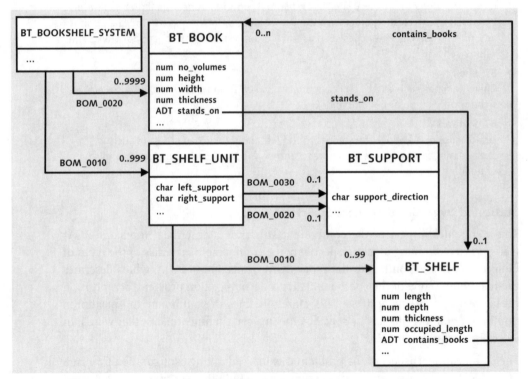

Figure 10.19 Component Structure for the Bookshelf System

486

The dynamic component structure is also taken into account in Figure 10.19. As previously outlined, each book is to be assigned to one shelf. The arrow labeled `stands_on` represents the unique, one-to-one relationship between a book and the shelf to which it is assigned, which is not mapped in the BOM. Conversely, the relationship between a shelf and its assigned books is represented by the arrow labeled `contains_books`. Several books can be assigned to one shelf, as indicated in the diagram by `0..n`.

The dynamic relationships between specific building blocks are described using ADT (*abstract data type*) characteristics. As the term *abstract data type* indicates, a reference from one building block to one or more building blocks (of the same type) also implies a certain behavior. An ADT characteristic is a typed reference. ADT characteristics with a single value are used for one-to-one relationships, whereas ADT characteristics with multiple values are used for one-to-many relationships. These ADT characteristics are included in the variant class of the referencing type of building block.

Behavior of Building Blocks

The behavior of types of building blocks is modeled using configuration rules. There are two types of configuration rules in SCE advanced mode:

▸ **Constraints**
Declarative rules are formulated as constraints, the modeling capabilities of which are greatly enhanced compared with LO-VC.

▸ **Rules**
Rules allow you to formulate algorithms that are processed using a procedural approach.

Monitoring rules are used to aggregate characteristic values, for example, to calculate the total thickness of all books assigned to one shelf in the shelving system. The balancing of supply and consumption values is also executed using monitoring rules, which are active in the background at all times. Reevaluating rules also exist, where the triggering of rule evaluation can be event-controlled, for example, based on a user action.

Constraint nets are an instrument for grouping configuration rules. For example, they can be used to group all configuration rules that define the behavior of one type of building block. For technical reasons, a constraint net contains either constraints or rules.

Some representative examples of constraints and rules are provided in the following, based on the composition problem presented by our bookshelf system.

[+] **Constraint for Value Restriction**

```
Objects:
   Shelf is_a (300) BT_SHELF.
Restrictions:
    Shelf.Domain depth >= Shelf.contains_books.width.
Inferences:
   Shelf.Domain depth.
```

The keyword Domain allows you to formulate explicit rules for the value range of a characteristic. The value range for the depth characteristic of the shelves is restricted in this example. Each shelf must at least be deep enough to accommodate the widest of the books assigned to that shelf. The books assigned to the shelf are represented by the contains_books ADT characteristic. The width characteristic of the relevant books is easily incorporated using the expression Shelf.contains_books.width. This type of expression is also called *running dot notation*.

[+] **Constraint for Assigning a Value to an ADT Characteristic**

```
Objects:
   Shelf is_a (300) BT_SHELF
   Book is_a (300) BT_BOOK.
Condition:
    Book.stands_on = Shelf.
Restrictions:
    Shelf.contains_books = Book.
Inferences:
   Shelf.contains_books.
```

Here, we assume that a book is always manually assigned to a shelf. In other words, the ID of one of the previously generated BT_Shelf building blocks is manually selected, for example from the input help for the stands_on ADT characteristic that is assigned to each book. The constraint then adds entries to the list of values for the multi-value contains_books characteristic for the relevant shelf.

[+] **Constraint for Building Block Instantiation**

```
Objects:
   Shelf_unit is_object (material) (300) (NR = BT_Shelf_unit).
Restrictions:
   has_part (Shelf_unit,
            pos_no = '0020' and pos_type = (300) BT_SUPPORT,
            with support_direction = 'right')
      if Shelf_unit.left_support = 'yes'.
```

Here the assumption is that for a shelf the user decides whether a lateral support is required by evaluating the characteristics `left_support` and `right_support`. This constraint then ensures that a `BT_Support` building block exists in BOM item '0020' for every shelf where the `left_support` characteristic is assigned the value 'yes' and that the `support_direction` characteristic is assigned the value 'right'. Otherwise, a `BT_Support` building block is generated by the constraint and assigned a relevant value.

Monitoring Rule for Calculating Shelf Capacity   **[+]**

```
Objects:
  Shelf is_a (300) BT_Shelf.
Then Increment:
  Shelf.occupied_length by
    Shelf.contains_books.no_volumes
  * Shelf.contains_books.thickness
```

This monitoring rule calculates the `occupied_length` aggregation characteristic of a shelf based on the total thickness of the books assigned to that shelf. The `no_volumes` characteristic is taken into account when calculating the thickness of multivolume editions.

Although the example of the bookshelf system described here only covers a very small part of the spectrum of rule diversity, it nonetheless provides some indication of how composition problems can be tackled in advanced mode in the SCE.

10.4   Summary

In this chapter, we discussed three topics that are sometimes viewed as being particularly challenging and therefore willingly handed over to a small group of experts. Having examined these in more detail here, you'll surely agree that there is, in fact, no reason to avoid these topics.

Our discussion of *performance optimization* clearly demonstrated that product configuration is one area where particular attention must be paid to the runtime behavior of software solutions. Critical bottlenecks may occur here, in particular in relation to interactive configuration. And the design and structure of your configuration models have a very significant influence on runtime behavior during the processing of a configuration task. So you're now aware of which aspects to consider when constructing your configuration models to ensure efficient runtime behavior and you understand that early stress testing in the planned application context represents an essential component of modeling. In addition, an accurate analysis of any bottlenecks detected is critical if you're going to be able to respond promptly by constructing more efficient alternative models.

Change management allows you to control and document the incorporation of planned changes to configuration models into an operational SAP ERP environment. Engineering Change Management (ECM) is used for this purpose, so we provided an example in this chapter of how this is done, and you should now be able to approach this task with confidence. Order Change Management (OCM) is provided in SAP ERP for making controlled changes to the configuration of a released sales order item, and so we provided an example of how OCM works based on the sales order change process.

In *system configurations*, multiple components are configured, and the interactions between these cannot simply be expressed in a super BOM. This chapter included detailed examples to help you distinguish between three classes of problems in this regard. These consist of products that require the inclusion of additional configurable components in an order BOM, interlinked configuration structures, where you often need to take account of extensive dependencies not only along but also across the hierarchical BOM structure, and composition systems, which only take shape during configuration with the gradual addition of configurable building blocks. The SAP Business Suite provides adequate support for order BOMs and interlinked configuration structures. For composition problems, you can use the advanced mode of the SCE as the core configuration tool in a problem-specific software solution.

At this point you should understand the diversity and complexity of product configuration, and know about the various tasks, tools, and practical examples relating to this topic. But who are the professional experts in this subject matter? What have they learned, and how can you benefit from their experiences? This chapter will answer these questions by introducing you to the Configuration Workgroup.

11 Configuration Workgroup

Many organizations have been set up to represent the interests of SAP software users. The best known of these are the *Americas' SAP Users' Group* (ASUG), established in 1991, and the *German-Speaking SAP User Group* or *Deutschsprachige SAP-Anwendergruppe* (DSAG), which has been in existence since 1997. For information about these user groups, both of which have many thousands of members, visit their websites at *www.asug.com* and *www.dsag.de*. Is there really a need then for a separate user group dedicated to product configuration with SAP software?

While many were mulling over this very question, a small working group originating in a software development project gradually evolved into an independent organization that has met with considerable approval from the configuration experts among SAP users, SAP partners, and SAP itself. This organization is known as the *Configuration Workgroup* (CWG).

11.1 Introduction to the CWG

The Configuration Workgroup is the international user group for product configuration based on SAP software. The working language of the CWG is English. Detailed information about the CWG is available on the web at *www.configuration-workgroup. com*. As an autonomous, independent organization, the CWG provides information for all SAP users who use the *Variant Configuration (LO-VC)* or *Internet Pricing and Configurator* (IPC) tool for any purpose.

Membership of the CWG is open to enterprises that use SAP software to process configuration tasks and partners that provide support for the implementation of the relevant software solutions from SAP, as well as to individuals who work within these contexts. The CWG carries out all of its activities on a voluntary basis, with many of the association's costs being covered by sponsors. For example, membership of the CWG has been free for many years. Most of its approximately 1,000 members are individuals who use LO-VC or the IPC on a daily basis. Some of these members are also heavily involved in voluntary activities for the CWG.

The most high-profile of these activities are the organization of the CWG conferences that are held twice a year and maintenance of the official CWG website.

▶ **CWG conferences**
The CWG conferences have always been based on very open communication, and the atmosphere is more casual than formal. The style and structure of the conferences are discussed in more detail in Section 11.5, CWG Conferences.

▶ **CWG Portal**
The CWG Portal is part of the official CWG website. It incorporates a document archive and many discussion forums and has become the second pillar of the CWG (see Section 11.6, CWG Portal), alongside the CWG conferences.

The CWG maintains a close working relationship with SAP. Many SAP employees, both from the development labs and from teams at SAP's international subsidiaries, can be found among the ranks of CWG members. In the past, the main focus of the user group was to facilitate the exchange of information between SAP software developers and end users. This interaction was, at times, clearly determined by the interests of the software provider. Over time, the user group has gradually redressed this imbalance. Its organizational structure now allows all activities to be planned and executed independently. Today, the exchange of experiences between users of SAP software has taken center stage (see Section 11.3, History).

11.2   Tasks and Objectives

The CWG has its own set of bylaws. For information about how these came into being, see Section 11.3, History. The CWG has enshrined the following three objectives in these bylaws:

▶ **Effective utilization of SAP software**
The CWG seeks to advance the effective utilization of the product configuration software developed and marketed by SAP by promoting the exchange of information and active dissemination of knowledge to the mutual benefit of its members.

▶ **Best practices**
The CWG provides a forum in which members can exchange ideas and information and advocates the correct use of SAP software for product configuration as tried and tested in many projects.

▶ **Influence**
In the interests of its members, the CWG seeks to exert an influence on the direction of the development activities, product strategies, and service offerings of SAP and SAP partners.

There are many ways in which the CWG pursues these objectives, and it has developed the following instruments for this purpose.

Exchange of Information

The main instrument used to pursue these goals is the exchange of information. In the early years of the CWG, the exchange of information was restricted to a very small number of users and a limited circle of representatives from the SAP development organization. This worked very well for a while. However, as the use of product configuration in SAP implementations became more widespread in enterprises of various sizes, the question arose as to whether a small working group could effectively understand and represent the interests of the large number of users outside of this group. For this reason, the CWG quickly set itself an additional task, which is not explicitly formulated in its bylaws, namely, to establish a network consisting of all professional users of product configuration in SAP solutions. The exchange of experiences and information requires a forum that unites individuals who possess the relevant experience and knowledge and those who can benefit from these.

Establishing a Network

By offering a number of specific and often very useful services, this nonprofit organization has gained numerous members in just a few years. Because all of the CWG's activities are carried out on a nonprofit basis, the services it offers are limited to those that are of mutual benefit to its members. If you present your work within the forum of the CWG, you're certain to find others with similar experiences. If you answer many questions or offer advice as a consultant, you can enhance your reputation as an expert.

The CWG began to attract increasing numbers of members when it refashioned its bi-annual meetings as conferences. Current development trends from the world of SAP development are presented at these conferences. However, most of the interest is normally generated by presentations concerning implementation projects, which aim to demonstrate how the software is actually used in practice. The original purpose of the CWG meetings to discuss specialized subjects among a body of experts has also been preserved. A number of focus groups meet periodically to work on very specific topics. For example, one of these focus groups formulated the requirements for future enhancements of LO-VC Variant Configuration and assigned various priorities to these. Another developed the *Knowledge Base Interchange Format* (KBIF) as an XML format for the exchange of configuration models. This was then proposed as the standard format at a CWG conference.

The establishment of the CWG Portal (see Section 11.6, CWG Portal) added a whole new dimension. This service is used by many individuals who use product configuration in SAP business process solutions on a daily basis but are unable to attend an international conference once or twice a year.

Overall, the high profile of the CWG, its independence, and the positive response it has received among users, SAP subsidiaries, and partners have established a sound basis for implementing its objectives as enshrined in its bylaws.

11.3 History

The Configuration Workgroup was established as a working group in the fall of 1993 under the original name *American Configuration Workgroup* (ACWG). Its objective was to facilitate collaboration between a number of renowned U.S. manufacturing enterprises to foster the development of a standardized configuration tool.

All members of the working group were vendors of configurable products that were manufactured discretely as components but were to be installed subsequently as larger systems. As a rule, these components were also sold as complete systems. The focus was therefore firmly on complex system configuration. Each of the members had already developed individual configuration tools and wanted to replace these with a sustainable standard solution. In addition, all were SAP customers and had standardized their processes to a large degree based on SAP business process solutions. It is hardly surprising, then, that some SAP consultants and developers became involved in the activities of the working group. However, the ACWG remained independent of SAP.

Its stated goal was to incorporate a new standard configuration tool into the SAP infrastructure. However, the question as to who would develop this tool remained open. Collaboration between the ACWG and SAP proved very fruitful and produced a range of truly innovative ideas, although, of course, there were also some challenges and tensions to be overcome.

Specialist Activities in the Early Years

The specification document produced by the ACWG in early 1994 contained the following key requirements, some of which were considered critical by SAP:

▶ The configuration tool should be integrated into the SAP business process solution but also be capable of running as a stand-alone solution, in other words, independently of a large backend system.

▶ An Application Programming Interface (API) should be provided to enable the creation of an application-specific graphical user interface.

▶ Functionality should be geared toward system configuration tasks.

▶ The creation of configuration models should be a very simple task. Reference was made to an easily accessible product modeling environment.

SAP accepted these challenges and, in the spring of 1994, collaborated closely with the working group to design the configurator that was later to be implemented as the Sales Configuration Engine (SCE). By this time, several variant manufacturers from Finland and Switzerland had joined the group, and the American Configuration Workgroup had been rebranded the Configuration Workgroup (CWG). The CWG was temporarily taken under the umbrella of the SAP High-Tech Industry Center of Expertise, which increased its membership from the high-tech industry. A small number of highly dedicated members of the CWG continued to work on the development and testing of the new SCE configurator, which was available as of summer 1997 and was deployed for the first time in a live environment in the spring of 1998 as part of a project solution. In the years that followed, the CWG collaborated on the development of a Java-based *Product Modeling Environment* (Java PME) and the *Internet Pricing and Configurator* (IPC), which became available in the summer of 1999 as the first component of the SAP CRM solution.

Evolution into an Organization

As a small customer user group, the CWG maintained close links over many years with the SAP development team responsible for the SCE configurator. CWG meetings were held twice a year, providing a forum in which technical and functional issues relating to complex system configuration could be discussed in-depth among focus groups. It soon became clear that the CWG would have to open its doors to a broader membership if it was to continue to exert influence over the development of the standard software. Its meetings were restructured so that, in addition to discussions within focus groups, time was also allotted to the presentation of new developments in the configuration tool and of experiences gained on implementation projects.

Initially, the main focus tended to be on SAP customers that used the IPC in a CRM scenario, such as Internet Sales. These were often manufacturers of configurable products that had enhanced their existing solution with additional sales channels based on the LO-VC Variant Configuration tool and SAP ERP. Because many variant manufacturers had experienced significant success on implementation projects with the Variant Configuration tool that was integrated into SAP ERP, this tool became the subject of increasing numbers of discussions and presentations at the CWG meetings.

The ranks of the CWG had risen to just under 300 by the time of its 10th anniversary in 2003. Its meetings were transformed into conferences with over 100 delegates. The challenge of organizing and hosting these conferences became increasingly complex because the CWG remained a loosely structured working group and therefore functioned as an organization without a defined legal form.

After many discussions, an association was established and registered in accordance with the German Association Law in the spring of 2005. This association and its bylaws regulate the now fully independent organization that is the CWG. In that same year, the CWG Portal was launched as part of the CWG website. This online platform now serves as an easily accessible and free-of-charge meeting place for over 2,500 users in the area of SAP product configuration.

11.4 Organizational Structure

The Configuration Workgroup is a registered association of the district court of Wiesloch in accordance with the German Association Law. The association gives the CWG a proper legal form. Its bylaws regulate the bodies and structures within the organization and set out its rules and regulations.

The CWG is open to the following types of members:

- *Member companies*, which can be SAP customers that use product configuration as part of an SAP solution or SAP partners that provide some form of support for product configuration solutions. The CWG also has associate member companies, which do not meet all of the requirements of a member company but are still considered on a par with these in accordance with CWG guidelines.

- *Representatives* are nominated by member companies. Each member company can designate one employee as their primary contact and one employee as their executive contact.

- *Individuals* can also apply for membership in the CWG, provided that they use SAP product configuration solutions in some way in a professional capacity.

Official membership is free of charge and is approved by the CWG Board of Directors. Application for membership is normally a very simple process. If you have not already joined the CWG, it is an option that is at least worth considering. For more information about membership, refer to the CWG website at *www.configuration-workgroup.com*.

The executive bodies of the CWG are the executive committee and the board of directors. Members of both bodies are elected each year by CWG members.

The executive committee is composed of the officers of the CWG, namely, the president, who represents the CWG both internally and externally, the vice-president and the immediate past president, both of whom provide additional support, the secretary, who manages business transactions and the list of members, and the treasurer, who handles financial matters. The board of directors consists of the executive committee plus between 5 and 16 CWG directors.

A new vice-president is elected each year. After serving their designated time in office, vice-presidents are automatically made president of the CWG. The current president then becomes the immediate past president, and the immediate past president he replaces retires from the executive committee. Because these terms of office each last one year, the usual term of membership of the executive committee is three years. This rotating system ensures a high level of stability and continuity but also continuous renewal at the executive level.

11.5   CWG Conferences

Whereas the CWG conferences, which are conducted in English, previously tended to be hosted in the SAP locations of Walldorf, Germany, and Philadelphia, they have, in recent years, convened in hotel settings in a variety of locations in Europe and the U.S. This development expresses the association's independence from SAP and has also significantly boosted membership. It has now also become tradition that the spring conference is held in Europe and the fall conference in the U.S. The fact that a conference fee is now charged as a contribution toward the expense of hosting the event does not present any real obstacle to attending, because it only represents a fraction of the overall travel expenses involved.

Since 2005, the three-day conferences have followed the same format. The first day typically consists of general-interest presentations. These usually provide information about new developments from SAP and its partners. A great deal of interest is always generated by presentations about implementation projects, which provide conference delegates with information about the various application scenarios of product configuration. Guest speakers on selected topics that touch on product configuration or who present market reports on configuration software complete the program for the first day.

The second day of the conference is normally dedicated to discussions in smaller focus groups. These provide a forum for presentations on specialist topics relating to product configuration. This provides an opportunity for detailed feedback discussions. Here, the opportunity to get to know the other delegates and exchange experiences is considered just as important as the specific topic under discussion. Dividing delegates into smaller groups also allows for hands-on experience of using and testing new applications. Very valuable insights can be gleaned and contacts forged in this context. It is part of the CWG's core mission to foster networking and provide its members with opportunities to establish and strengthen business contacts.

On the third day, plenary presentations are usually back on the agenda. However, unlike the first day's program, these are generally geared toward implementation teams and modeling experts. Issues relating to practical applications are discussed and tips and tricks are exchanged. A closing meeting with the CWG Board of Directors and a feedback questionnaire provide delegates with an opportunity to have input into the direction of future conferences.

11.6   CWG Portal

Since 2005, the CWG Portal has been an integral part of the CWG website (*www.configuration-workgroup.com*). All of the pages on the CWG website are in English. The website consists of a public area and the CWG Portal, where access is restricted.

There is no charge for using the portal, but online registration is required. You are required to provide certain contact details, and the name and company name you enter will be visible to other portal users. There are currently 2,500 registered portal users, and approximately 3,500 articles have been posted. To date, more than 100 users have actively participated in the discussions in the CWG Portal, which means that the vast majority of users (some 2,400) merely log on to read the information posted.

Currently, five portal functions are available to registered portal users:

▶ **CWG Blog**
In this section, you can read the articles posted by three bloggers relating to current issues in product configuration with SAP and CWG activities. Approximately 20 articles have been posted here in the past year. The bloggers' personal opinions frequently add flavor to their pithy articles. Readers of the blog often comment on what they have read, which also lends color to this section.

▶ **CWG Forum**
The discussion forum provides a space in which opinions and experiences can be exchanged and archived. The CWG Forum is currently divided into seven specialist forums and three administration forums. The most active by far of the specialist forums is devoted to the subject of product modeling. This forum has about 300 threads and a total of 1,200 posts. In many cases, these are questions that are

both posted and answered by modeling experts. A large community reads these posts and learns from them. The other specialist forums deal, for example, with product configuration issues in sales and service and in production. The CWG Forum also displays the portal users who are currently online.

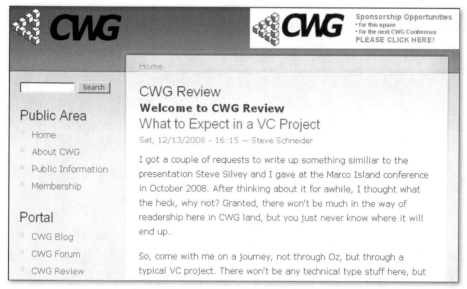

Figure 11.1 CWG Review: Online Magazine in the CWG Portal

▶ **CWG Review**
Whereas early editions of the CWG members' magazine were sent by email on a quarterly basis, it soon found a home in the CWG Portal, where it can be much more easily accessed as an online magazine. Between 2005 and 2008, 34 articles were published in its seven sections. In its most active section, "Configuration Contemplation," the editor discusses, for example, what project team members can expect when working on a typical SAP implementation project in the area of Variant Configuration (see Figure 11.1).

▶ **Document archive**
The CWG's document archive is one of its most valuable resources. Here you can access approximately 400 conference papers from over a dozen CWG conferences since 2002 in just a matter of seconds. The archive also serves as a space where regional groups can manage their documents. Currently, the only official regional branch of the CWG is the USA Midwest branch, with almost 50 members, which makes good use of this area of the Portal.

▶ **CWG Sandbox**
This area of the portal provides all of the information you require in relation to the SAP ERP system operated by the CWG for its members. This system is discussed in detail in Section 11.7, CWG Sandbox System.

11.7   CWG Sandbox System

Since late 2007, the CWG has been operating an SAP ERP system that can be accessed by its members subject to certain conditions, as seen in Figure 11.2.

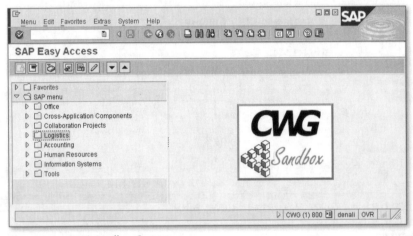

Figure 11.2   CWG Sandbox System

The SAP ERP solution, which is operated as a *sandbox* system or testing environment, has been used, for example, to set up a library of configuration models. Modeling experts can learn directly from one another in this environment, where examples of various modeling approaches can be compared and contrasted. Today, just one year after its launch, the sandbox has about 100 registered users, who have created more than 40 configuration models. These models can be presented and discussed with interactive system demos in the focus-group sessions at the CWG conferences.

In addition, partner companies can connect their supplementary software products to the CWG Sandbox to demonstrate their functionality to CWG members. This is a very useful function because it allows members to have some hands-on experience with a solution after it has been presented to them.

You'll find more information about the CWG sandbox system in the relevant area of the CWG Portal.

11.8 Summary

This chapter introduced you to the international user group for product configuration with SAP solutions. The core mission of this group is to facilitate the free exchange of information relating to all aspects of product configuration. Thanks to the voluntary efforts of many highly committed members, almost all services offered by the CWG are free to its community.

From its inception up to the mid-1990s, the CWG focused to a large degree on the development of the Sales Configuration Engine (SCE). As its membership grew, the group increasingly broadened its vision to incorporate other issues of interest. Today, the group represents all interests connected with product configuration with SAP in the broadest sense.

The CWG's bylaws lend a stable structure to the organization and ensure that it remains autonomous. The CWG conferences, which are held twice a year in Europe and the U.S., represent the lifeblood of the CWG's activities. The CWG Portal and the functions it provides have significantly increased the number of configuration experts within the CWG community and provide a basis for networking between members on a daily basis. The CWG Sandbox System, which is an SAP ERP implementation operated by the CWG for its members to promote the active exchange of information, is one of the most successful achievements of the CWG.

The previous chapters described how you can use product configuration with SAP solutions, which aspects you need to consider in your projects, and how you can benefit from other users' experiences. Now, we'll take a look at a new SAP solution, which is tailored to the requirements of medium-sized enterprises and pursues innovative approaches.

12   Outlook for SAP Business ByDesign

Variant Configuration with SAP is not new. All of the topics that were introduced earlier in this book deal with established components of the SAP enterprise solutions. If you use Variant Configuration LO-VC and the Internet Pricing and Configurator (IPC), you can rely on many years of experience. New software versions encompass the current functions of product configuration and provide incremental enhancements based on very stable and sophisticated business processing. This leads to a high level of protection of investment for numerous implementations of product configuration solutions based on SAP ERP and SAP CRM. The solutions that are now developed in comprehensive implementation projects will endure for many years.

This chapter discusses a brand new SAP solution that has been developed in the SAP research and development labs and is still in an early product launch phase. It's the new solution for medium-sized businesses: *SAP Business ByDesign*. This solution opens new doors and contains numerous essential innovations. This chapter deals with the product configuration concepts in this new solution for medium-sized businesses. You'll learn the differences between the configuration requirements of medium-sized enterprises and large enterprises. SAP Business ByDesign provides new options for the interaction between medium-sized enterprises and large enterprises. In the long run, some of the new concepts will also affect the further development of the already established SAP enterprise solutions.

12.1   SAP Business ByDesign

SAP Business ByDesign is the new business process solution from SAP, which is customized to the requirements of medium-sized enterprises. It is a complete solution that is integrated with all critical applications. It covers the following areas of use:

- ▸ Executive management for supporting managers
- ▸ Financial accounting for all financial processes
- ▸ Compliance management for ensuring the compliance with laws
- ▸ Customer relationship management for customer support
- ▸ Supplier relationship management for vendor support
- ▸ Supply chain management for planning, production, and stockholding
- ▸ Project management for processing the project-based business
- ▸ Product lifecycle management for product development
- ▸ Human resources management for personnel services

Accordingly, the range of functions covers all areas and is basically designed for enterprises with a wide and flexible range of products.

SAP Business ByDesign is completely based on a service-oriented architecture (SOA). All business functions are executed — also within the system — through defined service interfaces. These service interfaces are included in fully modeled business objects (BO). Examples of business objects are customer, product, sales order, and production order.

SAP Business ByDesign is operated in the on-demand mode. According to the *Software-as-a-Service* (SaaS) concept, the software is neither purchased nor installed and operated by the user. Moreover, the user accesses a centrally provided implementation via a network connection. Depending on the respective requirements, the operator provides the required resources, such as computers and storage capacities. All services that are required for the software use are leased as a whole. This allows for providing comprehensive functions at an affordable price.

12.2 Product Configuration in Medium-Sized Businesses

Product configuration is used in various forms in large enterprises. Sometimes a large set of product parameters is used. At least in manufacturing enterprises it often occurs that configurable products again have configurable components. Numerous configuration applications leverage a complex set of rules and comprehensive data-sets, for example, in variant tables. The option of covering a lot of variants with variant BOMs is an essential core function for many implementations.

So, what is the special feature of product configuration for medium-sized enterprises? Can you assume that smaller medium-sized enterprises generally have to solve less complex configuration tasks? There are lots of options to characterize medium-sized enterprises. However, simply structured variant products do not characterize medium-sized businesses. Without further entering the discussion about criteria for distinguishing between large enterprises and medium-sized enterprises, we would like to advert to the following two factors which are often part of the corporate philosophy in medium-sized enterprises:

▶ **Product flexibility**
The product flexibility is very high. If the customer has specific requirements for an offered product, the enterprise often does everything to fulfill these requirements. The motto is frequently, "What doesn't fit will be made to fit!"

▶ **Process flexibility**
The adaptation of products to individual customer requests is often a creative process. There are not only predefined product options that can be individually selected. Sometimes fulfilling special requirements involves pragmatic modifications to the logistics process. The entire process is based less on formally defined process steps and allows for manual modifications by processors who understand their daily business but also the customer's special requirements.

These requirements can also exist in large enterprises. The difference here is that smaller enterprises often couldn't survive without fulfilling these requirements. The following sections illustrate the requirements for product and process flexibility using an example.

Product and Process Flexibility: The Glass Door Example   [Ex]

A company that processes glass plates also provides glass doors or door leaves made of glass. Typical sales parameters are type of glass, glass thickness, and width and height of the door leaf.

To attach the hinges and locks, holes are cut into the glass plate. Some standard variants are available for the geometric dimensions of the holes; however, the dimensions for the locking device in particular often deviate from the standard. Furthermore, for about 5% of the orders, additional holes are requested as special requirements for design elements in the door leaf.

Figure 12.1 shows some examples of the geometry of the holes in the glass doors. The goal of modeling a respective configuration task would be to completely cover the largest possible number of orders with the configuration model. The modeling of the standard categories, *round hole* and *square hole* (as shown in Figure 12.1 on the left), already requires numerous product parameters and rules; for example, a minimum distance to the border must be ensured. If you also try to include the holes for design elements (additional holes, as shown in Figure 12.1 on the right) in the configuration model, you have to deal with a complex system configuration as introduced in Chapter 10, Challenges in Variant Configuration.

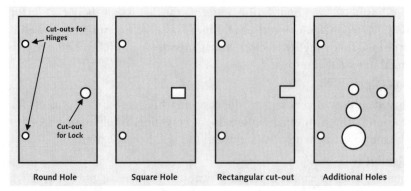

Figure 12.1 Door Leaves with Different Types of Holes

The decisive factor in this example is that a part of the individual specification can be easily modeled as a configuration task with formal product parameters. However, for the complete formal description in a product model covering all product variants that actually occur, the effort would be unacceptably high. This applies even more to a medium-sized enterprise in which perhaps one sales employee who can dedicate only a small proportion of his time to the product modeling carries out the modeling task.

A business process solution that takes into account this context must be able to support not fully modeled product options in logistics processing.

12.3 Make to Order in SAP Business ByDesign

SAP Business ByDesign supports various sales and production scenarios. The first scenario that supports lean product configuration during the product launch phase of the new solution for medium-sized businesses is the make-to-order (MTO) scenario. This section provides a brief overview of the core concepts used.

12.3.1 Extending the Product Concept

Usually, an entry in the product or material master describes the product. From quotation and sales order to planning, production, purchasing, and storage to delivery and invoicing, all logistics processing steps refer to the entry in the product master.

As you learned in the previous chapters, traditional MTO production refers to the product in logistics processing but also to the sales order or sales order item. For configurable products, the reference to the sales order item also ensures that the configuration results, that is, the description of the product options selected by the customer, are stored. As *make to order* implies, logistics processing focuses on the sales order. Combining multiple orders is also not possible if two orders refer to the same product with the same set of selected product options.

Specifiable Products [+]

Products that can be individually characterized are referred to as *specifiable products* in SAP Business ByDesign. It is assumed that the entry in the product master usually describes a standard version (that is, a specific variant) of the product. Deviations from the standard can be described with an additional product specification.

The combination of product and product specification then leads to a specified product.

Extending the product concept in the logistics processing of SAP Business ByDesign makes the handling of specified products more independent of the respective sales orders. In addition to the reference to the product master, which is always required, the product concept for specifiable products also envisions an optional product specification.

Product Specification [+]

A product instance that deviates from the standard is described in a specific business object, the *product specification*. In addition to the identifying product specification number, the business object also includes a reference to the product that is specified and the actual specification content.

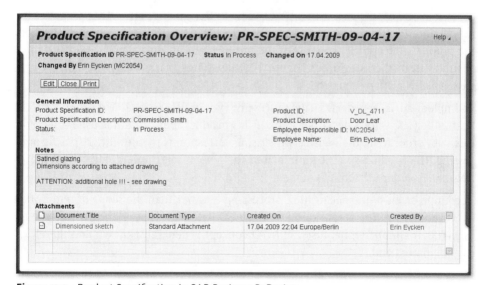

Figure 12.2 Product Specification in SAP Business ByDesign

Figure 12.2 shows the data sheet of a product specification. THE GENERAL INFORMA-TION section displays the PRODUCT SPECIFICATION ID that identifies the product specification and a PRODUCT SPECIFICATION DESCRIPTION. On the right, you can view the references to the product (PRODUCT ID and PRODUCT DESCRIPTION) that is specified in the product specification and to the employee that is responsible for the product specification. In the example illustrated in Figure 12.2, the actual specification content consists of a textual description in the NOTES section and a dimensioned sketch in the ATTACHMENTS section.

All logistics steps in the SAP Business ByDesign MTO scenario including the sales order processing refer to the product master entry and — if available — to the product specification. This enables you to work in a cross-order manner, if required. You can combine demand requirements that refer to the same product specification. If a sales order is canceled, you can post an already manufactured variant to the warehouse with reference to the product master entry and the product specification and sell it to someone else. If the finished product needs to be reworked for another sales order, you can post it to another warehouse stock with reference to another product specification. The major difference in logistics processing for standard products and specified products is the product concept, which has been extended with an optional product specification.

12.3.2 Make to Specification

Of course, the MTO process in SAP Business ByDesign supports sales-order-specific production. However, the infrastructure of the business process solution is based on the product concept that consists of the product and product specification and thus abstracts from the sales order. Figure 12.3 illustrates that, strictly speaking, it is a *make-to-specification* scenario.

To implement sales-order-specific processing, you must create a new product specification for a specific sales order item. If this product specification is used solely for this sales order item, the logistics processing corresponds to sales-order-specific processing. If a product specification is used for several sales orders, the processing is not merely order-related, of course.

The scope of make-to-specification processing is wider than the scope of traditional MTO processing. The scenario in SAP Business ByDesign is nevertheless referred to as MTO for simplicity, because the MTO concept is generally known and therefore easy to understand.

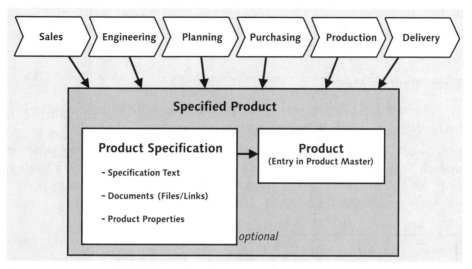

Figure 12.3 Concept of the Make-to-Specification Processing

12.3.3 Lightweight Product Variants

The product and product specification pair characterizes a specific product version. It thus corresponds to the concept that was introduced as product or material variant in the previous chapters.

An essential characteristic of the traditional material variant is that it can be put into stock, that is, planned and manufactured according to the make-to-stock procedure.

Usually, a product variant can also be subject of further master data. With the introduction of the MTO scenario in SAP Business ByDesign, you can also create BOMs with reference to a product and product specification. This is also referred to as lightweight product variants. By defining BOMs with respect to a certain product specification you can also implement order BOMs of an ETO scenario.

Additional master data references, such as specific sales prices for product variants that are identified via a product and product specification, are planned for future versions of SAP Business ByDesign.

12.4 Product Configuration in SAP Business ByDesign

Now that we've introduced the product specification as the central element of the MTO scenario in SAP Business ByDesign, in this section we'll discuss the concept on product configuration. You've probably noticed that the product specification is the one and only candidate for carrying configuration information. Still missing is the formal description of product options in a model.

12.4.1 Product Model

The product model business object enables you to combine information that you must always consider for the specification of products.

A product model can apply to one or several specifiable products. The specifiable products for which a product specification needs to be created based on a product model are assigned to this product model. Only one product model can exist for a specifiable product (see Figure 12.4).

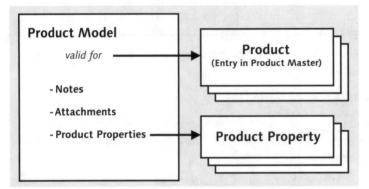

Figure 12.4 Product Model, Products, and Product Properties

The product model consequently serves as a template for creating a product specification. The following parts of the product model are used as a template for the product specification:

▶ **Notes**
For example, a text may contain notes that must be considered when creating a product specification. Questions that have to be answered in the specification can also be included in the notes.

▶ **Attachments**
You can attach various documents, which provide information for the specification, to the product model. These documents can be product brochures, questionnaires, or forms that need to be filled when creating the product specification.

▶ **Product properties**
Product options are formally modeled via product properties. You assign to the product model the properties that are relevant for the product that is supposed to be specified. You can restrict the allowed values of the reusable properties for the use in a product model.

> **From a Product To Be Specified to a Configurable Product**
>
> Defining the product model with assigned product properties converts a product to be specified into a configurable product. That means a standard product can easily become a configurable product. On the other hand, you can delete the properties from the product model. When you have deleted all properties from the product model or the assignment of the specifiable product to the product model, the product is no longer configurable.

In this way, SAP Business ByDesign provides a flexible transition from standard products to configurable products and vice versa.

12.4.2 Product Properties

SAP Business ByDesign enables you to define numerous product properties to describe product options. The following formats are distinguished here:

▶ **Codes**
To describe discrete options such as color or type of glass

▶ **Quantities**
For quantifiable figures such as width or volume

▶ **Boolean values**
For properties such as park heating or adverse weather lamp

▶ **Integers**
For countable properties such as number of hinges

▶ **Decimal numbers**
For figures without a unit, such as correction factor

▶ **User-defined free text**
For texts without defined value sets, such as first name or hobbies

Except for user-defined text properties, allowed values are defined for all product properties. You can restrict the allowed values if you use a product property in a product model. You can define a default value for each property in the product model. This default value is then transferred to the product specification as the initial value. For the property usage in the product model, you also define whether values must be assigned to a property in the product specification.

When you select the product in the product specification, the product properties are transferred to the product specification from the product model that is specified for the product. You can then assign values to the properties in the product specification. Then consistency and completeness checks are carried out. Figure 12.5 shows a product specification for a door leaf. This specification was created based on a product model with product properties. The TYPE OF GLASS code property and the properties with dimensions in millimeters for THICKNESS, WIDTH, and HEIGHT are defined for the DOOR LEAF product.

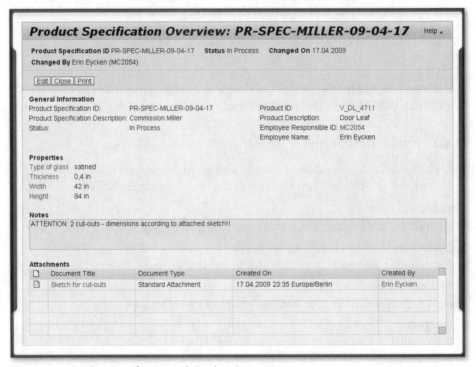

Figure 12.5   Product Specification with Product Properties

Consequently, all prerequisites for a lean product configuration in SAP Business ByDesign are met. For controlling dependencies between the properties of a configurable product a configurator can be integrated.

12.4.3   Integration of a Configurator

When entering property values in the product specification, you can process cross-property dependencies in the future using the *Core Constraint Engine* (CCE). The dependency rules are defined as a part of the product model. The set of rules is formulated with a declarative approach.

At the beginning variant tables for restricting allowed values will be supported. As you learned in the first two chapters, variant tables are a very powerful and flexible tool for the formulation of dependencies. This is the case for product options in particular, provided that they have been modeled via code properties. Code properties and value restrictions via variant tables enable you to fully model a large portion of use cases in product configuration. Moreover, the definition of variant tables is a quite intuitive method of expressing dependencies between product properties.

To define dependent numerical product properties, such as for the calculation of an area based on height and width, simple assignment formulas are very useful. A corresponding dependency type is planned, and further dependency types will be added.

Because configuration tasks in medium-sized enterprises don't differ significantly from configuration tasks in large enterprises regarding the required modeling capabilities in configuration rules, the set of rules as an integral part of the standard implementation will cover a wide range in the medium term. The product launch of the MTO scenario in SAP Business ByDesign, however, doesn't enable you to use the CCE yet.

There will also always be specific configuration problems in the long run, which a standard solution for product configuration can only cover to an unsatisfactory extent. This is particularly the case if a product-specific or industry-related user interface is required. For example, it is hardly possible to implement a graphical configuration of kitchens with a standard solution without project-specific enhancements.

Therefore, SAP Business ByDesign supports the concept of externally creating product specifications and copying them in SAP Business ByDesign via a message interface. The corresponding application-to-application (A2A) message is designed for the integration of third-party configurators with SAP Business ByDesign.

12.4.4 Process Automation

In the first step, the following business functions are linked to the usage of product properties in product models and product specifications:

▶ **Consistency and completeness check**
The consistency of the selected property values compared to the defined allowed values in the product model is ensured when you create a product specification. This check also includes a completeness check for the selected product options.

Only product specifications with a consistent and complete property value assignment can be released.

▶ **Search based on property values**
You can find product specifications by the property value assignment contained in them. A few property values are already very selective here.

The future versions of SAP Business ByDesign will probably contain additional business functions for product properties, for example, sales prices based on property values and configurable BOMs with selection conditions. The expansion of the product configuration to services and combinations of materials and services may also be added as a business function.

12.5   Summary

This chapter introduced the new business process solution for medium-sized enterprises, SAP Business ByDesign. The core of the solution's structure is completely based on SOA. The on-demand solution of software as a service allows for cost-effective operation and flexible growth according to the requirements of a dynamic enterprise.

Product configuration for medium-sized enterprises in particular requires high flexibility regarding the product portfolio and in the product provisioning process. A business process solution has to consider these central requirements. By extending the product concept with a product specification, the make-to-order scenario in SAP Business ByDesign supports a very wide range, from make-to-stock production to engineer-to-order processing.

The lean product configuration with SAP Business ByDesign is based on reusable product properties and product models, which can be used as a template for product specifications. To map cross-property dependencies, you can integrate a configurator. For logistics processing in SAP Business ByDesign, the Core Constraint Engine will be used as a configurator. The integration of third-party configurators will be implemented via an external creation of product specifications and a subsequent transfer to SAP Business ByDesign.

The simple but highly flexible product specification tools will also affect other business process solutions in the medium term. Just like you, we are curious about how the product configuration market will develop.

Appendices

A   Database Tables of Variant Configuration

This first section contains — virtually uncommented — database tables that play a role in the environment of Variant Configuration. These include both database tables that collect data from modeling and database tables that manage the configurations.

The upper half of Figure A.1 shows the database tables that collect master data from the classification system. The lower half shows the database tables for managing object dependencies. Here, the CUOB database table represents the assignments of object dependencies to the model's master data and is therefore the link between the object dependencies and the other databases of the model.

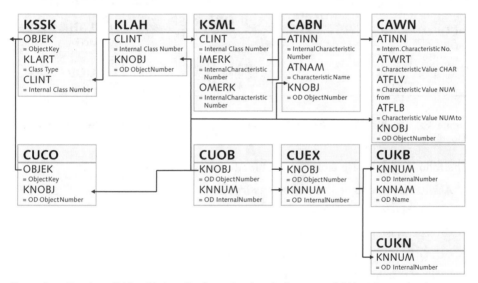

Figure A.1   Database Tables, Variant Configuration I — Assignment of Object Dependencies

Figure A.2, which is rather comprehensive, shows a list of additional database tables that contain master data of the model. This includes the material master (MARA, MARC), the bill of materials (MAST, KDST, STPO), the routing (PLPO, PLFH, PLFL — the three tables to which you can assign object dependencies), and the configuration profile (CUCO).Figure A.2 also shows the links to the two other figures that include database tables. For example, a link exists to the database tables of the classification system in Figure A.1. They've been supplemented by the storage of the actual classification (AUSP) and the conversion of object IDs (INOB). The database table of the sales document items is the last database table that was added to this figure.

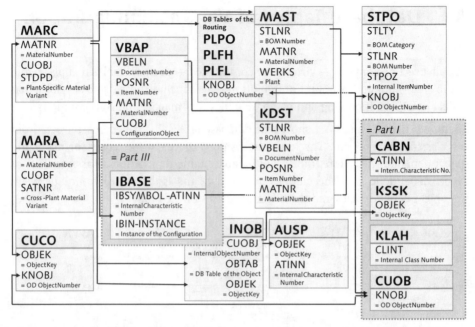

Figure A.2 Database Tables, Variant Configuration I — Links Between the Master Data of the Model

The following database tables are included in these two figures:

- AUSP: Values assigned to characteristics
- CABN: Master data characteristics
- CAWN: Master data characteristic values
- CUCO: Master data table for configuration profiles
- CUEX: Dependency compilation
- CUKB: Administrative information for dependency maintenance
- CUKN: Dependency source code based for variants/configuration
- CUOB: Assignment of object to dependency
- INOB: Assignment of an internal number to any object
- KDST: Link: sales order item — bill of materials
- KLAH: Master data class header
- KSML: Assignment: characteristics to classes
- KSSK: Assignment table: object to class
- MAST: Link: material — bill of materials

- MARA: General material data
- MARC: Plant data for material
- PLFH: Task lists — production resource/tool
- PLFL: Task lists — sequences
- PLPO: Task lists — operations
- STPO: BOM items
- VBAP: Item data for sales document

In addition to the tables and their links, Figure A.3 also shows a possible data analysis.

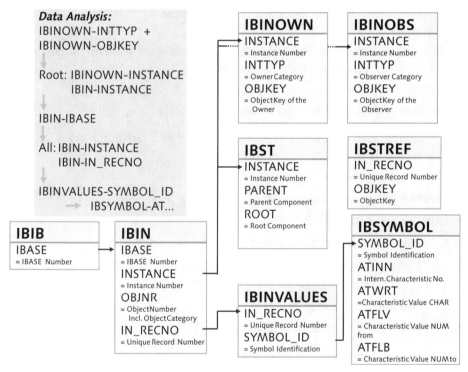

Figure A.3 Database Tables Variant Configuration I — Configuration Storage

You can access the IBINOWN table using the owner of the configuration, for instance, of a sales order item. Via the corresponding row in the IBINOWN table, you obtain the key of the *header instance*. You can use this key to access the IBIN table and to determine the key of IBASE, the entire configuration, for this instance. The IBASE, whose data is available in the IBIB table, can consist of multiple instances. This would be the case for multilevel configurations and would result in multiple entries in the IBIN table. You can determine the unique key, IN_RECNO, for each instance. You can use this key to also access the value assignment in two steps. The IBINVALUES table includes the symbol IDs of the values and the IBSYMBOL table the actual values.

- IBIB: Administrative data of IBASE
- IBIN: Data of an instance version
- IBINT: Description of the instances
- IBST : Root relationship and is-part-of relationship (*parent* and *root*)
- IBINVALUES: Assignment of a characteristic value to an instance
- IBINOWN: Owner of the root instance and consequently the IBASE (for instance, sales order item)
- IBINOBS: Instance observer (for instance, follow-up documents for the sales order, such as production order, and so on)
- IBSTREF: BOM reference to an is-part-of relationship/instance
- IBSYMBOL: Characteristic value (single value or interval)

More database tables exist for storing variant tables and variant functions. For the variant tables, these are:

- CUVTAB: Table header
- CUVTAB_ADM: Administrative data
- CUVTAB_TX: Texts
- CUVTAB_FLD: Characteristics and therefore columns of the table
- CUVTAB_IND: Value assignment alternatives
- CUVTLN: Rows of the table
- CUVTAB_VALC: Character values
- CUVTAB_VALN: Noncharacter values

For variant function, these are the following:

- CUVFUN: Function header
- CUVFUN_ADM: Administrative data
- CUVFUN_TX: Texts
- CUVFUN_PAR: Characteristics as parameters of the function
- CUVFUN_IND: Value assignment alternatives

B   APIs of Variant Configuration

SAP provides *application programming interfaces* (APIs) that application programs can use to communicate with other systems. These are function modules that satisfy fixed rules, for instance, the RFC capability (*remote function call*).

The complete functions of SAP Variant Configuration are available as smaller encapsulated function modules in the form of such APIs so that you can access the corresponding functionality in separate programs.

In addition, Transaction CAVC_TEST, which is shown in Figure B.1, provides a documentation and test environment for all APIs in the Variant Configuration environment. Here, you can find documentation and the source code of all APIs; you also have the option to start and therefore test them. If the APIs require user input, the system provides the corresponding fields. Moreover, the system lists all results after processing the APIs.

The naming convention for APIs followed a fixed scheme that is also shown in Figure B.1. CAVC means for Variant Configuration, which is followed by the description of the type, and the exact functionality which is kept as short as possible.

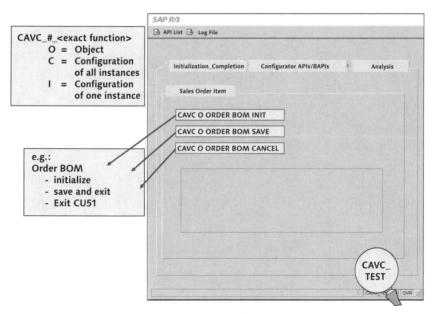

Figure B.1 APIs of Variant Configuration and Test Environment

C User Exits of Variant Configuration

User exits are a tool that enables you to integrate additional, enterprise-specific requirements with the standard processes. These user exits represent calls of customer-specific programs at a fixed time with a fixed interface. In this process, existing include commands access blank include files that can then be filled.

User exits exist as a special feature in Variant Configuration, which you can include as buttons in the value assignment screen to start them manually. A prerequisite is that you work with a user interface design as was discussed in Section 2.6.10, User Interface Design, in Chapter 2. There, you create buttons whose predefined names are CUSTOMER_PUSHBUTTON_###. You can select any name and pushbutton name. You don't require any further details. The user exits' names range between `Exit_SAPLCEIO_010` and `Exit_SAPLCEIO_019`. You can find them in the `LCEIOF01` program in the `EXECUTE_PUSHBUTTON_GROUP` form routine.

The following provides a small selection of additional user exits for Variant Configuration:

► **CCUX0000**
Additional check of the configuration that is executed during the final configuration check (contains the `Exit_SAPLCUKO_001` function module)

► **CCUX0001**
Load function for the configuration that controls whether loading can be done from an external source (contains the `Exit_SAPLCUDO_001` function module for loading from an external source using the `Exit_SAPLCUXC_001` function module)

► **CCUX0002**
Specification of a class node responds to inconsistencies during object search (contains the `Exit_SAPLCUDO_002` function module)

► **CCUX0800**
Controls the explosion depth for multilevel configurations; sets explosion of all or only configurable assemblies (contains the `Exit_SAPLCUKO_002` function module)

Transaction SMOD displays the full list of all current user exits in the Variant Configuration environment in the hierarchy display under SAP • LO • LO-VC • CU. In the components of the enhancement, you can also find the list of function modules, among other things.

D Comprehensive Examples of Variant Functions

Based on the descriptions provided in Section 2.6.9, Variant Functions, in Chapter 2, the following sections list examples of variant functions. This list also includes all additional modeling data that is relevant for the variant functions, including:

▶ Characteristics

▶ Messages in message classes

▶ The actual variant functions

▶ The associated function modules including interface parameters and source code

▶ Object dependencies when calling these variant functions

We hope this list of all data that is relevant in the environment of the variant function examples provides support and motivation for your first variant function.

Characteristics — Transaction CT04

TPC_09	PC application	Char 1 single v.
Values:	A: photo editing	
	B: text editing	
	C: data server	
TPC_10	Casing type (PC)	Char 1 single v.
Values:	X: customized	
	T: tower	
	M: minitower	
	D: desktop	
TPC_11	Number of CD drives	Num 1 0 single v.
Values:	0, 1, 2	
TPC_14	CPU	Char 1 single v.
Values:	A: slow	
	B: medium	
	C: fast	
TPC_31	PC package ID	Char 30 single v.
Values:	x-y-z	with x and z from (A, B, C)
		And y from (X, T, M, D)
TPC_32	Selection AWB	Char 1 single v.
No values		

TPC_34   Message number   Num 2 0 single v.
No values

Message Class — Transaction SE91

Message class: ZTPCSET
In development class: TBIN
Description: Messages for T-PCSET
Messages:
001   Caution: Hard disk capacity too small for this amount of software!
002   Caution: If w = 20 cm, d = 40 cm, h = 30 cm, better use minitower!
003   Caution: If w = 20 cm, d = 40 cm, h = 60 cm, better use tower!
004   Caution: If w = 50 cm, d = 40 cm, h = 10 cm, better use desktop!

Variant Functions — Transaction CU67

You need to provide the following interface parameters for the function module that has the same name as the variant function:

- Import: parameter name: GLOBALS
  associated type: CUOV_00

- Export: no entries

- Changing: no entries

- Tables: parameter names: QUERY and MATCH
  associated type: CUOV_01 (for both)

- Exceptions: exception: FAIL and INTERNAL_ERROR

Variant function: Z_PCSET_NOTES

Function for calling warning messages

     Characteristics: TPC_32   TPC_34
     Key: X
     Source text:
Function module:

```
DATA: MLD(3) TYPE C,
      LS_T100 TYPE T100.
****************************************************************
  READ TABLE QUERY WITH KEY VARNAM = 'TPC_34'.
  CHECK SY-SUBRC IS INITIAL AND
      NOT QUERY-ATWRT IS INITIAL.
  MLD = QUERY-ATWRT.
```

```
    SELECT SINGLE * FROM T100 INTO LS_T100
                              WHERE ARBGB = 'ZTPCSET'
                              AND SPRSL = SY-LANGU
                              AND MSGNR = MLD.

    CALL FUNCTION 'POPUP_TO_CONFIRM_MSG_WITH_CALL'
         EXPORTING
              TXT01                          = LS_T100-TEXT
              TITLE                          = TEXT-400
              LENGTH                         = 40
         EXCEPTIONS
              FUNCTION_MODULE_MISSED         = 1
              TEXT_SECOND_PUSHBUTTON_MISSED  = 2
              OTHERS                         = 99.
```

Variant function: Z_PCSET_ID

Generate a string from three parts

	Characteristics:	TPC_09	TPC_10	TPC_14	TPC_31
	Key:				X
	Source text:				

Function module:

```
*"----------------------------------------------------------
*"*"Local interface:
*"  IMPORTING
*"     REFERENCE(GLOBALS) LIKE  CUOV_00 STRUCTURE  CUOV_00
*"  TABLES
*"     QUERY STRUCTURE  CUOV_01
*"     MATCH STRUCTURE  CUOV_01
*"  EXCEPTIONS
*"     FAIL
*"     INTERNAL_ERROR
*"----------------------------------------------------------

DATA: id_char  LIKE cuov_01-atwrt,
      PC_ID    LIKE cuov_01-atwrt,
      USAGE    LIKE cuov_01-atwrt,
      CASE     LIKE cuov_01-atwrt,
      CPU      LIKE cuov_01-atwrt,
      dash(1)  TYPE c VALUE '-'.

*..initialize table with export parameters.................
```

```
      REFRESH match.

*..get value of input characteristic tpc_31 (PC ID)
   CALL FUNCTION 'CUOV_GET_FUNCTION_ARGUMENT'
        EXPORTING
             argument      = 'TPC_31'
        IMPORTING
             sym_val       =  PC_ID
        TABLES
             query         = query
        EXCEPTIONS
             arg_not_found = 01.


   IF sy-subrc <> 0.
     RAISE internal_error.
   ENDIF.

* Split string
   USAGE = PC_ID(1).
   CASE  = PC_ID+2(1).
   CPU   = PC_ID+4(1).


*..add result TPC_09 (application)
   CALL FUNCTION 'CUOV_SET_FUNCTION_ARGUMENT'
        EXPORTING
             argument              = 'TPC_09'
             vtype                 = 'CHAR'
             sym_val               = USAGE
        TABLES
             match                 = match
        EXCEPTIONS
             existing_value_replaced = 01.


*..add result TPC_10 (casing)
   CALL FUNCTION 'CUOV_SET_FUNCTION_ARGUMENT'
        EXPORTING
             argument              = 'TPC_10'
             vtype                 = 'CHAR'
             sym_val               = CASE
        TABLES
             match                 = match
```

```
        EXCEPTIONS
            existing_value_replaced = 01.

*..add result TPC_14 (CPU)
   CALL FUNCTION 'CUOV_SET_FUNCTION_ARGUMENT'
        EXPORTING
            argument                = 'TPC_14'
            vtype                   = 'CHAR'
            sym_val                 = CPU
        TABLES
            match                   = match
        EXCEPTIONS
            existing_value_replaced = 01.
```

Object Dependencies for the Configuration Profile — Transactions CU02 and CU42

T_PC_FUNCTION_NOTES1
Description: Warning message in case of hard disk problem
Syntax:

```
function Z_PCSET_NOTES
(tpc_34  =   '002'   ,
 tpc_32  = $self.tpc_32)
* Output message 1 as warning message
if $self.tpc_29 > 50,
* if at least 50 % of hard disk occupied by software
```

T_PC_FUNCTION_ID
Description: PC-ID generated via function
Syntax:

```
* SCREEN_DEP-NOINPUT
$self.p_24 = 'tpc_31'        if $self.tpc_09 specified
                             or $self.tpc_10 specified
                             or $self.tpc_14 specified,

( $self.p_24 = 'tpc_09' ,
  $self.p_24 = 'tpc_10' ,
  $self.p_24 = 'tpc_14' )  if $self.tpc_31 specified,

* Function
```

```
function   Z_PCSET_ID
       ( tpc_09  =    $self.tpc_09 ,
         tpc_10  =    $self.tpc_10 ,
         tpc_14  =    $self.tpc_14 ,
         tpc_31  =    $self.tpc_31 ) if $self.tpc_31 specified,

* String Operation
$self.tpc_31 = $self.tpc_09 || '-' ||
               $self.tpc_10 || '-' ||
               $self.tpc_14   if $self.tpc_09 specified
                              and $self.tpc_10 specified
                              and $self.tpc_14 specified
```

T_PC_FUNCTION_NOTES2

Description: Warning message in case of casing problem

Syntax:

```
function Z_PCSET_NOTES
(tpc_34  =   '002'   ,
 tpc_32  = $self.tpc_32)
* Output message 2 as warning message
if $self.tpc_21=20 and $self.tpc_20=40 and $self.tpc_22=30
* if width= 20 cm, depth = 40 cm and height = 30 cm
* then minitower
* -------------------------------------------------------------
function Z_PCSET_NOTES
(tpc_34  =   '003'   ,
 tpc_32  = $self.tpc_32)
* Output message 3 as warning message
if $self.tpc_21=20 and $self.tpc_20=40 and $self.tpc_22=60
* if width = 20 cm, depth = 40 cm and height = 60 cm
* then tower
* -------------------------------------------------------------
function Z_PCSET_NOTES
(tpc_34  =   '004'   ,
 tpc_32  = $self.tpc_32)
* Output message 4 as warning message
if $self.tpc_21=50 and $self.tpc_20=40 and $self.tpc_22=10
* if width = 50 cm, depth = 40 cm and height = 10 cm
* then desktop
```

E   The Authors

Dr. Uwe Blumöhr received a Ph.D. in Mathematics and a Master's degree for Teaching. He works as a training consultant in the area of customer and partner trainings at SAP Deutschland AG & Co. KG, the German subsidiary of SAP. His work focuses on the areas of Variant Configuration and Lifecycle Data Management. Since 1996, Uwe Blumöhr has been responsible for all of SAP's international training developments in the Variant Configuration area; he develops most of SAP's training material on this topic himself.

Dr. Manfred Münch received a Ph.D. in Aerospace Engineering. He works as process architect in the software development area at the SAP Labs in Walldorf, Germany. In 1988 he became involved with product configuration at Heidelberger Druckmaschinen AG, a market leader for printing presses. Since 1999, he has been responsible for product configuration in the product management and development of SAP AG. From 2001 to 2004 he was secretary of the Configuration Workgroup (CWG) and managed the business of the SAP user group for product configuration.

Marin Ukalovic has a Master's degree in mechanical engineering and initially worked as a design engineer in mechanical engineering and plant construction. In 1999, he joined SAP's sales team and supported customers of discrete manufacturing in the logistics area. In 2003, as solution architect, he assumed responsibility for the paper and furniture industry at SAP Deutschland AG & Co. KG, the German subsidiary of SAP. In 2006 he joined the Industry Business Development EMEA area, where he is now responsible for mechanical engineering and plant construction in Europe. SAP Variant Configuration has been the central topic throughout his career at SAP. He has supported the Configuration Workgroup (CWG) as Account Manager since 2006, and he's a member of the CWG board of directors.

Index

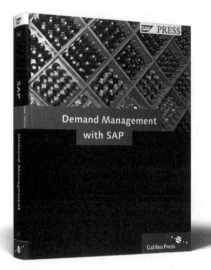

Teaches what demand management is
and how to use it effectively

Describes how to configure SAP solutions,
such as APO, SNC, DSR, etc., to support
demand management

Provides case studies, real-world
examples, and best practices throughout

Up-to-date for ERP 6.0

Christopher Foti, Jessie Chimni

Demand Management with SAP

This is the first resource that managers and business leaders need to
understand and effectively use SAP's demand management tools. Each
chapter begins with a general description of a business strategy or a
process in Demand Management from the SAP perspective, and then
teaches how SAP's solution is designed to work and teaches readers how
to configure and customize the solutions. After reading this book, users
will understand demand management, and will know how to configure
and use the SAP solutions.

approx. 450 pp., 69,95 Euro / US$ 69.95
ISBN 978-1-59229-267-7, Oct 2009

>> www.sap-press.com

Discover how to use SAP MM to meet your business process needs

Learn step-by-step how to configure SAP MM

Explore case studies, real-world examples, and best practices

Up to date for SAP ERP 6.0

Akash Agrawal

Customizing Materials Management Processes in SAP ERP Operations

This is a complete resource to teach consultants and managers how to configure SAP MM according to their company's business processes. After reading this book, users will have a thorough understanding of the strategies and processes for materials management, and know how SAP MM can be configured to support these business processes. Case studies and best-practice research are presented throughout to help readers develop a strong and thorough understanding of why and how to configure SAP MM.

approx. 395 pp., 69,95 Euro / US$ 69.95
ISBN 978-1-59229-280-6, Aug 2009

>> www.sap-press.com

Interested in reading more?

Please visit our Web site for all
new book releases from SAP PRESS.

www.sap-press.com